# PI IN THE SKY

*Counting, Thinking, and Being*

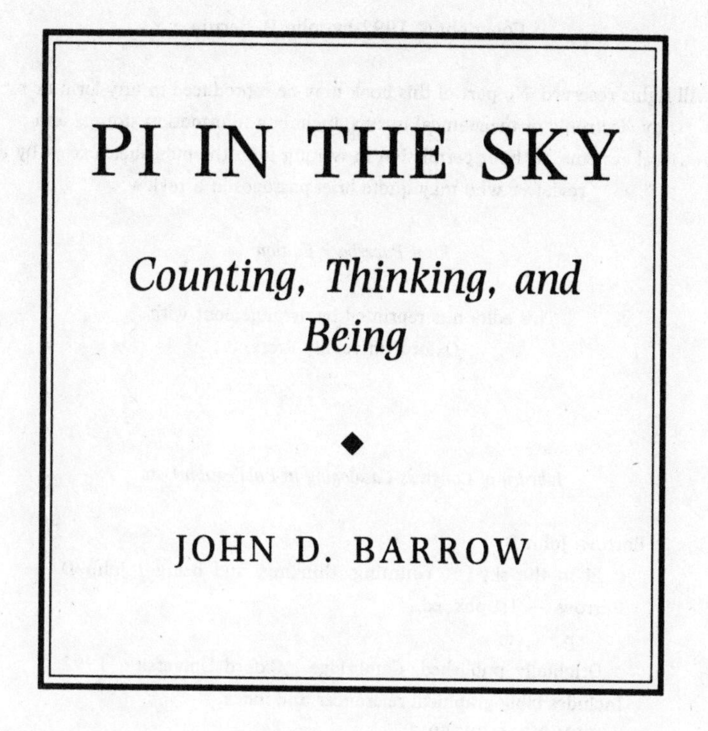

# PI IN THE SKY

*Counting, Thinking, and Being*

◆

JOHN D. BARROW

LITTLE, BROWN AND COMPANY

*Boston New York Toronto London*

*First Paperback Edition*

This edition is reprinted by arrangement with Oxford University Press.

*Library of Congress Cataloging-in-Publication Data*

Barrow, John D.
Pi in the sky : counting, thinking, and being / John D. Barrow. — 1st pbk. ed.
p. cm.
Originally published: Cambridge : Oxford University, 1992.
Includes bibliographical references and index.
ISBN 0-316-08259-7
1. Mathematics. I. Title.
QA36.B37 1994
510 — dc20 93-24402

10 9 8 7 6 5 4 3 2 1

RRD-VA

Published simultaneously in Canada by Little, Brown & Company (Canada) Limited

*Printed in the United States of America*

*From the moment I picked up your book until I laid it down, I was convulsed with laughter. Some day I intend reading it.*

*Groucho Marx*

*To David,*
*Who now sees why people read books,*
*but not why they write them*

# PREFACE

There is safety in numbers. By translating the actual into the numerical we have found the secret to the structure and workings of the Universe. For some mysterious reason mathematics has proved itself a reliable guide to the world in which we live and of which we are a part. Mathematics works; as a result we have been tempted to equate understanding of the world with its mathematical encapsulization. At the root of every claim we make to possess an understanding of the physical world sits a mathematical truism. But upon what does this faith in mathematics rest? How can our inky squiggles on pieces of paper possibly tell us how the world goes round? Why is the world found to be so unerringly mathematical? These are questions that scientists rarely ask. For they lead us into greater mysteries still: What is mathematics? Do we invent it? Do we discover it? Could it be merely a handy summary of our experiences of the world or could it be something immaterial and other-worldly that exists in the absence of mathematicians? Maybe we should regard the presence of order in the world as a prerequisite for the existence of things as complex as readers of books published by the Oxford University Press. Seen in this light, the existence of mathematics would be no more, but no less, surprising than that of life itself.

In the chapters to follow I have tried to introduce the reader to the mystery of the utility and universality of mathematics in order to explore the possible meanings of mathematics. As a result, this is a book about mathematics without being a book of mathematics. Mindful of the definition of a mathematician as the person you don't want to meet at a party, I have tried to avoid writing a history of mathematics and I have assiduously avoided doing any mathematics. Rather, our journey will take us first to the most ancient and diverse anthropological evidences of counting, and the origins of mathematical intuition, with a view to learning how and why it germinated and grew. Did it arise spontaneously and independently in many cultures, or was it the fruit of a peculiar insight that spread from one culture to others, riding the tides of trade and commerce carried by a wind of words? The fragility of this process will become clear. The unique features of its success are impressed most forcefully upon us by the fact that we can conceive of so many ways in which it could have failed.

We look at the principal interpretations of mathematics, the strange events and individuals that created them, and the challenges they face from the latest discoveries in mathematics, computation, and physics. The meaning of mathematics will emerge as a key question that must eventually be answered in any quest for a fundamental understanding of the physical world. At present no embryonic Theory of Everything takes wing and flies without an unquestioning faith in the reliability of mathematics. If we seek a Theory of Everything, or even a Theory of Anything, then it is an article of faith that this theory will be a mathematical theory. Without a deep understanding of the meaning and possible limits of mathematics we run the risk of building our house upon the sand.

Philosophy is a steadily evolving subject; but that part of it that deals with the nature of mathematics is stuck in a time-warp. The philosophy of mathematics has hardly progressed at all when compared with the development of philosophy or of mathematics. Now is the time to rejuvenate a study of the meaning of mathematics amongst philosophers. New developments in fundamental science challenge us to choose between a paradigm for the Universe as a great symmetry or as a great computation. They move us towards seeing mathematics in strange new ways. They lead us into a deeper appreciation of the connection between the computational process and the applicability of mathematics to the workings of the natural world.

But mathematics is not only about numbers. And in our story we shall meet many strange characters and unexpected events that make plain the humanity of those engaged in the search for number and its meaning. These events inspired great tides of change in human thinking about our Universe and the existence of other immaterial worlds beyond the reach of our senses. It gave flesh to the Platonic yearning for another perfect world where 'pi' really was in the sky.

To succeed, even a little, writers can't afford to be just one person; they've got to be a whole lot of people trying hard to be one person. That process is made possible, and even pleasant, by the contributions of others. For that reason I am grateful to those whose comments and contributions have helped to bring this book into its present state. Whether that state is one which prevents them from putting it down after they have picked it up, or prevents them from picking it up after they have put it down, I cannot say. Either way, I would like to thank David Bailin, Margaret Boden, Laura Brown,

Terry Bristol, John Casti, Paul Davies, Wim Drees, George Ellis, Giulio Giorello, John Harrison, John Hays, John Lucas, John Manger, Leon Mestel, Stephen Metcalf, Arthur Miller, Sir William McCrea, Olaf Pedersen, Michael Redhead, Aaron Sloman, John Maynard Smith, Donald Winch, and Gavin Wraith for their contributions and also Anjali and David Bailin for access to their collection of unusual pictures. Very special thanks are due to Elizabeth for all her help. Finally, younger family members, David, Roger, and Louise, have shown a lively interest in what they thought was going to be a cookery book and their wariness of all things mathematical appeals to no less an authority than Sherlock Holmes himself who taught them that Professor Moriarty was 'a professor of mathematics at a provincial university'. But alas, it was not possible to dedicate this work to Geoffrey as they had hoped.

*Brighton* J.D.B.
May 1992

# CONTENTS

1 From mystery to history 1

A mystery within an enigma 1
Illusions of certainty 2
The secret society 6
Non-Euclideanism 8
Logics—To Be or Not To Be 15
The Rashomon effect 19
The analogy that never breaks down? 21
Tinkling symbols 23
Thinking about thinking 24

2 The counter culture 26

By the pricking of my thumbs 26
The bare bones of history 28
Creation or evolution 33
The ordinals versus the cardinals 36
Counting without counting 41
Fingers and toes 45
Baser methods 49
Counting with base 2 51
The neo-2 system of counting 56
Counting in fives 60
What's so special about sixty? 64
The spread of the decimal system 68
The dance of the seven veils 72
Ritual geometry 73
The place-value system and the invention of zero 81
A final accounting 101

3 With form but void 106

Numerology 106
The very opposite 108
Hilbert's scheme 112
Kurt Gödel 117
More surprises 124
Thinking by numbers 127
Bourbachique mathématique 129
Arithmetic in chaos 134
Science friction 137
Mathematicians off form 140

4 The mothers of inventionism 147

Mind from matter 147
Shadowlands 149
Trap-door functions 150
Mathematical creation 154
Marxist mathematics 156
Complexity and simplicity 159
Maths as psychology 165
Pre-established mental harmony? 171
Self-discovery 176

5 Intuitionism: the immaculate construction 178

Mathematicians from outer space 178
Ramanujan 181
Intuitionism and three-valued logic 185
A very peculiar practice 188
A closer look at Brouwer 192
What is 'intuition'? 196
The tragedy of Cantor and Kronecker 198
Cantor and infinity 205
The comedy of Hilbert and Brouwer 216
The Four-Colour Conjecture 227
Transhuman mathematics 234
New-age mathematics 236
Paradigms 243
Computability, compressibility, and utility 245

6 Platonic heavens above and within 249

The growth of abstraction 249
Footsteps through Plato's footnotes 251
The platonic world of mathematics 258
Far away and long ago 265
The presence of the past 268
The unreasonable effectiveness of mathematics 270
Difficulties with platonic relationships 272
Seance or science? 273
Revel without a cause 276
A computer ontological argument 280
A speculative anthropic interpretation of mathematics 284
Maths and mysticism 292
Supernatural numbers? 294

*Further reading* 298

*Index* 311

# CHAPTER ONE
# *From mystery to history*

> *If we would discover the little backstairs door that for any age serves as the secret entranceway to knowledge, we will do well to look for certain unobtrusive words with uncertain meanings that are permitted to slip off the tongue or the pen without fear and without research; words which, having from constant repetition lost their metaphorical significance, are unconsciously mistaken for objective realities.*
>
> CARL BECKER

## A MYSTERY WITHIN AN ENIGMA

> *Ask a philosopher 'What is philosophy?' or a historian 'What is history?', and they will have no difficulty in giving an answer. Neither of them, in fact, can pursue his own discipline without knowing what he is searching for. But ask a mathematican 'What is mathematics?' and he may justifiably reply that he does not know the answer but that this does not stop him from doing mathematics.*
>
> FRANÇOIS LASSERRE

A mystery lurks beneath the magic carpet of science, something that scientists have not been telling, something too shocking to mention except in rather esoterically refined circles: that at the root of the success of twentieth-century science there lies a deeply 'religious' belief—a belief in an unseen and perfect transcendental world that controls us in an unexplained way, yet upon which we seem to exert no influence whatsoever. What this world is, where it is, and what it is to us is what this book is about.

This sounds more than a trifle shocking to any audience that watches and applauds the theatre of science. Once there was magic and mysticism, but we have been taught that the march of human progress has gone in step with our scientific understanding of the natural world and the erosion of that part of reality which we are willing to parcel up and label 'unknowable'. This enterprise has been founded upon the certainty that comes from speaking the language of science, a symbolic language that banishes ambiguity and doubt, the only language with a built-in logic which enables an intimate communion with the innermost workings of Nature to be established and underpinned by thought and action: this language is mathematics.

As science has progressed, it has become more mathematical in its expression and more unified in its structure. Scientists believe there to be one Universe with a single universal legislation from which all the diverse subdivisions of science ultimately receive their marching orders. In recent years, the search for this single 'Theory of Everything' has become the new Grail of fundamental science. If found, its content will be a piece of logically consistent mathematics. But what *is* mathematics and why do we entrust it with the secret of the Universe? Why do we look to mathematics for answers to ultimate questions about the nature of physical reality? What is the foundation upon which this magical mathematics rests? Indeed, what is mathematics and why does it work? If we cannot answer these questions our scientific explanations of the Universe are based ultimately upon things we do not understand, upon the intangible mysteries that lie behind the impregnable battlements of a castle in the air.

## ILLUSIONS OF CERTAINTY

> *Petite, attractive, intelligent WSF, 30, fond of music, theatre, books, travel, seeks warm, affectionate, fun-loving man to share life's pleasures with view to lasting relationship. Send photograph. Please no biochemists.*
>
> PERSONAL AD, NEW YORK REVIEW OF BOOKS

History is full of people who thought they were right—absolutely right, completely right, without a shadow of a doubt. And because history never seems like history when you are living through it, it is tempting for us to think the same. Cocooned within this illusion it is easy to look into the past and smile at the touching naïvety of the ancients inbred by all manner of notions of the mystical and the fantastic. Smiles might turn to despair as we detect such notions, not merely as intellectual decoration but the very fabric of their view of reality; strange immaterial imaginings about gods and ghosts who steer the world in its course: the things that are seen rest confidently upon the things that are not. We return confidently to an image of the inexorable march of progress. A march that has brought us to a new realism, a mature perspective on the world and its workings that has divested it of the metaphysical and the mystical. This realism is called something like the 'scientific attitude' and even if we know nothing *of* it we certainly know *about* it.

Science surrounds us with devices and machines without whose help our lives would come to a grinding halt. We have become so used to the routine success of everyday technology that even its spectacular failures seem only a temporary aberration attributable to 'human error'. Such an equation confirms the unnoticed prejudice that human intervention is ultimately unwanted as well as unnecessary. The future beckons with a vision of ever-growing competence and a deeper understanding of the world we live in, built upon the certain foundation of pragmatic experience and objective investigation which follows the recipes of those philosophers of science who stress the rigid doctrines of falsification and verification by observation and experiment. In this way the sun of human understanding is imagined to rise, evaporating the mists of mysticism, illuminating the way through the clouds of unknowing.

Such a grandiloquent view of things, if not entirely universal, is prevalent enough to dictate thinking across a wide spectrum of human activity. If we look to the most technologically advanced societies we find the greatest thirst to discover the latest accounts of the progress of science, whilst the most backward human societies are dominated by superstition, wedded to the inflexible ways of the past. If we look closer still at the technologically successful societies, we find them founded upon the language of mathematics. It lies at the heart of the scientific enterprise and at the root of the computer revolution. No practical progress can flourish without its sustaining presence.

The development of our understanding of the natural world around us and within us has been a long and painful process. At first we were impressed by the irregularities and vagaries of Nature: catastrophes and calamities—the out-of-the-ordinary events that were so devastating or striking that they assumed a place in our histories, legends, and memories. From this appears to have grown gradually an appreciation that the 'out-of-the-ordinary' requires an 'ordinary'. And what characterizes the ordinary is its reliability, its predictability, and its exploitability. It is that part of our world which permits us to predict its future from its history. So powerful were the effects of this reorientation and so great were the fruits that followed from it that it became a defining line between the simple and the sophisticated ancient society. The systemization of these predictable aspects of the world came to form an activity that is rewarding and rewarded. It was what we now call 'science': it is simply the replacement of our observations of the world by abbreviations which retain some or all of the information instantiated in the world.

As we look back at the growth of this activity, we find that it is closely linked to number: to measuring, counting, surveying, dividing, and making patterns. If we look at the activity of science today we find that, for the outsider, it is wellnigh indistinguishable from mathematics. There is a reason for this. Mathematics 'works' as a description of the world and the things that occur within it. The little squiggles that we make on pieces of paper seem to tell us about the ultimate structure of matter, about the motions of the stars and planets and the workings of our own minds. But while this is indeed the reason, it is not an explanation. Why should mathematics 'work'? Why does the world dance to a mathematical tune? Why do things keep following the path mapped out by a sequence of numbers that issue from an equation on a piece of paper? Is there some secret connection between them; is it just a coincidence; or is there just no other way that things could be? Most of the other things we do here on Earth are at best only partially successful. The systems of thought and practice that we codify and seek to follow rarely work out as planned. Reality has a habit of diverging from the path planned for it by human imaginings. Mathematics seems to be the extraordinary exception that proves the rule. But what exactly is that rule?

From these simple questions about the ubiquity and utility of mathematics others follow in overwhelming profusion. What *is* mathematics? What are numbers and how did we come to discover them? But, did we really discover them—perhaps we merely invented them? Most people find mathematics difficult; is it a natural activity for the human mind or just a curious skill that is possessed by a few? At the other end of the spectrum to these simple questions, there might be queries of a more esoteric nature. How did our ancestors pass from the mundane activity of counting to the concept of 'number' in the abstract, devoid of any particular collection of objects to label? And did this graduation launch us into an immaterial world of mathematical truths which we can tune into because of some unusual propensity of the human mind? If so, does this mean that mathematics is really a religion, albeit a rather austere and difficult one? If we contacted some alien intelligence in the Universe, would it possess the same type of mathematical knowledge that we do? Would it too think it interesting to write books exploring the puzzle of 'why the world is mathematical'?

These questions, and others like them, are what motivate the story we are going to tell in this book. They expose a frayed edge to the tapestry of meaning that our contemplation of the world is designed to create and embroider. We see interwoven there the relation between the reality around us and the images of it that we create in our minds. We find also the tension between what our minds have abstracted from the environment in which they have developed and what may have been imprinted upon them by the very genetic signature that makes us human. We find ourselves being drawn towards deep questions which were once the sole preserve of the theologians. Down the centuries there have been those who saw in mathematics the closest approach we have to absolute truth—the furthest that our minds can go from subjectivity. Its very structure forms a model for all other searches after absolute truth.

The fact that such unexpected connections can be made with other questions leads us to suspect that there is more to a number than meets the eye. Throughout our culture we come across vestigial remnants of a bygone age when numbers had deeper and stranger meanings than they needed. Who has not thought they had a lucky number? Or at least an unlucky one? Most religions possess curious numerical eccentricities that wed them to the cultures in which they developed. Whether it be '666' or '7' makes little difference. Numbers possessed an intrinsic meaning for some. If only they could fathom that meaning then some little part of the secret of the Universe would be revealed.

Numbers dominate much of our everyday lives in diverse and barely perceptible ways. Parents are anxious to see their children attain good grades in mathematics, assured that this is a passport to success and security. The government stipulates that mathematical studies are compulsory for most of a child's time at school. Intelligence tests contain particular ingredients to test the candidates' ability to reason mathematically. Advertisements promising challenging and well-paid employment invite us to seek further details if we can successfully spot the next number in puzzling sequences. Students change subjects when they discover the mathematical component of their course is too difficult to master. Young children turn out to be extraordinarily talented in mathematics with abilities outstripping children twice their age, yet their performance in other school subjects is unremarkable. Only in art and music is such precociousness so impressive. Looking out into the wider world, we see the cogs of the Western world turned by the engine of mathematical understanding. Computers programmed with inexorable logic control our economies and money markets, our aeroplanes and trains. Our own minds are seen as a fallible 'natural' form of intelligence that these computers will ultimately perfect in an embodiment of 'artificial' intelligence that is nothing more than a complicated mathematical algorithm that can be implemented electronically far faster and less fallibly than by our own feeble brains. In our universities there is an unspoken suspicion that the further a discipline lies from mathematics and the smaller the body of mathematical statements at its core the less rigorous and intellectually respectable it is.

These examples are all rather predictable. For most people mathematics lives in classrooms, impenetrable books, income tax returns, and university lecture halls. Mathematicians are the people you don't want to meet at cocktail parties. But this is not the whole truth. There are curious but very significant examples of mathematics breaking out of these cloistered constraints to exert completely unsuspected influences upon all sorts of other areas of human activity.

## THE SECRET SOCIETY

*The different branches of Arithmetic—Ambition, Distraction, Uglification and Derision.*
LEWIS CARROLL

There have always been those who saw the Universe as a vast cryptogram whose meaning could be unlocked by finding the right combination. The secret, once found, would guarantee the *cognoscenti* power and domination over the uninitiated. This Hermetic tradition has always retained some place in Western thought. Even today this delusion persists in eccentric circles. Quite frequently I receive letters from well-meaning members of the public who claim to have unravelled the ultimate structure of the Universe by strange jugglings of numbers. Sometimes the numbers involve the dimensions of particular objects, like the Great Pyramid, which have a traditional place in such numerological gymnastics. On other occasions they involve endowing other symbols, perhaps letters or names, with numerological equivalents. These are strange echo from the past.

Pythagoras was a native of Samos, a Mediterranean island off the coast near Ephesus, now part of Turkey. During the sixth century BC it was a thriving economy which rivalled Miletus, Ephesus, and other coastal towns. Samos was ruled by the despot Polycrates who seems to have amassed most of his wealth by piracy, and attempted to keep it by swopping alliances at opportune times when more powerful neighbours threatened to descend upon him. Eventually his untrustworthiness caught up with him and he was lured into a trap by the Persians, captured, and finally crucified in 522 BC. Pythagoras' early life is more of a mystery. Tradition has it that he left Samos for Miletus and there became a student of Thales, before travelling on to Phoenicia and studying for twenty-two years in Egypt until Cambyses conquered Egypt in 525 BC, whereupon Pythagoras followed his army back to Babylon where he immersed himself in the study of arithmetic, music, and other matters of the mind. After further Mediterranean travels he comes to rest in Croton on what is now the southern coast of Italy, but which in Pythagoras' time had been a Greek colony since about 710 BC. Whether he really pursued such a peripatetic existence is hard to ascertain. Because of what happened later there grew up many romantic traditions about Pythagoras' early life designed to increase his prestige and authority. A period of study in Egypt followed by a sojourn in the advanced culture of Babylon was just the sort of curriculum vitae that would provide a respectable apprenticeship for any great philosopher.

Fortunately, what occurred at Croton is better known. It appears that the region had experienced a religious revival of sorts during the sixth century BC leading to a plethora of quasi-religious communities united by a particular set of views about the afterlife and a shared appreciation of a roster of taboos and rituals. By forming a communal existence they were able to a create a 'support group' to act as a source of encouragement and insulation from the larger pagan culture within which they had embedded themselves. Distinct dress, distinct diet, and a secret way of life reinforced their separateness. Within one such micro-society Pythagoras nurtured a cult based upon numbers that was to become a weird mixture of the sublime and the ridiculous.

Pythagoras and his followers saw numbers as the true essence of things. They saw the numerical as a symbolic picture of the true meaning of the Universe. Whereas we are used to regarding numbers as descriptions of collections of things or relationships between things, the Pythagoreans were drawn deep into an older mystical belief about the significance of the numbers themselves. Thus a particular number, like 4, would have both a symbolic representation and a symbolic meaning. Pythagoras and his disciples turned mathematics into a mystical religion in which the numbers themselves were signs and symbols of an occult knowledge behind the world of appearances which could be found only by special discernment and interpretation. Eventually, a bizarre crisis occurred which shook the foundations of the sect. They had believed all numbers to be of two sorts, either whole numbers (like 1,2,3,..., and so on) or fractions (like $^1/_2$, $^4/_5$, $^2/_7$,..... and so on), which were made by dividing any whole number by another. These numbers were called 'rational' numbers. But the famous theorem about triangles that bears Pythagoras' name revealed that if one drew a square whose sides were each one unit in length then the length of the diagonal joining any two opposing corners of the square was equal to a quantity that we call the square root of 2. Try as they may, using their well-tried systematic procedures, they could not express this number as a ratio of two whole numbers, that is, as a fraction. Eventually it was proved that the square root of 2 could not be expressed as the ratio of any two whole numbers: it was a new and peculiar type of number that came to be known at first by the Pythagoreans as *arrheton*, meaning that it was inexpressible as a ratio. Later these perplexing numbers became known as 'irrational numbers' and this term was meant to reflect the same idea about the inexpressibility of the number. Numbers like the square root of 2 were shocking to them because they could not be measured out precisely with a measuring device. This was a challenge to the Pythagoreans' faith in the ultimate power of numbers to rule the Universe. If they failed to capture something as mundane as the diagonal of a square, their entire religion was under threat. They were plagued by this result because from it one could construct a never-ending list of other irrational numbers beginning with the square roots of 3, 5, 6, and 7. It also opened up a rift between the school of arithmetic, which could create these strange 'irrational' numbers, and the school of geometry, which could not measure them. Tradition has it that, when these irrational numbers were first discovered, the fact was kept a secret by the Brotherhood lest the news should spread that there were numbers that challenged their doctrines. When Hippasus committed the sin of violating his oath of secrecy to reveal this terrible truth, he was drowned. According to a later commentator, 'the unutterable and formless were to be concealed; those who uncovered and touched this image of life were instantly destroyed and shall remain the play of the eternal waves'. One suspects, however, that like many secret societies, the Pythagoreans took themselves rather more seriously than they were taken by outsiders. Epicharmos, an aptly named comic writer of the fifth century BC, gives us this dialogue about some of the peculiar consequences of considering oneself to be nothing more than one's vital statistics:

'When you have an even number or, for all I care, an odd number, and someone adds a pebble or takes one away, is the original unchanged?'

'May the gods forbid it!'

'Now if some one adds to a length or cuts off a piece, will it measure the same as before?'

'Why, of course not.'

'But look at people. One may grow taller, another may lose weight ... Then by your argument, you, I, and others are not the same people we were yesterday and we shall be still different individuals in the future.'

The world is not static but the Pythagoreans' numbers were. Later, we shall see to what lengths we must go to avoid objections like these.

## NON-EUCLIDEANISM

*Geometry, throughout the 17th and 18th centuries, remained, in the war against empiricism, an impregnable fortress of the idealists. Those who held—as was generally held on the Continent—that certain knowledge, independent of experience, was possible about the real world, had only to point to Geometry: none but a madman, they said, would throw doubt on its validity, and none but a fool would deny its objective reference.*

BERTRAND RUSSELL

Lest one think the excesses manifested by the Pythagorean numerologists of the past are merely the symptoms of looking too close to the dawn of history in a place where the noonday sun is too high overhead, it is instructive to look at a more recent example which influences much of our modern system of values.

One of the cornerstones of human thinking from the early Greeks until the nineteenth century had been the certainty offered by Euclid's study of geometry. It was rigorous and definite. Sure theorems about lines and triangles, circles and squares, following with unimpeachable logic from clearly stated assumptions called 'axioms'. Indeed, this whole edifice provided the model for the construction and application of logic to sets of assumptions. As such, it was taught in European schools in a manner that remained unchanged for more than a hundred years. Indeed, the wording of Euclid's works created for generations of schoolchildren a curious droll stylized language, every bit as distinctive in its own way as that of the Authorized Version of the Bible which they listened to in morning assembly. The effect is caricatured most memorably in Stephen Leacock's piece entitled *Boarding-House Geometry*, penned in 1910:

*Definitions and Axioms*

All boarding-houses are the same boarding-house.
Boarders in the same boarding-house and on the same flat are equal to one another.
A single room is that which has no parts and no magnitude.
The landlady of a boarding-house is a parallelogram—that is an oblong angular figure, which cannot be described, but which is equal to anything.
A wrangle is the disinclination for each other of two boarders that meet together but are not in the same line.
All the other rooms being taken, a single room is said to be a double room.

*Postulates and Propositions*

A pie may be produced any number of times.
The landlady can be reduced to her lowest terms by a series of propositions.
A bee-line may be made from any boarding-house to any other boarding-house.
The clothes of a boarding-house bed, though produced ever so far both ways, will not meet.
Any two meals at a boarding-house are together less than two square meals.
If from the opposite ends of a boarding-house a line be

> drawn passing through all rooms in turn, then the stove-pipe which warms the boarders will lie within that line.
>
> On the same bill and on the same side of it there should not be two charges for the same thing.
>
> If there be two boarders on the same flat, and the amount of side of the one be equal to the amount of side of the other, each to each, and the wrangle between one boarder and the landlady be equal to the wrangle between the landlady and the other, then shall the weekly bills of the two boarders be equal also, each to each.
>
> For if not, let one bill be the greater.
>
> Then the other bill is less than it might have been — which is absurd.

But at the beginning of the nineteenth century Euclidean geometry was a good deal more than a game to be played with pencil, ruler, and compasses in schoolchildren's exercise books. It was a description of how the world really was. Euclid derived his intuition about geometrical truths by drawing figures in the sand and inspecting the relationships between lengths and angles and shapes. The self-evident 'truths' of what he saw on the flat ground in front of him he idealized into postulates which were to underpin his reasonings about what it might in the future be possible for him to draw in the sand. What, in retrospect, characterizes Euclid's geometrical edifice most singularly is the postulate that parallel lines never meet. This truth seems self-evident. All attempts down the centuries to derive it as a consequence of the other basic assumptions adopted by Euclid had failed.

Euclid's pristine geometry had subtle influences upon other areas of human thinking. It undergirded all architecture and artistic composition, all navigation and astronomy. In the sciences it lay at the foundation of all Newton's work on motion and gravitation. His famous *Principia*, which appeared three hundred years ago, appears to the casual observer to resemble a vast treatise on geometry, because Newton was a master at the application of geometry to the description of Nature. Such mastery was the hallmark of a seventeenth-century mathematician. Geometry stood for absolute logical certainty. It was the way the world was. And the determinism of the Newtonian clockwork world derived its certainty from its geometrical underwriting. This certainty had been abstracted into many other realms during the seventeenth and eighteenth centuries. There were Newtonian models of government and human behaviour that appealed to the certainty of its mathematics. There were arguments for the existence of God founded upon the mathematical certainty of the geometrical laws of Nature that Newton had revealed. Geometry provided its students with a system of thought that was absolutely true because it employed flawless logical reasoning from premisses which were statements of how the world was seen to be. It had withstood all misguided attempts at its disproof. It witnessed to the attainability of ultimate truth and personified the Almighty as the Great Geometer and Architect of Nature. It stood as a rock of certainty in the midst of the buffeting seas of human speculation.

Its special status can be discerned from its treatment by Kant in his philosophical system during the eighteenth century. His system of thought was wedded to the inevitability of Euclidean geometry. He gave it as an example of synthetic *a priori* knowledge, that is, something that is necessarily true. For Kant, this necessity arose because of the nature of human modes of thinking. The way in which the human brain was constructed ensured that we must find the mathematical truths of geometry to hold.

In Britain, where Kant's sceptical influence was small during the eighteenth and nineteenth centuries, Euclidean geometry was the exemplar of a peculiarly 'British' way of thinking. There, the common view was that space has a structure independent of our minds and that in elucidating the rules of Euclidean geometry we had uncovered the essence of this reality with complete certainty. In other areas of experience we had so far failed to probe as deeply into the mind of God and our knowledge was partial to a degree that reflected the extent to which it failed to appeal to the truths of geometry.

The discovery that Euclidean geometry was not a unique, inevitable, and absolute truth about the world came therefore as a stunning blow. Its impact was far-reaching and irreversible. It undermined absolutist views about human knowledge across a vast spectrum of human thinking. Although its demise was resisted by mathematicians for a long while, those seeking to overthrow traditional Euclidean certainties seized upon it as a signal that relativism was the rule. The prefix 'non-Euclidean' came to signify something more general than what happened to lines in space.

Lobachevski, Gauss, and Bolyai, in a complicated and disputed sequence of events, realized that Euclidean geometry was but one of several possibilities. On any curved surface, for example the surface of a sphere, one can lay down analogous postulates and rules of reasoning to those of Euclid. The result is a logically impeccable system that differs from that of Euclid. If straight lines are still taken to be the shortest path that can be taken between two points on a surface, then a triangle can be drawn on the surface of a sphere but the three internal angles of the triangle do not sum to 180°. The famous fifth 'parallel' postulate of Euclid—that parallel lines never meet—is not true on such a curved surface and is not one of the postulates used. In retrospect it seems strange that it took so long for this idea to emerge. It was known that the Earth was not flat; there were innumerable curved surfaces in front of our noses. Artists and navigators were well used to drawing the shortest paths between points on such surfaces. More persuasive still should have been the fact that if we look at the world through a distorting lens or in a reflecting mirror then we see a skewed geometry that is not that of Euclid. But the unique and determined nature of the distortion of the image of the Euclidean geometry means that there must exist rules governing the distorted geometry which are merely the distortions of the rules of Euclid. Eventually, Bernhard Riemann showed how to systematize the study of all possible geometries within a very wide class (which includes Euclid's as the simplest case) in terms of the changes that are made to Pythagoras' famous theorem regarding the relations between the lengths of the sides of right-angled triangles.

There was considerable resistance to the promulgation of these new discoveries, not least amongst mathematicians. The discovery of non-Euclidean geometries was regarded for a long time as a sort of logical curiosity, probably flawed in some way. Although curved surfaces clearly did exist, almost everyone believed that the physical space in which we move and have our being really was Euclidean. It was the fixed background stage on which the events of the universe were played out.

At this time mathematicians did not think of axiomatic systems—like the rules and postulates defining Euclidean geometry in particular—as being entities that one was at liberty to create at will subject only to the stricture of self-consistency. Axiomatic systems, of which arithmetic and geometry were the only familiar examples, were aspects of the how things were in physical reality, not mere paper creations.

The discovery that Euclidean geometry was not a unique God-given attribute of the world caused all manner of similar prejudices and beliefs to be questioned in the late nineteenth century and the growing trend towards cultural relativism gained momentum through the fortuitous resonance with other revolutionary ideas like Charles Darwin's theory of evolution through natural selection and Simon de Laplace's nebular hypothesis to explain the origin of the solar system from a swirling primordial cloud of gas. In all these realms the actual had been revealed to be but one of many possibilities. The revolution that occurred was not like many so called 'scientific revolutions', which would have seen non-Euclidean geometry replacing Euclidean geometry as *the* description of the world. Euclidean geometry at first remained unchallenged as the description of the space we lived in; what changed was its status as an exemplar of absolute truth. Other geometries could exist and possessed a logical status no less secure than that of Euclid's world. The latter could now be distinguished only by the claim that it was the geometry that was employed in Nature but no reason could be given as to why it should be this one and no other. The effect upon mathematics itself was important. The idea of axioms was for the first time divorced from physical reality and the doors were open for mathematics to branch off into innumerable logical 'paper worlds' of its own construction. The divide between pure and applied mathematics was born. Mathematics could be viewed as a tool for describing how things worked or as an open catalogue of logical interconnections. In time, as we shall see in subsequent chapters, this would lead to radical new ideas as to the nature of mathematics itself. Later, in 1915 there was to be a dramatic dénouement to this story when Einstein founded his new theory of gravitation upon the premiss that our physical space possesses a non-Euclidean geometry that is created by the presence of mass and energy in the Universe.*

Non-Euclidean geometries no longer existed solely as logical systems on pieces of paper, distinguished from the Euclidean case only by the brute fact that God had chosen the latter in his architecture of the Universe. Observations confirmed the predictions of Einstein's non-Euclidean theory of space. The real world was not Euclidean after all. The deviations from Euclid's geometry are very small, little more than one part in a hundred thousand over the dimensions of our solar system, but their presence was undeniably confirmed by observation exactly as Einstein had predicted.

The demise of Euclid's ancient assumption found a parallel in the development of new algebras which gave up another deeply-held assumption: that when action *A* was followed by action *B*, the result was identical with what followed from first doing *B* and then *A*. The addition or multiplication of numbers possesses this familiar 'commutative' property: 1 + 2 equals 2 + 1. But in 1843 the Irish physicist and mathematician William Hamilton constructed a logically complete and consistent algebra for objects (which he called 'quaternions') defined by a

* Some years earlier, in 1900, the German astronomer Karl Schwarzschild had written a little-known scientific paper in which he produced a model of the Universe with a non-Euclidean geometry of space, governed by Newtonian gravitational forces, and as early as 1850 Bernhard Riemann had speculated that the geometrical form of the Universe should be dictated by the forces of Nature—a speculation that was developed further by the English mathematician William Clifford as a result of becoming involved in the translation of some of Riemann's work into English.

set of rules which were not commutative. Non-commutative operations are not really very rare—the operations of putting on one's shoes and putting on one's socks are examples where different orders of so doing produce very different results. Hamilton's development marked the beginning of a period in which mathematicians freely created systems of symbols using prescribed self-consistent rules governing their combination one with another giving no thought for whether such systems described anything in the real world.

In the early part of the nineteenth century the absolutist belief in liberal democracy as the best form of human government was widespread. Its establishment was seen as the culmination of human history. The most famous manifestation of this spirit of the age is probably to be found in the words of the American Declaration of Independence. As the century neared its end, there could be found a universal trend towards a new type of relativism, with which the downfall of Euclidean certainty struck a resonant chord. No longer was it universally believed that there was an optimal system of ethics, or code of laws, or norms of human behaviour. The example of non-Euclidean geometry showed that there could be axioms which would lead to a set of values counter to the accepted authority in any sphere of human affairs. Examples sprung up in political thought, in ethics, in law, and in sociology. The term 'non-Euclidean' came to be employed in non-mathematical fields as a label for unconventional, non-traditional, or radical thinking, as the message got through that logical self-consistency was necessary but not sufficient to demand adherence to an accepted way of doing things. We find articles with titles like '*Soundings in non-Euclidean economics*' and studies paralleling traditional Christian ethics with Euclideanism and its radical alternatives with the non-Euclidean paradigm. Anthropologists reassessed their attitude to the development of non-Western civilizations. Where once they had measured their merit by the closeness of their approach to the absolute 'ideal' displayed by the evolution of Western culture, now they recognized more forcefully the dubious nature of any such hierarchy of values. In the New World some anti-government commentators liked to refer to the American political and legal system as a 'Euclidean theory' because of its appeal to the 'self-evident' truths upon which it was founded.

Prior to the coming of non-Euclidean geometry, there was a unity, a confidence, and a certainty to our knowledge of the world. Afterwards, it was not enough to know that God is a geometer. The one unassailable truth about the nature of the physical world had been eroded and so, along with it, had centuries of confidence in the existence and knowability of unassailable truths about the Universe. How are the mighty fallen.

## LOGICS—TO BE OR NOT TO BE

*Logic is invincible because in order to combat logic it is necessary to use logic.*
PIERRE BOUTROUX

If Euclid's geometry was a cornerstone of the Universe, it was built upon that foundation stone of rationality which is logic itself. From Aristotle's first systematization of logical argument two thousand years ago, the laws of logic had been equated with the 'laws of thought'. This was never questioned, because the traditional laws governing the machinery of logical deduction which had been laid down by Aristotle are seemingly uncontentious. First, there is the law of *identity*: everything is what it is; that is, given some entity labelled by *A*, we have that *A* is *A*—'a spade is a spade'. Second, there is the law of *non-contradiction*: that is, both *A* and its negation, not-*A*, cannot be true—'you cannot be both dead and alive'. Finally, there is the 'principle of the excluded middle': that every statement is either true or not true—'you are either reading this book or you are not'. Just as Euclidean geometry had been distinguished from others by the assumption of its exact applicability to reality, so these rules were believed to be the description of the way of the world.*

But the transmogrification of Euclidean geometry into but one of many possible systems of geometry each distinguished by a particular set of defining axioms whose validity required only self-consistency undermined the status of simple logic as well. Even if experience did follow those three laws of simple logic, this did not endow them with an inviolate and special status in the eyes of the mathematician. Logic, like geometry, was to become divorced from physical reality: a complex suite of games played with symbols that created paper worlds of their own.

All of the three 'laws of thought' were challenged by logicians exploring new systems. But it was the third, the so called 'principle of the excluded middle', that came under the closest scrutiny. This assumption made simple logic a two-valued logic because every statement has two possible truth values: it is either true or false. This assumption lies at the root

* In a non-Western culture like that of the Jains in ancient India one finds a more sophisticated attitude towards the truth status of statements. The possibility that a statement might be indeterminate is admitted as well as the possibility that uncertainties exist in our analysis. These would correspond to statistical statements in which we simply give the likelihood that a certain statement is true or false. Jainian logic admits seven categories for a statement which reflect both its intrinsic uncertainty and the incompleteness of our knowledge of it: (1) maybe it is; (2) maybe it is not; (3) maybe it is, but it is not; (4) maybe it is indeterminate; (5) maybe it is but is indeterminate; (6) maybe it is not but is indeterminate; (7) maybe it is and it is not and is also indeterminate.

of many of the most famous mathematical proofs created by the early Greeks. It underwrites the technique of proof by contradiction (the *reductio ad absurdum*), in which one assumes that what one seeks to prove true is in fact false. From this one derives a logical contradiction—for example that 1 = 2—and hence we conclude that the original assumption one made cannot have been false. If the original statement is either true or false, then, since we have shown that it cannot be false without contradictions arising, it must be true. The essential role played by the law of the excluded middle is obvious. Without it, this means of proof fails. Such a method of proof has been compared to a gambit in the game of chess where a piece is risked in order to establish a greater advantage. The mathematician G.H. Hardy saw the *reductio ad absurdum* as 'a far finer gambit than any chess gambit: a chess player may offer the sacrifice of a pawn or even a piece, but a mathematician offers the *game*'.

In the early 1920s logicians like Jan Lukasiewicz, Emil Post, and Alfred Tarski showed that there could exist perfectly consistent logics which are not two-valued. They do not assume the validity of the law of the excluded middle. They were called three-valued logics (there are actually 3072 possible versions of them) because they allowed a statement to be either *true* or *false* or *undecided*.* Subsequently, higher-valued logics with more than three alternative truth statuses for any statement were shown to be consistent as well. And indeed, one could even invent logics in which every proposition may take on any one of an *infinite* number of possible truth values. These demonstrations showed, in the words of one of their creators that 'the law of the excluded middle is not writ in the heavens'. In 1932, as the dust settled, the American logician Alonzo Church drew out the significance of the creation of these new logics and the parallels with the creation of new geometries; it encapsulates the new attitude of mathematicians to the relativism of logics:

> We do not attach any character of uniqueness or absolute truth to any particular system of logic ... We may draw the analogy of a three dimensional geometry used in describing physical space ... there may be, and actually are, more than one geometry whose use is feasible in describing physical space. Similarly, there exists, undoubtedly more than one formal system whose use as a logic is feasible, and of these systems one may be more pleasing or more convenient than another, but it cannot be said that one is right and the other is wrong.

* Such a possibility seems to have been considered earlier but less systematically by William of Ockham in the fourteenth century and by the German philosopher Georg Hegel in the nineteenth century.

Yet the situation with regard to the undermining of the status of classical Aristotelian logic differs in interesting ways from the erosion of the status of Euclidean geometry. The demonstration that neither were necessary truths about the Universe did away with the only plausible candidates available to philosophers and theologians as examples of absolute truths about the Universe which our minds could discover. But nonetheless one could see that locally the world did obey the axioms of Euclidean geometry. Even though astronomical observations spectacularly confirmed Einstein's assumption that the geometry of space and time was a non-Euclidean one, it was still the case that as one surveyed smaller and smaller dimensions, the geometry of the world approximated that of Euclid to greater and greater accuracy. The situation with logics was not quite so transparent. Rather, it is positively incestuous. For while we do not need to assume any particular geometry applies to the real world in order to discuss the issue, we do need logic to talk about logic. All our discussions about which logic one might employ, for example whether one has a special status over all others or whether we are justified in assuming the law of the excluded middle is true, take place using the two-valued logic of ordinary parlance, which assumes it to be either true or false. Although some legal systems permit three different statuses for statements, all the legal argument *about* them employs traditional two-valued logic. Moreover, as we shall see later on, it is impossible to demonstrate the non-

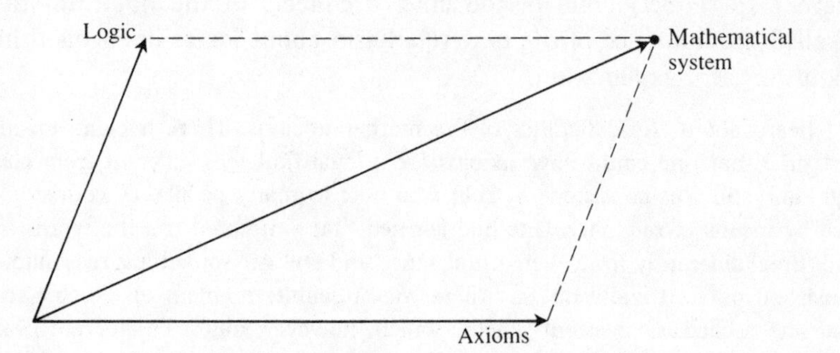


**Figure 1.1** We can envisage a mathematical system to be built upon two foundations: 'axioms' which are things that are assumed to be true, and 'logic' which is the rules by which one is allowed to deduce new truths from the axioms. We can think of the axioms and logic as two sides of a parallelogram and the resulting mathematical system as the resultant diagonal. If one alters either the side marked 'axioms' or the side marked 'logic', then a new parallelogram is produced with a different resultant diagonal. Moreover, we see that it is possible for the same diagonal to arise from quite different combinations of logic and axioms.

contradictoriness of a logical system using only the devices of the system itself. There does not appear to exist some overarching superlogic within which all the possible logics can be seen as particular or limiting cases.

There is an illuminating way of representing the impact of the changes that occurred at the roots of geometry and logic in a diagram which is displayed in Fig. 1.1. The picture shows how we can regard a mathematical theory as constructed from a collection of stated principles, or 'axioms', together with a logic which tells us how we generate further truths from these axioms. We can think of the axioms and logic adopted as being sides of a parallelogram which generate a particular mathematical theory represented by the diagonal of the parallelogram. With regard to the mathematical description of geometry, mathematicians at first thought the sides of our parallelogram were both fixed; then they considered changing the nature of the 'axioms' side. Only much later did they realize that one could tinker with the 'logic' side as well. We notice also that any particular line can be the diagonal for parallelograms with quite different sides.

The psychological effects of the realization that one could invent all manner of different, self-consistent forms of logical reasoning were deep and wide. They had a liberating effect upon thinkers struggling with problems that seemed to defy traditional forms of argument for their resolution. During the early 1920s physicists were wrestling with the first glimpses of the quantum character of the atomic world. Many years later, Werner Heisenberg, one of the chief architects of the quantum theory, recalled the influence of the new relativism about logics upon his thinking about physical explanation.

> I heard about the difficulties of the mathematicians. There it came up for the first time that one could have axioms for a logic that was different from classical logic and still was consistent ... That was new to many people. Of course it came also by means of relativity. One had learned that you could use the words 'space' and 'time' differently from their usual sense and still get something reasonable and consistent out ... I could not say there was a definite moment at which I realized that one needed a consistent scheme which, however, might be different from the axiomatics of Newtonian physics. It was not as simple as that. Only gradually, I think, in the minds of many physicists developed the idea that we can scarcely describe nature without having something consistent, but we may be forced to describe nature by means of an axiomatic system which was thoroughly different from the old classical physics and even a logical system which was different from the old one.

In the 1930s a further blow to human confidence in its mental certainties was dealt by Kurt Gödel's work. He showed that the traditional method of reasoning from consistent axioms according to specified rules of deduction was unexpectedly limited. If the axioms of the system were only moderately complicated, so as to include the possibility of doing ordinary arithmetic at the very least, then there would always be statements which one could make about the quantities defined by the axioms whose truth or falsity could never be demonstrated using its rules of logical reasoning. One says that the system is 'incomplete' and so necessarily contains undecidable statements. If one attempts to cure this incompleteness by adding some extra axioms to enable a particular undecidability to be resolved, then one always creates new undecidable statements. This result is now firmly established in our folklore and it is interesting to reflect on how it has led to a general feeling that it somehow establishes an intrinsic human fallibility. Before this result was established in the 1930s there was great confidence amongst mathematicians that one would be able to decide the truth or falsity of all statements by a systematic procedure which might even be mechanized. After Gödel's demonstration there was at first surprise by mathematicians that truths could lie beyond the reach of logical deduction. A long period then ensued until the early 1960s during which this result had little impact upon other areas of mathematics and science. Since that time it has been cited repeatedly, in different guises, as a fundamental barrier to human understanding of the Universe. A favourite target has been the quest to create a form of 'artificial intelligence'. First John Lucas, and more recently and in a different way Roger Penrose, have tried to argue that the theorems proved by Gödel limit the capability of any axiomatically programmed machine to reproduce the judgements and processes of the human mind. At this point we are not interested in the truth or otherwise of these arguments merely in the fact that they have been made. One finds a growing literature which draws from Gödel's results support for an intuitive belief (or hope?) that there must always be gaps in our knowledge of the physical world. In this vein it has been suggested that if we were to define a religion to be a system of thought which contains unprovable statements, so it contains an element of faith, then Gödel has taught us that not only is mathematics a religion but it is the only religion able to prove itself to be one.

## THE RASHOMON EFFECT

*A great truth is a truth whose opposite is also a great truth.*

CHRISTOPHER MORLEY

Akira Kurosawa's most famous film, *Rashomon*, presents a sequence of events which at first seems straightforward. A young woman is attacked as she travels through the woods with a party of friends and relatives. What follows is a series of accounts of events by all the participants. Each is different—very different—yet each is an eyewitness account of the same events. One expects these contradictions to be resolved in the end, leaving the story with a simple and single account of the truth—the absolute truth of what took place. But no such dénouement appears, and as the film ends one is left pondering the way in which our desire for an unambiguous view of the world is so strongly felt and so seldom satisfied. One of the ways in which we have tried to create such a simple view of reality is to represent the diverse aspects of reality in terms of symbols. We can choose as few symbols as we wish and associate only particular aspects of a complex phenomenon with some symbol. In this way very different things can be directly compared by reference to the different symbols that have been employed. The use of symbols, for example a national flag, is both a way of conveying rather precise information and a means of conjuring up all manner of other sympathies and associations in the minds of people who see it. Again, there is a widely held view that such diversity of interpretation is merely the hallmark of things that are rather vague and pre-scientific. If they can't be measured, then they don't exist. It is a consequence of a poor choice of symbols. A good choice, an unambiguous choice, needs to be made and when that is done we have certainty and rigour where before we had doubt and confusion.

Over a period of thousands of years different human civilizations have come to appreciate the utility of symbols to represent quantities. The 'numbers', as we now call them, served a long apprenticeship during which they were used for nothing more exciting than the tallying of objects or the recording of the days and the seasons and the passage of the time of day. Later, in very few places, something strange happened. These symbols were found to have a life of their own that dictated how they should be manipulated. And the ways in which they were used enabled new facts about the world to be predicted. The growth of Western civilization has proceeded hand in hand with the availability of this type of knowledge. It made possible the growth of science as we know it. Today we look back and see the effectiveness of 'number' as the symbolic language of Nature—the analogy that never breaks down—and we start asking: Why is the world like this and what is this strange language of mathematics that enables us to unpick and predict the workings of Nature? Is it merely a succinct summary of what actually happens, that we invent, or is it a part of reality that we discover? And if we do discover it, then does this mean that it exists independently of ourselves? If so, 'where' does it exist? Certainly not in the world of space and time in which we live, because that is itself describable by that very same mathematics. But if it exists in another realm, how is it that we make contact with this 'other' world of mathematical forms?

## THE ANALOGY THAT NEVER BREAKS DOWN?

*Mathematics is the handwriting on the human consciousness of the very Spirit of Life itself.* CLAUDE BRAGDON

When scientists attempt to explain their work to the general public they are urged to simplify the ideas with which they work, to remove unnecessary technical language and generally make contact with the fund of everyday concepts and experience that the average layperson shares with them. Invariably this leads to an attempt to explain esoteric ideas by means of analogies. Thus, to explain how elementary particles of matter interact with one another we might describe them in terms of billiard balls colliding with each other.* Near the turn of the century this practice provoked the French mathematician, Henri Poincaré, to criticize it as a very 'English' approach to the study of Nature, typified by the work of Lord Kelvin and his collaborators, who were never content with an abstract mathematical theory of the world but instead sought always to reduce that mathematical abstraction to a picture involving simple mechanical concepts with which they had an intuitive familiarity. Kelvin's preference for simple mechanical pictures of what the equations were saying, in terms of rolling wheels, strings, and pulleys, has come to characterize the popularization of science in the English language. But if we look more closely at what scientists do, it is possible to see their descriptions as the search for analogies that differ from those used as a popularizing device only by the degree of sophistication and precision with which they can be endowed. Just as our picture of the most elementary particles of matter as little billiard balls, or atoms as mini solar systems, breaks down if pushed far enough, so our more sophisticated scientific description in terms of particles, fields, or strings may well break down as well if pushed too far.

Mathematics is also seen by many as an analogy. But it is implicitly assumed to be the analogy that never breaks down. Our experience of the world has failed to reveal any physical phenomenon that cannot be described mathematically. That is not to say that there are not things for which such a description is wholly inappropriate or pointless. Rather, there has yet to be found any system in Nature so unusual that it cannot be fitted into one of the strait-jackets that mathematics provides.

This state of affairs leads us to the overwhelming question: Is mathematics just an analogy or is it the real stuff of which the physical realities are but particular reflections?

* One Hungarian physicist once remarked in the course of writing a textbook that although he would often be referring to the motions and collisions of billiard balls to illustrate the laws of mechanics, he had neither seen nor played this game and his knowledge of it was derived entirely from the study of physics books.

This leads us to our first glimpse of the mysterious foundation of modern science. It uses and trusts the language of mathematics as an infallible guide to the way the world works without a satisfactory understanding of what mathematics actually is and why the world dances to a mathematical tune. If you question a variety of scientists about their trust in mathematics, you probably won't get a very convincing answer. They will not have given it a moment's thought. But in general they work as though they discover mathematical truths about the world rather than invent them or simply organize them into a pattern that fits the available catalogue of mathematical notions. Mathematicians, likewise, will typically have ignored this awkward question unless they happen to work on those areas of pure mathematics that wander perilously close to the subjects of logic and philosophy. If you pressed the mathematician harder, you would get a more considered response than you did from the physicist. You would find him working from Monday to Friday as if mathematical ideas really existed independently of himself waiting to be discovered. But, questioned at the weekend, reclining more thoughtfully in his armchair he would find this view very difficult to defend and most likely plump for regarding mathematics as a sort of logical game which you explore using the stated

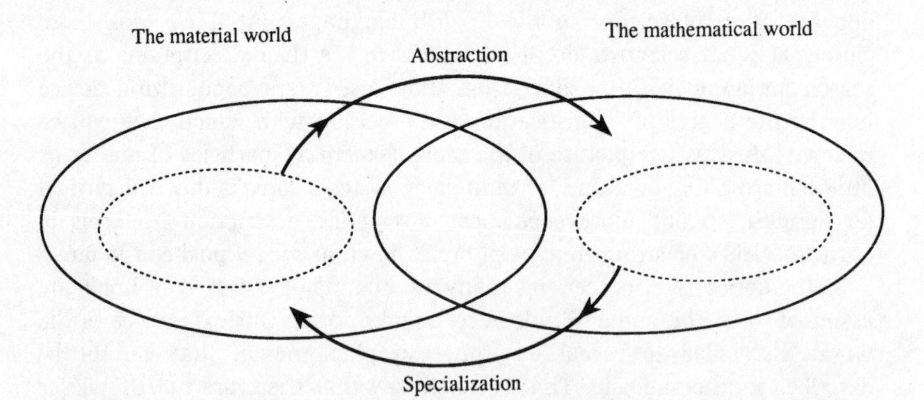


**Figure 1.2** The mysterious interrelationship between the material world of particular things and the mathematical world of abstract relationships, geometries, numbers, and logics. The things we see around us in the material world can act as a source of structures that can be abstracted into the mathematical world. Mathematical structures and relationships in the mathematical world can be specialized to particular examples of things and events in the real world. It is tempting to ask whether there are things in the physical world which cannot be abstracted in the mathematical world and if there may not be inhabitants of the mathematical world which have no concrete manifestations in our physical Universe.

rules. The fact that it works as a description of the world he would regard as a side issue that doesn't affect what he does. Nonetheless he would stress that he works as if he is discovering new truths. He acts as if mathematical objects exist already and perhaps believes that mathematics would surely exist in the absence of any mathematicians at all.

The dilemma can be represented in a simplified form by means of a diagram shown in Fig. 1.2. We recognize a 'real' world of particular objects and the interrelationships between them. We have been extraordinarily successful in understanding that world by use of what we recognize as an abstract mathematical 'world' of equations, numbers, geometries, algebras, and much, much more.

Parts of each world can be shifted to the other: from the real world to the mathematical world by the process of abstraction—as when we see a real pattern woven in the carpet and consider the collection of all possible patterns of similar type—and in the opposite direction by the process of specialization—when we take all possible geometries and pick out one that describes geometry on a flat surface. But our picture creates two interesting questions: Are there any parts of the real world which cannot be abstracted into the mathematical world, and conversely, are there any parts of the mathematical world which do not have any particular instantiation in the real world around us?

## TINKLING SYMBOLS

> *We can hardly imagine a state of mind in which all material objects were regarded as symbols of spiritual truths or episodes in sacred history. Yet, unless we make this effort of imagination, medieval art is largely incomprehensible.*
>
> KENNETH CLARK

A strange thing about people is that they are never content to take things as they come. Human consciousness acts as a buffer which cushions us from raw reality. It creates symbols or pictures of reality for us. This ability to represent the world in a succinct way is necessary for our evolution and survival. In order to learn from experience, we need to have ways of gathering and storing information about the environment in which we live. The fact that those ways and means are extraordinarily complicated is a reflection of the complexity of the environment in which we find ourselves.

As we observe and reflect upon the environment around us, so we find a use for symbolism. When we regard something as a symbol, we endow it with a plurality of meanings. For the initiated it acts as a store of information about things other than itself. This explains the usefulness and ubiquity of symbols of many sorts. When confronted with the bare facts of experience, our minds search for patterns and abbreviations of the information on offer. If the sequence of experiences is not random, then it will possess some pattern which permits the information within it to be represented in an abbreviated form without any degradation of its content. If no such mnemonic is available, the sequence of experiences would be regarded as random and it could not be communicated in any way except by a complete repetition. Science is the search for these abbreviations, or 'compressions', of the fruits of experience. This quest is remarkably successful, a fact that bears witness to the extraordinary compressibility of the Universe in which we live. Indeed, perhaps we could only survive in a universe that possessed a fair measure of compressibility which our brains could take advantage of to store information about the environment for future use. We can see why any such compression of experience relies upon those things that we call 'symbols'. Compressions seek to store lots of pieces of information in a single mark or representation. Symbols are the result of such a desire and mathematical symbols have proved to be the simplest and most succinct way of storing information.

In our everyday experience we encounter another type of 'symbolism' which seems at variance with the sense in which we have just presented it. Instead of using symbols, as mathematicians do, to make the complex simple, we see the use of symbols in art and religion, where the aim is to create a diversity of suggestive ideas by elaboration and extrapolation from a single seed. This contrast reminds us of a conversation that once took place between two great physicists. Paul Dirac, one of the creators of quantum mechanics, was famous for his prosaic manner. He was brief and perfectly logical, never using two words where one would do. J. Robert Oppenheimer was, by contrast, a gregarious polymath with a wide knowledge of literature and music. Indeed, T.S. Eliot wrote his play 'The Cocktail Party' whilst a visitor of Oppenheimer at his Institute in Princeton. Once Dirac asked Oppenheimer what he would have done in life had he not become a physicist. Oppenheimer replied immediately that he would have become a poet. Dirac was appalled. Finding this an inconceivable alternative, he replied, 'As a physicist I take what is complicated and make it simple. But the poet does the very opposite.'

## THINKING ABOUT THINKING

*An intellectual is a man who doesn't know how to park a bike.*

SPIRO AGNEW

Human development has two stages: in the first we think about things; in the second we begin to think about thinking. When we recognize that what we perceive is a representation of reality that could differ from it in essential ways we have invented philosophy.

Ever since ancient times philosophers have wrestled with the problem of the relation between the world as it really is and our perception of it. We have seen that mathematics is a peculiar part of the world in which we live but its exact nature is tantalizingly different from just about everything else we encounter. In the chapters that follow we shall explore some of the ideas that have been proposed to explain what mathematics is and why it describes the way of the world so widely and so successfully. We shall find reflected there many of the traditional alternatives that human minds have grasped when coming to grips with the problems of the nature and meaning of the world around us. There are those who seek to explain the problem away, concentrating instead upon strait-jacketing things within a rigid logical system which isolates them from the values and meanings in the outside world. Then, there are the 'doubting Thomases' who wish never to stray too far from what they can simply see and touch. They believe only in what they can make or measure. There are those who perceive only the fruits of their own minds; their world is of their own making. And then there are those who believe the ultimate nature of things resides in another world, a world whose contents we can by some means discover like explorers in the dark: for them 'pi' really is in the sky. Before we see what these searchers think about mathematical thinking, we need to look into the past to see just how 'natural' a human activity counting really was, and is; where and how it began and spread. To do this we must begin to uncover something of the origins of numbers in the dim and distant days before there were any mathematicians at all.

## CHAPTER TWO

# The counter culture

*Data without generalization is just gossip.* ROBERT PIRSIG

### BY THE PRICKING OF MY THUMBS

*The reasonable man adapts himself to the world; the unreasonable one persists in trying to adapt the world to himself. Therefore all progress depends on the unreasonable man.* GEORGE BERNARD SHAW

During the Second World War, in India, a young Indian girl found herself having to introduce one of her oriental friends to an Englishman who appeared at her home. The problem was that her girlfriend was Japanese and would have been immediately arrested were this to become known. So, she sought to disguise her nationality by telling her English visitor that her friend was Chinese. He was somewhat suspicious and then surprised them both by asking the oriental girl to do something very odd: 'Count with your fingers! Count to five!' he demanded. The Indian girl was shocked. Was this man out of his mind, she wondered, or was this another manifestation of the English sense of humour? Yet the oriental girl seemed unperturbed, and raised her hand to count out on her fingers, one, two, three, four, five. The man let out a cry of triumph, 'You see she is not Chinese; she is Japanese! Didn't you see how she did it? She began with her hand open and bent her fingers in one by one. Did you ever see a Chinese do such a thing? Never! The Chinese count like the English. Beginning with the fist closed, opening the fingers out one by one!'

We have all counted on our fingers, but perhaps, as our story reveals, there is more to be discerned from such dexterity than the limitations of childhood? This is how many of us begin to count when we are very young. Was it perhaps the way in which all counting originated?

Scientists are fascinated by 'origins'—the origin of life, the origin of the Universe, the origin of the solar system. In a sense this fascination is a mature version of those irritating sequences of 'why?' questions that inquisitive young children persist in asking. But the search for origins usually has another motivation besides the desire to trace everything back to the earliest possible state. Our discovery that the world operates according to a predictable pattern of 'causes' yielding definite consequences leads us to believe that a knowledge of how things might have originated will shed considerable light upon how they are now. So perhaps by looking at the possible explanations for the invention, discovery, or whatever, of numbers and other primitive mathematical notions, we might learn something interesting about the nature of these things. We are going to be interested only in the very simplest aspect of what is now called mathematics—counting and number. We want to look at how these notions may have arisen and evolved in human societies. One of the things that this may reveal is whether the common idea that the human mind possesses some universal natural intuition for number as part of its innate construction is a credible one. Alternatively, we might discover that very particular practical problems that mankind needed to solve resulted in the creation of a notion and method of counting that was attributable to the nature of those challenges rather than to anything intrinsic to the mind or deep properties of the Universe. We must try to decide whether the notion of number has grown out of our experience of the world or has merely been plucked from the recesses of the primitive human mind where it lay dormant, waiting to be nurtured by use.

One of the oddities about the origins of numbers and of counting is how little we find out about such things in the very places we would expect to learn most about them. No course of mathematics would ever consider the question. No history book delves into such matters. The origins of the Western world, the beginnings of civilization, or of culture and anatomy, of star-gazing and agriculture, or architecture and art, of all these things we find volume upon volume of evidence, speculation, and detail. But of the origin of counting there is next to nothing. We are going to take a look at the collection of facts that have been gathered up by anthropologists. Such a search may shed some light upon how the notion of number and the process of counting originated and spread.

In dealing with a problem like the origin and diversity of counting, one must be aware of some Victorian prejudices. In the mid-nineteenth century, anthropologists were content with a belief in the psychic unity of the human race and started from the premiss that, on the average, all human beings shared similar mental capacities. But this view was influentially challenged by the French anthropologist Lucien Lévy-Brühl, primarily through the publication of his book, *How natives think*, in 1910, where he argued for a sharp divide between the logical and the 'prelogical' mind. Native cultures were imagined to be universally prelogical and non-literate and unable to grasp simple logical connectives or develop the concept of number in the modern sense. Subsequent anthropologists steadily undermined such a simplistic picture which appears to over-emphasize the significance of technology as the hallmark of intellectual capability and was too ready to use the contrasting capabilities of many primitive peoples as an implicit argument for the superiority of modern European cultures. Yet despite Lévy-Brühl's prejudices his ideas were very influential in early studies of counting numeracy in native cultures and one must be aware of taking on board his assumptions without a more critical appraisal. We should stress that we are not studying the origins of counting as a means of assessing the capabilities of individual cultures; primitive or otherwise. Counting is but one activity of a society; the extent to which it was developed depends upon the needs and circumstances of its perpetrators, not simply on the power of their imaginations. Moreover, when any culture does something for the first time, whether it be the making of tools or the manipulation of numbers, it is going to produce a result that appears clumsy or simple-minded when viewed from the vantage point of the future.

## THE BARE BONES OF HISTORY

*The world must be getting old, I think; it dresses so very soberly now.*

JEROME K. JEROME

We are about to take a tour through the ancient world to seek out what is known about the origins of counting; to determine what can be deduced about people's natural, or unnatural, aptitude for things numerical. Whilst modern mathematics in general, and those parts of it which find felicitous application in the realms of science in particular, are considerably more exotic than the process of mere counting, counting is undeniably a necessary precursor to these more sophisticated enterprises of the mind. But before we embark upon this quest, we should plant a few markers to give some cultural context to what follows.

Mankind is the only living species on planet Earth that has a collective conscious memory. This memory, our history, as we have come to call it, is both the hallmark of civilization and the means by which we come to know of it. The earliest human ancestors have been on Earth for more than two million years, but our own species, *Homo sapiens*, appears to have been around for a mere hundred thousand of those years. The earliest traces of fortified buildings stretch back to about 7000 BC in Jericho, but if we seek evidence of 'civilization' in the form of social organization at a fixed location involving some form of central administration, rules, and rulers, that amounted to something more than just large numbers of people living in close association then there are a very small number of ancient centres. Within these centres the most significant

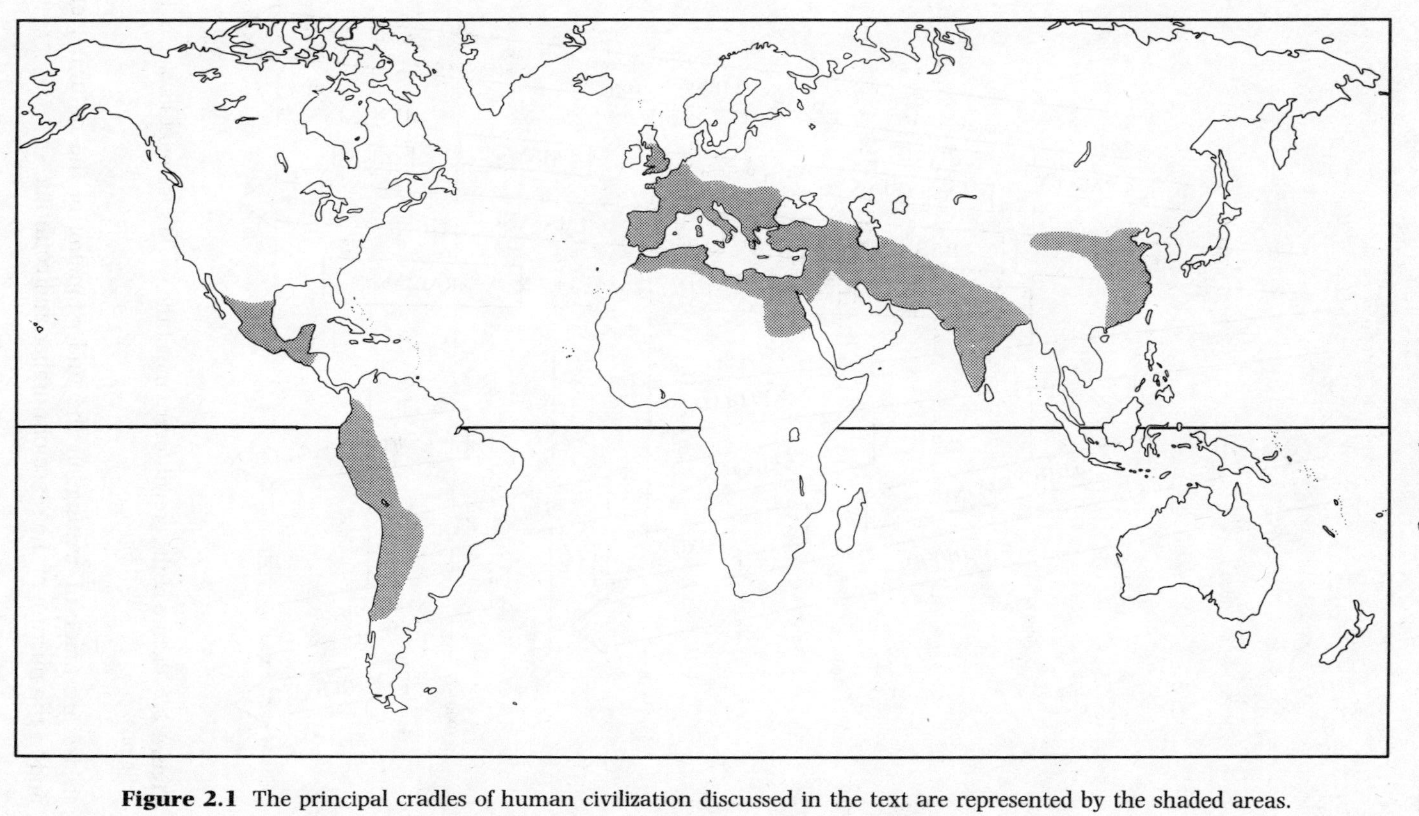

**Figure 2.1** The principal cradles of human civilization discussed in the text are represented by the shaded areas.

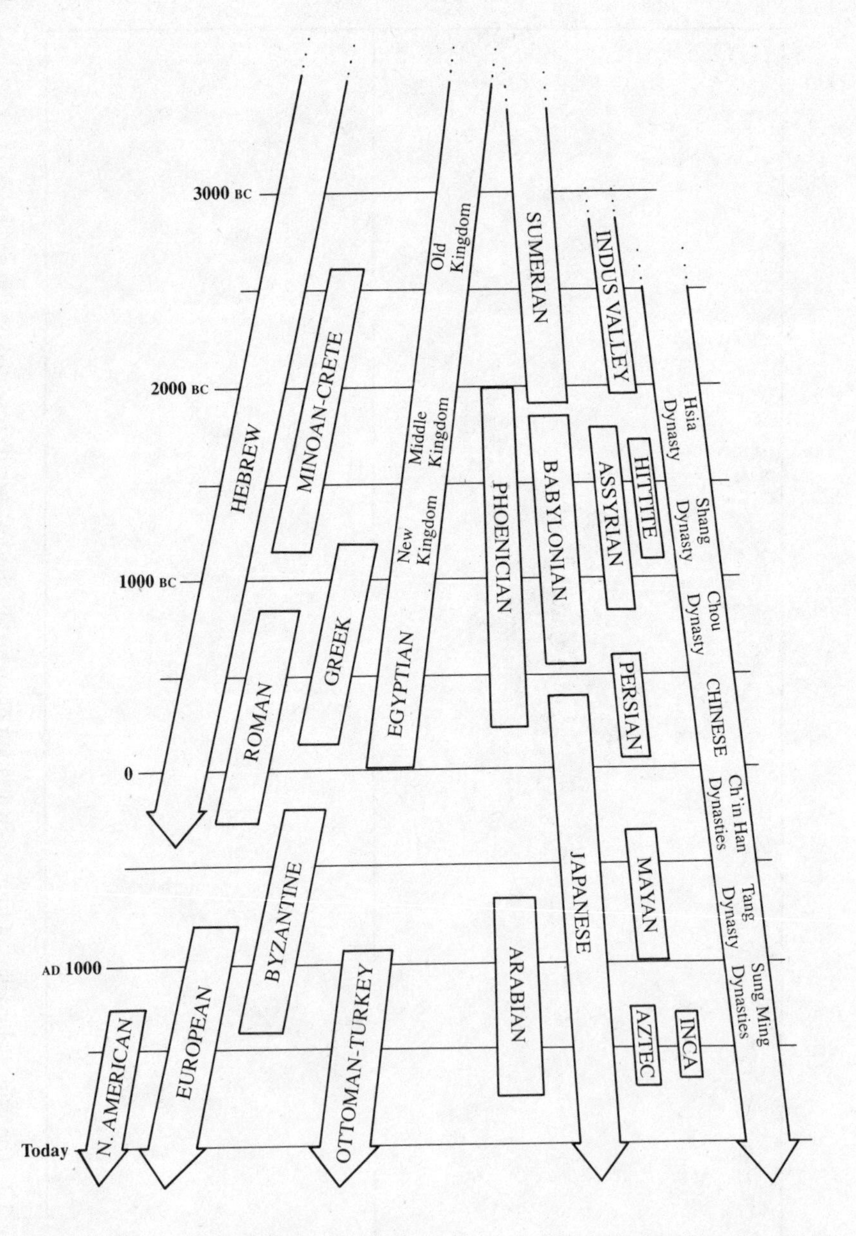


**Figure 2.2** Picture of the growth over time of the world's foremost ideologies and religions.

strides are made in writing, in the understanding of the natural world, and in the nurture of the sense of number and counting.

The oldest of these cultural centres are those which appear to have emerged around 3000 BC in Egypt and Sumer, some in the Indus valley region that we now call India around 2800 BC, those on the island of Crete, where the Minoans were to be found from 2000 BC, and the Shang civilization of China which predates 1500 BC. In the Americas things began later with the Olmecs flourishing in Mexico around 1000 BC, and the Chavin of Peru in 900 BC. The geographical distribution and development of the principal human civilizations is shown in Fig. 2.1 and 2.2. But the most ancient evidence of human counting is to be found in the remains of smaller groups of hunters and gatherers who existed long before any of these great centres of civilization.

In 1937 the bone of a young wolf was discovered at Věstonice in Czechoslovakia (see Fig. 2.3). It dates from about 30 000 BC and is now kept in the Moravské Museum in Czechoslovakia. It is about seven inches long and displays a line of fifty-seven deep notches. The first twenty-five are possibly grouped in fives and are all the same length. This set is ended by a single notch that is twice as long as the others. A new series of thirty short notches then begins preceded by another long notch. This extraordinary relic from palaeolithic times displays systematic tallying and perhaps some appreciation of grouping in fives, inspired no doubt by the fingers on the hunter's hand. Presumably the tally was a record of the hunter's kills. We know very little else about the civilization that created this recording device except that they did make artistic images. A small carved ivory head of a woman was found at the same site. But what these intuitions of art and counting amounted to we shall probably never know.

In Africa similar artefacts of far greater antiquity have been unearthed. In the Lebembo Mountains bordering Swaziland the small part of a baboon's thigh bone, dating from about 35 000 BC, has been discovered with twenty-nine notches engraved upon it. This is the earliest known tallying device. But it is far from being the most intriguing. That distinction must surely fall upon the 'Ishango bone' found at Ishango

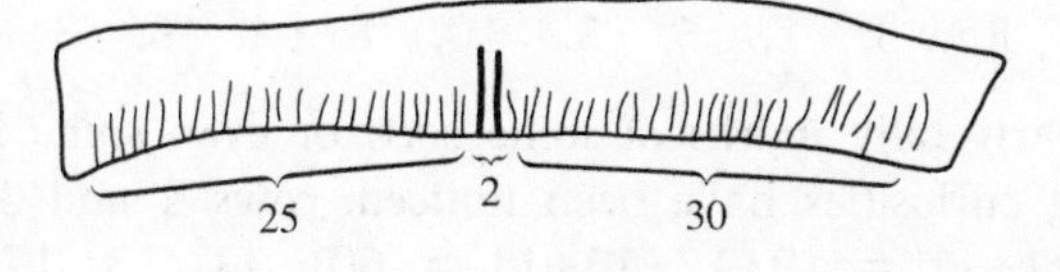


**Figure 2.3** Pattern of incisions found on the palaeolithic wolf bone discovered at Věstonice by Karl Absolon in 1937. They were made in about 30 000 BC.

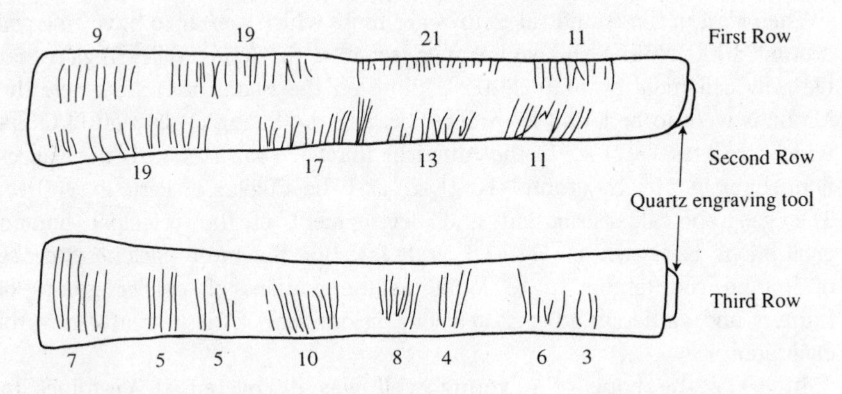


**Figure 2.4** Views from each side of the bone tool handle found by Jean de Heinzelin at Ishango near Lake Edward in Africa. The right-hand end would originally have held a larger quartz tool. Markings are found in suggestive groups in three rows. They were made in about 9000 BC.

by Lake Edward in the mountains near the borders of Zaire and Uganda by the Belgian archaeologist Jean de Heinzelin. For a period of just a few hundred years around 9000 BC the shores of Lake Edward supported a small community who farmed and fished. Eventually, they were eclipsed by a volcanic eruption. Their legacy is a collection of remains displaying their tools and fishing harpoons. Among the remains found at Ishango was a bone handle into the end of which was tightly fixed a quartz tool-head, perhaps used for engraving or cutting. The bone is now discoloured and solidified by the chemical effects of thousands of years of exposure, but there remain three rows of notches cut down different sides of the bone. Sketches of the bone from photographs of the three side views are shown in Fig. 2.4. The notches are grouped in tantalizingly suggestive ways that have provoked those anxious to support an African intellectual genesis to make elaborate speculations. The facts are as follows. The notches are arranged along each row in groups:

Row 1: 9, 19, 21, 11
Row 2: 19, 17, 13, 11
Row 3: 7, 5, 5, 10, 8, 4, 6, 3

These are clearly tally marks of some sort, or even some form of game. The following curiosities have been noticed: rows 2 and 3 both sum to 60, (9+19+21+11 = 19+17+13+11 = 60); 11, 13, 17, and 19 are sequential prime numbers (as indeed are 5 and 7 if one reads around the end of row 3 on the bone on to row 2)! The first row displays the counting pattern 10+1, 20+1, 20–1, 10–1. In the third row the close grouping of 3 and 6, 4 and 8, and 10 and 5 seems to indicate some notion of doubling or duplication. On this evidence de Heinzelin concludes that these people had rudimentary notions of prime numbers, counting based upon 10 and 20 (no doubt taken from counting their fingers and toes), and a system of doubling. This seems a little far-fetched. A more interesting line of enquiry is to take the hint supplied by the two 60's and believe them to denote two lunar months. We know that the Ishango environment undergoes very strong seasonal variations, which would have induced its original inhabitants to become semi-nomadic, moving between the high and low ground as the lake's waters rose and fell with the dry and rainy seasons. It is a reasonable supposition that these people would have shared mankind's universal fascination with the heavens and have taken note of the phases of the Moon. However, the sum of the remaining row of tallies is not 60; it is 48. This seems to scotch the idea that the tallies were of lunar phases; but some scholars who have subjected the bone handle to microscopic analysis have claimed that there are further indentations on the bone that are now only visible under the microscope which render it possible to link the entire sequence of gradations with a record of the phases of the moon. But the evidence is not very compelling.

## CREATION OR EVOLUTION

*History is a nightmare from which I am trying to awake.* JAMES JOYCE

Is counting instinctive? It is such a familiar and all-encompassing aspect of human thinking that there is a tendency for us to regard it as an ability so simple and obvious that it must have been invented by every civilization the world has ever known. If they wanted to do anything at all—to farm, to trade, to check that no one had stolen one of their cattle—surely they would need to count. This line of reasoning nudges one towards the view that counting was an activity that was invented over and over again in each and every primitive society. The alternative view that we might take is that counting is a rather difficult concept to hit upon and was only invented in a small number of great cultural centres and then spread from those into other less-advanced cultures who interacted with them. If the idea of counting diffused in this way from one or two sources, rather than springing into being all over the world, it alters our judgement regarding the ubiquity of human mathematical intuition about numbers, which some mathematicians have taken as a basis for the entire philosophy of mathematics.

Despite our feeling that counting is the most elementary of intuitions that we assimilate effortlessly just by being in the world, there are many primitive human groups who cannot count beyond two and have no developed number-sense at all.* Some Australian aboriginal tribes only possess words for the quantities 'one' and 'two'. All greater quantities are expressed by a word with the sense of 'many'. Some South African tribes exhibit a similar language structure for number words. In fact, there is even a vestigial remnant of it in modern Indo-European languages where we find that the original root word for 'three' has a meaning of something like 'over' 'beyond', or 'afar', indicating that the early sense was merely that of something more than the particular words for one and two. In Latin, this affinity exists between *trans* meaning 'beyond' and *tres* meaning 'three'; the same relationship is manifest in French where we find *très* for 'very' and *trois* for the number 'three'. Presumably, only later did words emerge to distinguish the various different varieties of 'many'. Some languages, for example Arabic, also retain a threefold treatment of quantity with differentiation between singular, dual (for two only), and plural (for more than two). Some Oceanic tribes even began conjugating and declining words in the singular, the dual, threefold, fourfold, and finally in a plural for all larger quantities.

Early anthropological studies of the Kalahari Bushmen in the first half of this century found an interesting state of affairs that was a little further advanced. Words existed for quantities up to five but not beyond, and those questioned were unable to describe quantities greater than five. In New Guinea, parts of South America, and Africa this limited system has been developed slightly by the device of repetition, so that, whilst there are no words for numbers greater than two, larger numbers can be expressed by juxtaposition as 'two-one', 'two-two', 'two-two-one'. The Aranda aboriginals in Australia combine this device with the one-two-many structure, using only two basic number words, *ninta* for 'one' and *tara* for 'two', so 'three' is denoted by *tara-ma-ninta* (i.e. 'two-and-one') and 'four' by *tara-ma-tara* ('two-and-two'). All greater quantities are referred to as 'many.' A further nuance is provided by the Botocoudo Indians of Brazil, who have words for numbers up to four, but then use a word meaning 'a great many' and point to the hairs of their head when they say this word to stress the notion that beyond the number four things become as practically innumerable as the hairs on their heads.

* Australian friends tell me that, while there still exist tribal groups with little or no well-developed numerical sense or tradition, nonetheless on occasions when children from these groups receive modern educational training their mathematical aptitude often displays no significant differences from that of other children with more sophisticated cultural backgrounds.

This method of counting is called the '2-system'. One should compare it to that which we use today which is founded upon the base 10, so that we have distinct words for numbers up to and including ten and then we compose ten-one (which we term 'eleven'), ten-two (twelve), ten-three (thirteen), ten-four (fourteen), up to ten-nine (nineteen) and ten-ten (twenty), before continuing with ten-ten-one which we call 'twenty-one'.

One might at first wonder why, having adopted the device of joining 'one-two' and 'two-two' to capture the ideas of 'threeness' and 'fourness', the perpetrators did not simply continue indefinitely, say, expressing five as two-two-one, six as two-two-two, and so on. But that would require a big leap. The quantities 'two-one' and 'two-two' are seen merely as pairings of separate things, not as the whole numbers three and four in themselves as we view them. When only singlets and pairs are identified, it is not possible to compose threefold combinations like 'two-two-two' without a major conceptual leap which changes one's picture of what all these compounded quantities really are. They have to become more than mere combinations before combination is unlimited.

But those cultures which count only up to some small number like two, three, or five cannot really be said to be counting at all. All they have are adjectives which describe certain states of affairs. They have no systematic way of describing an ascending set of quantities because they see no sequential relationship between the quantities involved. This is a systemless description of quantities using number-words and is rather different to the process of *counting* which implies the recognition of a relationship between different quantities. This recognition must be present whether one elects to count in ones, twos, or some other sequence of increments.

To obtain some feeling for what life is like if one has a conception of 'oneness' and 'twoness' but nothing more, one might consider the following vivid account of the world of the Damaras in Southern Africa which is taken from Francis Galton's turn-of-the-century account of his early contacts with these people (it also reveals something of their ingenuity in overcoming their conceptual limitations):

In practice whatever they may possess in their language, they certainly use no greater number than three. When they wish to express four they take to their fingers, which are to them formidable instruments of calculation as a sliding rule* is to an English schoolboy. They puzzle very much after five, because no spare

* Young readers who do not know what a 'sliding rule' or 'slide rule' is should ask someone over the age of thirty-five.

hand remains to grasp and secure the fingers that are required for units. Yet they seldom lose oxen; the way in which they discover the loss of one is not by the number of the herd being diminished, but by the absence of a face they know. When bartering is going on each sheep must be paid for separately. Thus suppose two sticks of tobacco to be the rate of exchange for one sheep, it would sorely puzzle a Damara to take two sheep and give him four sticks. I have done so, and seen a man take two of the sticks apart and take a sight over them at one of the sheep he was about to sell. Having satisfied himself that one was honestly paid for, and finding to his surprise that exactly two sticks remained in hand to settle the account for the other sheep, he would be afflicted with doubts; the transaction seemed to him to come too 'pat' to be correct, and he would refer back to the first couple of sticks, and then his mind got hazy and confused, and wandered from one sheep to the other, and he broke off the transaction until two sticks were placed in his hand and one sheep driven away, and then the other two sticks given him and the second sheep driven away.

## THE ORDINALS VERSUS THE CARDINALS

*All are lunatics, but he who can analyze his delusion is called a philosopher.*

AMBROSE BIERCE

The extent to which numbers are used as labels rather than seen as items in a sequence that can be created by a process of addition is revealed by an examination of the words used to express cardinal and ordinal aspects of numbers. The cardinal aspect refers to the total number in some collection or set of things, for example 'one', 'two', and so on. The ordinal aspect refers to the order of rank, for example 'first', 'second', and so on. We often use a symbol which can mean either of these interchangeably. If we pick up a pack of computer disks with '6' printed on it then we understand that to mean there are six disks in the collection, whereas if my aeroplane seat has the number '6' on it this refers to only one seat—the sixth seat in a sequence of seats in the plane. In the first case we see the cardinal aspect of number; in the latter, we see the ordinal aspect. The recognition of this distinction is a fundamental one in the evolution of human thinking about enumeration—the first step towards an abstract view of the essence of counting as the relationship *between* quantities rather than merely a listing *of* quantities. It is interesting to search for any linguistic residue of an early failure to see numbers in both an ordinal and a cardinal context. We can do this by looking at the words used for 'one' and 'first', and so on, in several languages to see if, as in English, the words are etymologically quite distinct. Here is the situation in English, French, German, and Italian,

English: one/first two/second three/third four/fourth
French: un/premier deux/second or deuxième trois/troisième quatre/quatrième
German: ein/erste zwei/ander or zweite drei/dritter vier/vierte
Italian: uno/primo due/secondo tre/terzo quattro/quarto

In each of these four languages the words for 'one' and 'first' are quite distinct in form and emphasize the distinction between solitariness (one) and priority (being first). In Italian and the more old-fashioned German and French usage of *ander* and *second*, there is also a clear difference between the words used for 'two' and 'second', just as there is in English. This reflects the Latin root sense in English, French, and Italian of being second, that is, coming next in line, and this does not necessarily have an immediate association with a total of two quantities. But when we get to three and beyond, there is a clear and simple relationship between the cardinal and ordinal words. Presumably this indicates that the dual aspect of number was appreciated by the time the concepts of 'threeness' and 'fourness' had emerged linguistically, following a period when only words describing 'oneness' and 'twoness' existed with greater quantities described by joining those words together as we described above.

In all the known languages of Indo-European origin, numbers larger than four are never treated as adjectives, changing their form according to the thing they are describing. But, numbers up to and including four are: we say they are 'inflected'. This linguistic phenomenon indicates a very close connection between the noun and the adjective. It is also a rather antiquated structure that barely survives in the modern forms of many Indo-European languages. For example, in French we find two words *un* and *une* corresponding to the English 'one' and they are used according to the gender of what is being counted. An analogous feature of language that certainly survives in English is the way in which different adjectives are associated with the same quantities of different things. We speak of a pair of shoes, a brace of pheasants, a yoke of oxen, or a couple of people, but we would never speak of a brace of chickens or couple of shoes. Numbers are also linked with the forms of the things being counted as we speak of one 'woman' but of two 'women'.

We have seen that the distinction between cardinal and ordinal aspects of number and the use of inflected adjectives is clear up to the number four but conflated beyond that.* Can we understand why this might be so?

* In Finnish there are still two kinds of plural, as in classical Greek, Biblical Hebrew and Arabic: one for two things and another for more than two. Also interesting in this respect is

Why did the divide occur at 4 and not 6 or 3? One proposal is that it is simply a reflection of the presence of four fingers on our hand. The thumb was clearly different in appearance and position and was not regarded as being equivalent to the other four fingers. This idea is supported by the existence of a measure called a 'handsbreadth', equal to 4 fingers, in many ancient civilizations. The Greeks and Egyptians compounded it into the 'ell', which was equal to six handsbreaths, twenty-four fingers; the Roman foot was composed of four 'hands'; this was in turn equal to $4 \times 4 = 16$ *digiti*, or fingers, from which we get our word for digit, which is still used to mean either a finger or a single-figure number between 0 and 9. In some cultures where 4-counting is found, it turns out that they did not begin by counting the fingers of their hands but rather counted the spaces between the fingers. Moreover, in some languages, the word for a number like 6 or 8 has a dual ending, indicating that it was conceived originally as 'two-threes' or 'two-fours' rather than as a single entity like the words for the quantities from one up to four. Although counting to the base 4 is not common today, there are at least two known cases amongst the Afudu and Huku tribes of Africa, and some others in South America. A curious speculation arises from these cases and others which evolved from them to give special status to the number 8—the total number of fingers excluding the thumbs—that many known languages originally possessed a base-8 system[†] (which they later replaced by something better), because the word for the number 'nine' appears closely related to the word for 'new' suggesting that nine was a new number added to a traditional system. There are about twenty examples of this link, including Sanskrit, Persian, and the more familiar Latin, where we can see *novus* = 'new' and *novem* = 'nine'.

These linguistic relics remind us of an early stage in our cultural development when number words and number sense were contextual. There are many primitive languages where we find a variety of words for, say, five things. There is a word for five flat stones, another for five smooth stones, another for five fingers, and another still for five people. Whilst the affinity between different stones, or between different people is clearly seen, the affinity between different collections of fives is not.

the fact that there is no connection between the words for '2' and '$\frac{1}{2}$' in the Romance and Slavic languages (nor in Hungarian which is not an Indo-European language) but in all the European languages the words for '3' and '$\frac{1}{3}$', '4' and '$\frac{1}{4}$' and so on, are closely related, just as they are in English. This may indicate that the concept of a fraction, or the relation between a number and the concept of a ratio, only emerged after counting beyond 'two'.

[†] The Cumus Indians of California count to 16 using base 4 but then adopt base 20 for counting larger numbers. The Yuki Indians extended a base 8 system derived from the number of gaps between the fingers of both hands by placing groups of twigs between the gaps. In this way their system was used to count to 64.

| No. | Oral Counting | Flat objects | Round objects | Men | Long objects | Canoes | Measures |
|---|---|---|---|---|---|---|---|
| 1 | gyak | gak | g'erel | k'al | k'awutskan | k'amaet | k'al |
| 2 | t'epqat | t'epqat | goupel | t'epqadal | gaopskan | g'alpēeltk | gulbel |
| 3 | guant | guant | gutle | gulal | galtskan | galtskantk | guleont |
| 4 | tqalpq | tqalpq | tqalpq | tqalpqdal | tqaapskan | tqalpqsk | tqalpqalont |
| 5 | ketōne | ketōne | ketōne | keenecal | k'etoentskan | tetōonsk | ketonsilont |
| 6 | k'alt | k'alt | k'alt | k'aldal | k'aoltskan | k'altk | k'aldelont |
| 7 | t'epqalt | t'epqalt | t'epqalt | t'epqaldal | t'epqaltskan | t'epqaltk | t'epqaldelont |
| 8 | guandalt | yuktalt | yuktalt | yuktleadal | ek'tlaedskan | yuktaltk | yuktaldelont |
| 9 | ketemac | ketemac | ketemac | ketemacal | ketemaetskan | ketemack | ketemasilont |
| 10 | gy'ap | gy'ap | kpēel | kpal | kpēetskan | gy'apsk | kpeont |

**Figure 2.5** Table of number words used by the Indian tribes in British Columbia, Canada, who spoke the Tsimshian language. The information was gathered by the American anthropologist Franz Boas and was first published in 1881.

Franz Boas, one of the pioneers of systematic anthropology, presented the results of his studies of the northwestern tribes of Canada to the British Association for the Advancement of Science in 1889 and claimed that the Tsimshian language of British Colombia contained no less than seven distinct systems of number words to be used according to the nature of the things being counted. These are shown in Figure 2.5. The first column displays the words used when no particular object is being referred to, or when numbers are talked about. The remaining columns give the words used for counting inanimate objects of different shapes and sizes where one finds several hybridizations of *kan*, the word for 'tree'. In the last two columns there are words for the counting of canoes and for measures where, in the latter case, one finds adaptation of *anon*, the word for 'hand'. This is an example of an ancient language which possesses words for individual members of collections, or for collections of particular things, but not words for the general idea of a collection, a set, or a quantity.

We have distinguished between mere number sense and counting. At first we might think that one could not have the first without the second; but our minds clearly allow us to entertain an intuitive feeling for small quantities divorced from any conscious notion of counting. We know that if we glance at collections of one, two, three, four, or even five objects then we can instantly apprehend the number present. But when there are more objects present we lose this facility and have to resort consciously to 'counting' the things present if we want to know how many there are. As an exercise the reader might care to glance repeatedly at Fig. 2.6. You will find that you can immediately sense how many objects are present in the smaller collections, but not in the larger ones.

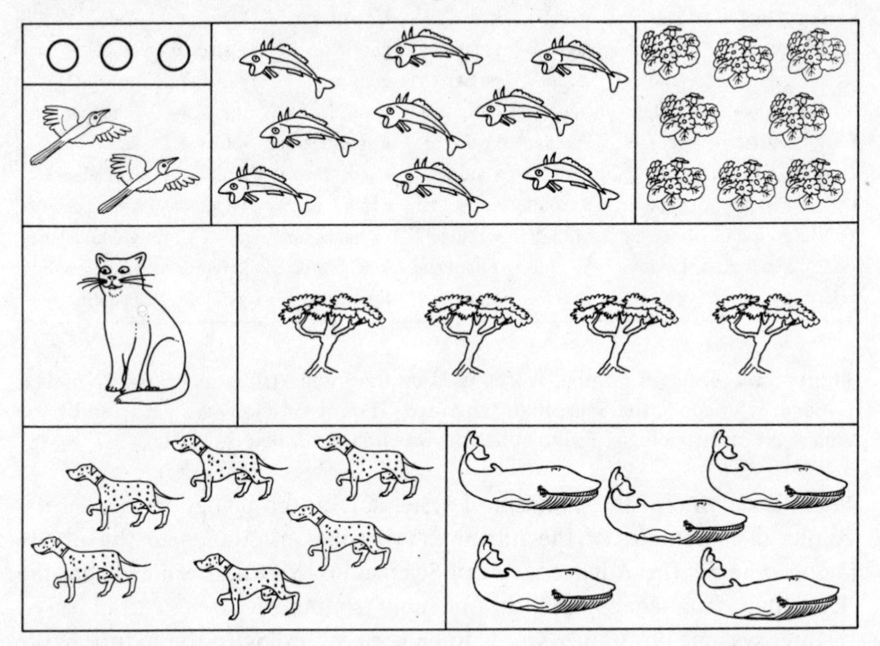

**Figure 2.6** Some small and large collections of objects. If you glance quickly at this picture and turn away at once you will find that you are generally able to remember how many objects there are in the collections of 1,2,3,4, or 5 objects. Beyond 5 the mind does not seem able to register the quantity of objects unless one makes a conscious effort to count, or the collections are arranged in convenient subsets. A similar threshold is found with the number sense of some animals and birds.

Small animals share this 'feeling' for number and will readily detect if one of a sufficiently small number of pieces of food or members of their young are missing. The process appears to be characterized by an act of comparison between two collections so that any difference is recognized. This is an important realization because it enables us to understand the way in which this sense then graduated to become something more systematic in several innovative cultures. We find the process of mental comparison becoming increasingly organized. There are examples of many different means by which this can occur. Some turn out to be unwieldy and impractical in the long run, whilst others turn out to be easily generalizable in such a way that they encourage their practitioners to take the greatest step of all: the recognition of an abstract concept of number divorced from collections of particular objects.

This focus upon our intuitive grasp of quantity provokes us to turn the spotlight upon ourselves a little more searchingly. It is easy to regard our own number sense as superior to those of the ancients because we are familiar with very large numbers. We think nothing of dealing with millions or billions of dollars or people. But just because our system of recording numbers allows us to express numbers of unlimited size very efficiently, this does not mean we have any intuitive comprehension of these large quantities. Very few people, if any, have a natural intuitive feeling for the difference between a million and ten million. The only way that sense arises is by thinking of it as a smaller number of visualizable collections, like populations of a typical city, audiences at a sports stadium, or mourners at an Indian public funeral. Upon investigating the natural numerical sense of the average twentieth-century person, one discovers that lurking behind the inherited scheme for writing large numbers is a very limited feeling for what the number symbols are conveying about quantity.

## COUNTING WITHOUT COUNTING

*Lecturing is the transfer of information from the notes of the lecturer to the notes of the student without passing through the minds of either.*

GENERATIONS OF STUDENTS

We can compare two collections of things, like the seats on a bus and the people who want to sit on them, and determine whether they are equal without resorting to counting. We do this by a correspondence principle which assigns a seat to a person until one or other collection is exhausted. This procedure can be found at the root of the first counting technique used by most primitive cultures who developed a practical number sense. They kept track of quantity by matching one collection of things with another. On returning home from the fields in the evening, shepherds could check that they had the same number of sheep as they set out with in the morning by setting aside one stone for each sheep. On returning in the evening, they could retrieve one stone for each sheep that returned safely to the fold. If any stones remained unassigned, then some sheep had got lost. In this way a large flock can be tallied without any sense of large numbers or systematic counting being available. Unfortunately, whilst this procedure gives one a sense of more, less, and equal, it is of no help if one wants to know how many sheep one has. If you want to discover if someone else has more sheep than you do, then one would have to compare physically the sheep or the collection of tallied stones with his sheep or his stones. This is very cumbersome and stones are even easier to lose than sheep. One way in which you could make it more efficient is to create some special collections for comparison. Suppose one picked on three-leafed clovers for this purpose and tallied a collection of fifteen sheep as five clovers. Other model collections are available, but it is hard to find any occurring naturally which have more than four or five members. So, while this strategy produces some reduction in the quantities to be tallied, this only helps up to a point; if one wishes to build up a procedure for tallying much larger numbers, some more flexible form of tallying* must be found.

Before we look at what that was, it is interesting to see again how a residue of these things remains in some of our modern words. The use of stones for tallying is seen in the root of our word 'calculate' which derives from the Latin *calculus* meaning a 'pebble'. Even in comparatively modern times one finds the very simple use of stones to enumerate large collections. Georges Ifrah writes about a Madagascan military roll call:

> In Madagascar, not so long ago, military leaders used a practical method to determine how many soldiers they had: they made them walk through a narrow passage in single file, and each time a soldier passed, a pebble was dropped on the ground. When there was a pile of ten pebbles, it was replaced with one pebble that was the beginning of a pile representing tens. The process was repeated till the tens pile contained ten pebbles, then another pile was begun for hundreds; and this continued till all the soldiers had been counted.

There are other curious associations between counting, ritual, and the use of stones in some cultures. Many ceremonies exist in which the participants carry stones or present them for counting. Other ancient records tell of the practice of piling up stones as a memorial or 'cairn' (which is Gaelic for a 'heap of stones'). In the book of Genesis we read of Jacob and Laban's covenant which was marked by the gathering and heaping of stones by Jacob and his brothers. There are creation rituals in which creation is symbolized by the casting and breaking of stones. There are several Central American legends which look to the origin of Mankind in stones. These mysterious associations are clearly deep and wide because

* Our word 'tally' derives from the old French *taille* meaning a cut or a notch; it has evolved into the modern French *tailler* meaning to cut, from which we obtain *tailleur* and hence our own cutter of cloth, the 'tailor'.

they are reflected in the Greek word for people, *laoi*, which is derived from *laas*, the word for 'stone'. From this we obtain the English word 'laity' which is used to describe people in the Church who are not members of the clergy.

In 1929 a remarkable archaeological discovery was made in the ruins of the palace of Nuzi, a city of the fifteenth century BC, southwest of Mosul in modern Iraq. A small rounded clay container was found with a cuneiform inscription on the outside, which was translated to read:

> Objects concerning sheep and goats
> 4 male lambs
> 21 ewes that have lambed   6 female goats that have kidded
> 6 female lambs   1 male goat
> 8 adult rams   2 kids

The total is therefore 4+21+6+6+1+8+2 = 48 animals. When the container was unsealed it was found to contain forty-eight clay balls. Later, other evidence came to light which revealed the purpose of the clay balls. One of the servants attached to an expedition in the region was sent to market to buy some chickens. He returned and put them in the poultry pen, but without counting them first. The servant was uneducated and did not know how to count, so he could not say how many chickens he had bought. But he needed to be reimbursed for their purchase and so he showed the expedition leader a collection of pebbles that he had set aside at the market: one for each chicken.

Study of the ancient container, which is now in the Harvard Semitic Museum, indicates that it belonged to an accountant who was responsible for his master's animals. He gave them to the shepherds—who unlike him could neither read nor write—to take out to pasture. The inscription on the container was a record for him of what animals they had taken and the clay balls served the shepherds as a check on the number of animals they had to return. The object was thus a form of receipt: its inscription for the owner, its contents for the shepherds.

The Pueblo Indians in South America exhibit a more sophisticated approach to tallying large quantities by using different objects in different orientations or patterns to denote different quantities. This creates a virtually unlimited potential for extension of the counting system but, again, places unreasonable strains upon the memory if large numbers are to be employed. Moreover, many of the positional nuances exploited in such systems are very slight, relying upon different positions of the hand or the fingers and can easily be mistaken or obliterated by an individual's manual quirks.

The ubiquity of tallying means that we might expect many languages to have employed words for numbers which describe the objects or collections of objects which enshrine the number being denoted. But whilst such traces can be found in primitive and ancient languages, no traces appear to remain in the modern languages of Indo-European origin. The only possible exception is the case of the word for 'five', which in some languages is still traceable to words for the 'hand': in Russian one sees this association between *piat* meaning 'five', and the word for an outstretched hand, *piast*. This lack of linguistic memory of the original tallies is a surprising state of affairs because the words employed for numbers in Indo-European languages have remained extremely stable over long periods of time, so one cannot argue that the original number words have simply got replaced by other words which carry no obvious link to the original entities used to signify the numbers. If that had happened it would have occurred in quite separate ways in different languages and one would expect to have rosters of number words in different languages today which bear little or no relation to one another. This is far from being the case. In the table below we can see the number words of several Indo-European languages. The close interrelationship is remarkable and indicates the persistence of the original words. The only other explanation one might offer for this unusual state of affairs is that, whilst the number words have remained very stable because of their importance in daily life and commerce, the words for tallying objects to which they originally referred need have no such enduring status. Those words could have undergone considerable change, so that the words now used to label these things bear no relation to the number words with which they were originally associated.

| Number | Greek | Latin | German | English | French | Russian |
|---|---|---|---|---|---|---|
| 1 | hen | unus | eins | one | un | odyn |
| 2 | duo | duo | zwei | two | deux | dva |
| 3 | treis | tres | drei | three | trois | tri |
| 4 | tettares | quattuor | vier | four | quatre | chetyre |
| 5 | pente | quinque | fünf | five | cinq | piat |
| 6 | hex | sex | sechs | six | six | shest |
| 7 | hepta | septem | sieben | seven | sept | sem |
| 8 | okto | octo | acht | eight | huit | vosem |
| 9 | ennea | novem | neun | nine | neuf | deviat |
| 10 | deka | decem | zehn | ten | dix | desiat |

## FINGERS AND TOES

> *In Samoa, when elementary schools were first established, the natives developed an absolute craze for arithmetical calculations. They laid aside their weapons and were to be seen going about armed with slate and pencil, setting sums and problems to one another and to European visitors. The Honourable Frederick Walpole declares that his visit to the beautiful island was positively embittered by ceaseless multiplication and division.*
>
> R. BRIFFAULT

One special set of tallies, our fingers, made it possible for tallying to graduate to large numbers and to take the step from cardinal to ordinal numbers. It is impossible to determine when or how this method of counting began but Mankind's earliest methods of communication probably rested upon gesture as much as upon language and it is not difficult to imagine that the gesticulation required to get across the notion of a particular quantity would most readily draw upon the waving of a collection of fingers. The historian Tobias Dantzig has gone so far as to claim that

> Wherever a counting technique, worthy of the name, exists at all, finger counting has been found to either precede it or accompany it.

These tallies are universal across cultural barriers and do not require every individual who employs them to have fashioned a collection of similar sticks or stones. Moreover they possess the attractive feature of enabling the user to tally larger numbers than with other naturally occurring reckoners, and extend them further if toes are used as well. In fact, things need not end there and there exist whole counting cultures which are based entirely upon the human anatomy. One of the most elaborate examples was that practised by the Torres Strait Islanders as late as the nineteenth century. Starting on the right side of their body, they touched first five fingers for the numbers 1 to 5, then the wrist (6), elbow (7), shoulder (8), chest (9), left shoulder (10), left elbow (11), and so on, down to the left fingers which took them to 17. Then they continued with the left little-toe (18), through to the left big-toe which takes them to 22, left ankle (23), left knee (24), left hip (25), right hip (26), and then back down the right side in the same way, finishing with the right little toe (33). One imagines that maths lessons might have resembled a form of St Vitus's dance.

This scheme was far from unique in general form; many other Pacific islanders display similar strategies for tallying using, for example, fingers, eyes, nose, ears, and mouth as anatomical reference points. An essential part of the mechanics of counting by such techniques is the suite of associated gestures by which the various parts of the anatomy are identified. This is not performed by naming them or using other forms of oral communication. Consequently, one expects that the anatomical associations preceded the development of number words. There are many nice examples of number words in primitive cultures which have very plain anatomical meanings. A typical structure is that displayed by the Zuni Indians who had two root words—for 'one' and 'two' only—and the ensuing number words were created from them using different displays of the fingers. They end up with a set of number words whose meanings are easy for us to visualize:

1. töp-in-te = taken to start with
2. kwil-li - raised with the one before
3. ha'-i = dividing finger
4. a-wi-te = all fingers raised except one
5. öp-te = notched off
6. to-pa-lĭ-k'ya = another added to what has already been counted
7. kwil-li-lĭ-k'ya = two brought and held up with the rest
8. ha-i-lĭ-k'ya = all but two held up with the rest
9. ten-a-lĭ-k'ya = all but one held up with the rest
10. äs-tem-'thla = all of the fingers
11. äs-tem-'thla-to-pa-yä'thi-to-na = all fingers and one more held up.

This scheme continues in similar fashion up to 20 where the word employed means 'two times all of the fingers'. One significant feature of this method of counting which distinguishes it from the use of other fixed collections of tallies is the fact that parts of the body remain in fixed positions relative to one another (unlike stones rattling around in a bag), and so it is possible to register the last part of the body that is used in an emumeration. It will always be the same when the same quantity is enumerated. This enables one to remember the size of the collection being counted simply by remembering the last part of the body referred to. So, if we were counting half a dozen sheep on our fingers and always followed the same order, we would associate the first finger of the second hand with the number 6; we no longer need to associate them with the whole set of six fingers that have been tallied off. This is an important step because it permits the complementary aspects of the cardinal and the ordinal aspects of numbers to become transparent. When you tally with stones, this does not happen. Any one of your collection of stones might be the sixth one counted each time you count out sets of six.

It has always been a matter of conjecture how we might link our modern Indo-European number words to terms arising from finger counting, but time seems to have eroded the connections as the words have evolved and changed. There are a few traces remaining though. The German *zehn*, for 'ten', derived from the Old German form *zehan*, resembles *zwei hand* meaning 'two hands';* while our English 'twenty' may derive from 'twains of tens'.

Soon, we shall see that the use of fingers and toes is closely related to the base that was employed by the early counters. But, first, there is another curious feature of finger counting that is of interest, because it sheds light upon the extent to which different cultures may have inherited their counting habits from others rather than invented them independently. Despite the ubiquity of finger counting, there is nonetheless a curious variation in the sequence of finger movements which accompany such counting. This is shown in Fig. 2.7. It was this variation that enabled the English gentleman we introduced at the start of the chapter to discover the nationality of the young oriental girl. By tracking similarities in finger counting technique one can trace the spread of counting practices from one culture to another.

British children today would count on their fingers by opening their fingers one by one from a closed fist beginning with the thumb and ending with the little finger before doing the same with the other hand. Yet one finds that there exist interconnected parts of Australia, America, Asia, and Africa in which finger counting begins with the left little finger; elsewhere it habitually begins with the right little finger; whilst in regions of central Africa and the middle of South America one finds a procedure in which the right index finger is used to start counting. Today, Westerners generally finger-count by extending out their fingers one by one from a clenched fist, but the Japanese begin with an open hand and then close their fingers one by one. There are American Indian tribes, like the Dene-Dinje, whose number words describe just this sequence of finger-closing. The meanings of their words for the numbers from one to five are:

1 = 'the end is bent' (i.e. the little finger is bent in half)
2 = 'it is bent once more' (now the ring finger is bent)
3 = 'the middle is bent' (now the middle finger is bent)
4 = 'only one remains' (now bend the index finger so only the thumb remains)
5 = 'my hand is finished'.

Despite the plethora of examples of finger counting from primitive and ancient cultures, it is by no means confined to such cases. Finger counting was fairly common in Europe right up until the sixteenth century when it was displaced by the written numbers of Hindu–Arabic origin which eventually evolved into the symbols 0,1,2,3,4,5,6,7,8,9 that we employ today. Fingers were rather useful as a *lingua franca* for foreign traders, and common reckoning devices like the abacus are easily combined with them to

* However, it is curious that *zehn* is also close to *zehen* the German for 'toes' which might even indicate some predecessors who counted to ten on their toes!

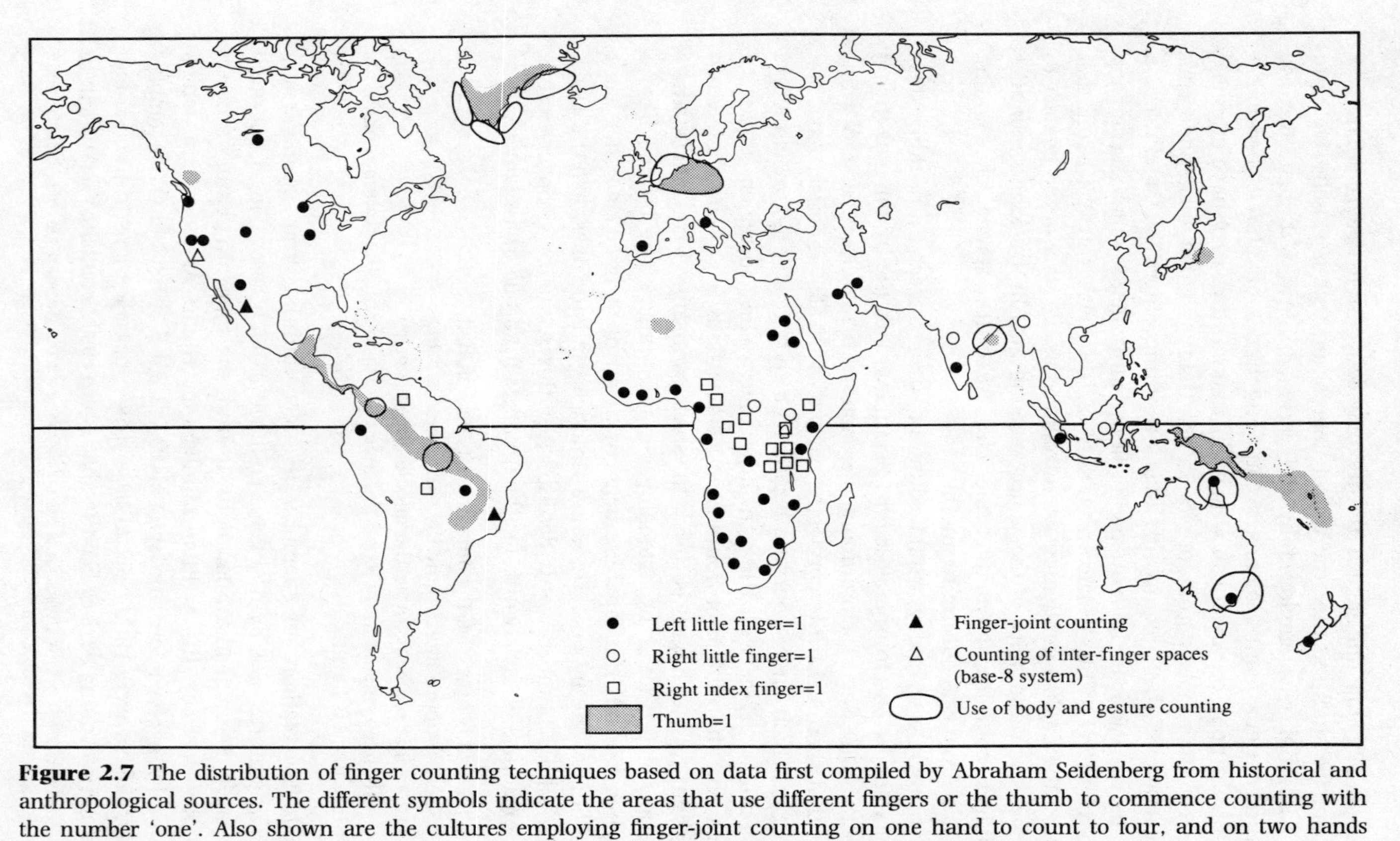


**Figure 2.7** The distribution of finger counting techniques based on data first compiled by Abraham Seidenberg from historical and anthropological sources. The different symbols indicate the areas that use different fingers or the thumb to commence counting with the number 'one'. Also shown are the cultures employing finger-joint counting on one hand to count to four, and on two hands count to eight. Extensions of finger counting which employ other parts of the body to count beyond ten are also marked.

calculation that has many stages. The most famous surviving account of the 'theory' of finger counting in early Europe is that of the Venerable Bede, who died in 735. Here we see the development of finger symbolism, first introduced by the early Greeks and Romans, which enabled very large numerals to be recorded by fingers. This method seems to have been propagated around the known world by the Roman legions during the expansion of their Empire and the establishment of its trading links. Bede's treatise 'On the Reckoning of Times' is dedicated to the question of computing the date of Easter and other important times and dates. The first chapter, entitled 'Concerning counting, or the speech of the fingers', is devoted to a description of finger counting which seems to have remained the standard reference on the subject for over a thousand years. It begins with the statement:

> We have considered it necessary (with God's help) to write about the measurement of time, and first of all to explain briefly the useful and convenient skill of bending the fingers. . .

Bede gets up to 9000 simply using the fingers in different configurations, but then continues on up to multiples of a hundred thousand by using gestures in which the fingers are positioned in different ways on different parts of the body. The Romans, by contrast, had extended finger counting only up to ten thousand. A later illustration of Bede's scheme is shown in Fig. 2.8. You notice that Bede's scheme differs from the usual one we might practice. It starts with the hand open and bends fingers in half, beginning with the left little finger. Also, the fingers are only bent at the middle knuckle. Perhaps the only places where we could now find such fine numerical nuances expressed through hand signals is by bookies at a racecourse or by traders on the dealing floor of an old-fashioned (pre-electronic 'big bang') stockmarket. Bookmakers and dealers are using elaborate hand signals to communicate very precise numerical information in a deafening environment that makes oral communication inefficient and prone to misinterpretation.

## BASER METHODS

> *Sir, allow me to ask you one question. If the church should say to you, 'Two and three make ten', what would you do? 'Sir,' said he, 'I should believe it, and I would count like this: one, two, three, four, ten.' I was now fully satisfied.*
>
> JAMES BOSWELL (OF DR JOHNSON)

The technique of finger counting is especially interesting because it appears to be linked in some way to the 'base' that is employed for doing arithmetic. Although one can count for ever using a different word for each new number, this is a cumbersome and impractical system that soon taxes one's

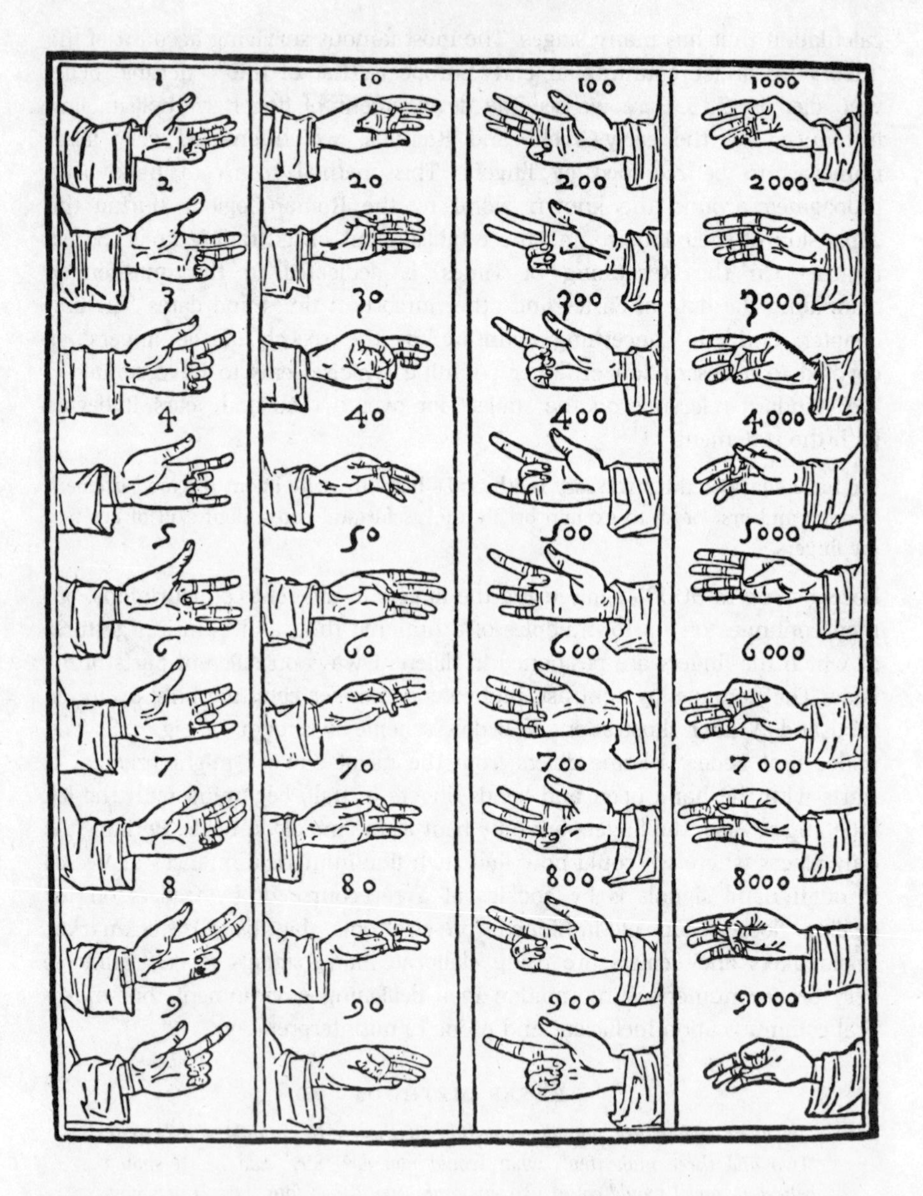


**Figure 2.8.** An illustration of 'The art of finger reckoning': the finger-counting scheme of the Venerable Bede drawn a thousand years after Bede by Jacob Leupold and published in Germany in 1727. You notice that this sequence follows the method of bending the fingers of the left hand, rather than our more familiar practice of unfolding fingers from a clenched fist.

powers of memory beyond their limit. It is more effective to adopt some quantity as a collective unit. So one might count five sheep on one's hand, but then start counting again on that hand whilst keeping a record of how many fives you have got using the fingers of the other hand. This idea is fundamental to all number systems. The quantity that defines the size of the collective unit is called the *base* of the counting system. Our own 'decimal' system uses 10 as its base. This, as we shall see, is a very common choice and derives from the availability of ten fingers for preliminary reckoning before one needs to keep a further count of how many lots of the collective unit one has tallied. But not everyone counted using the base–10 system. We have already described some cases where words only existed for the numbers one and two, and larger quantities were built up by juxtaposing those two words. These cultures were employing a base–2 or 'binary' system.

As we look around the ancient civilizations and tribal cultures of the ancient world we find a surprising range of counting systems. Their primary difference is in the base chosen for the counting system. With virtually no exceptions (except the isolated 4 and 8 counters we mentioned earlier) one finds counting *systems* derived only from the use of the bases 2, 5, 10, 20, and 60.

## COUNTING WITH BASE 2

*The animals went in two by two*
*Hurrah, hurrah*

POPULAR SONG

Counting in twos is the simplest possible system of counting that one could employ beyond the mere labelling of every quantity by a different unrelated word. There is evidence that it was once fairly widespread around the Earth, but in most places it was superseded by more effective numbering systems. The present distribution of pure 2-counters is charted in the map in Fig. 2.9. What one sees from this map is that pure counting by pairs, in which the counters have words for 'one' and 'two' only and then build up greater quantities by amalgamating these words, is a peripheral and relatively rare phenomenon. In Africa it is found only among the Bushmen, who have one of the oldest African cultures. Beyond 1 and 2 their composition of numbers continues up to 10 via the verbal constructions that we would denote by the combinations 2+2, 2+2+1, 2+2+2, and so on, up to 2+2+2+2+2. In eastern Australia and New Guinea one finds many 2-counting societies, all of whom have influenced one another and speak related languages. Amongst the South American Indians there is another region of linked languages and peoples. What is striking about these isolated and primitive cultures is the strong similarity of the linguistic structures that

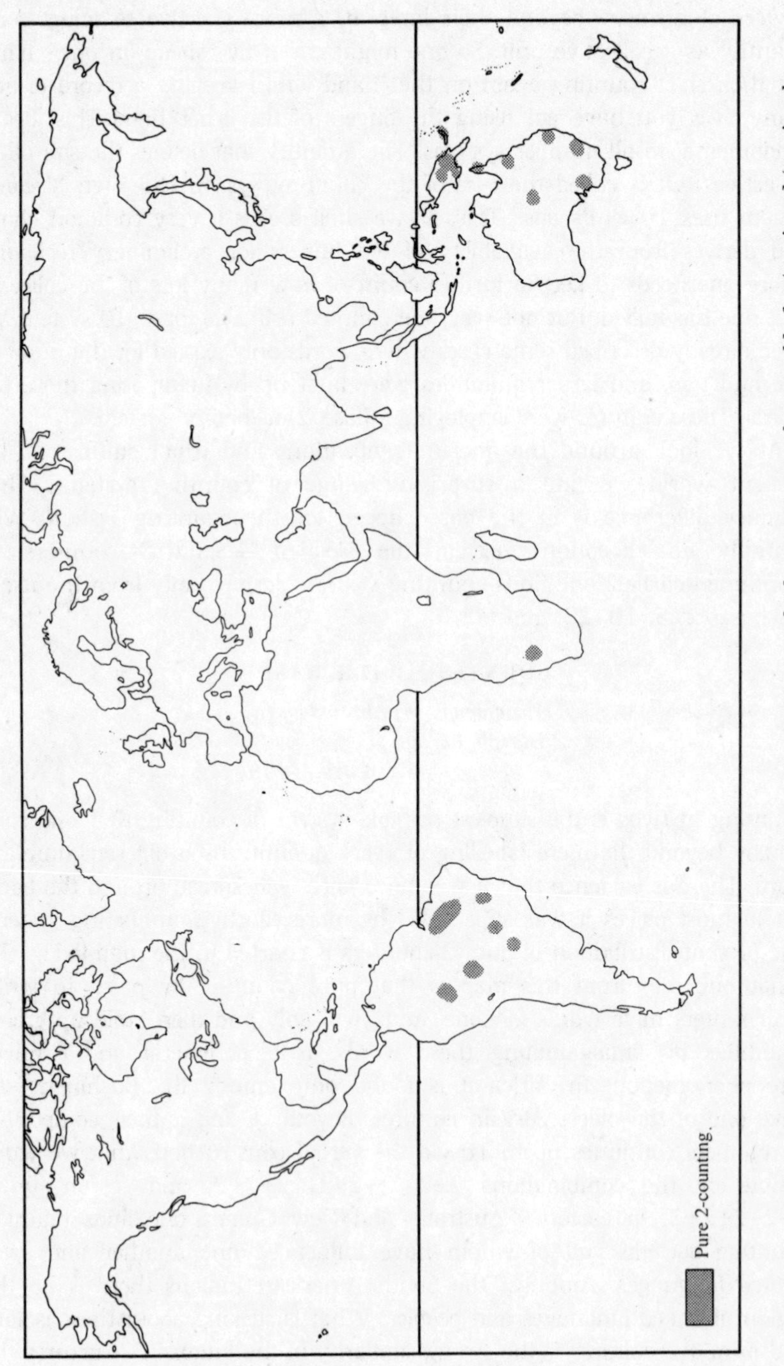

**Figure 2.9.** The distribution of existing pure 2-counting systems which contain number worlds for 'one' and 'two' only.

they employ to manifest the 2-counting system. The table below shows a comparison, collated by Abraham Seidenberg, of the way in which typical representatives from these three different continental centres of 2-counting build up their rosters of number words for 'one' and 'two':

| Gumulgal (Australia) | Bakairi (South America) | Bushmen (Africa) |
|---|---|---|
| 1 urapon | tokale | xa |
| 2 ukasar | ahage | t'oa |
| 3 ukasar-urapon | ahage tokale | 'quo |
| 4 ukasar-ukasar | ahage ahage | t'oa-t'oa |
| 5 ukasar-ukasar-urapon | ahage ahage tokale | t'oa-t'oa-t'a |
| 6 ukasar-ukasar-ukasar | ahage ahage ahage | t'oa-t'oa-t'oa |

We see at once the manner in which the word for two plays a pivotal role. Each system escalates as 'one', 'two', 'two-one', 'two-two' (or 'two-again'). The structure shown in the three geographically diverse examples above is so similar, even to the extent that the larger number precedes the smaller in the conjunctions, the peoples involved are so primitive in other ways and their locations are so much at the extremities of civilization, that it invites a speculation. The independent invention of this counting system by such tribal groups seems most unlikely. If this had been the case, one might expect to find far more similar counting systems spread randomly about, but there are virtually none. A more plausible explanation for the similarity of such primitive systems is that these far-flung outposts of 2-counting are the only remnants of what was the first widespread system of counting which arose from a particular centre of ancient culture. The existing two-counters are the only ones who never went on to higher things.

We know that 2-counting was known in Sumeria in 3000 BC with a three number-word sequence meaning 'man', 'woman', and 'many', so indicating the source of the idea of pairing in this case in human sexual distinction and fertility. With the passage of time the advanced cultures developed better systems of number words and adopted more powerful number systems. For while the 2-counting system can in principle continue to describe as large a number as one wishes, it is very cumberome to do this, and a system with a larger base enables one to count without exercising prodigious feats of memory. If one looks through the records of all the individual tribes displayed on the map above who are true 2-counters then in South America the largest composite number that one finds is 2+2+2+2+1 of the Zamuco Indians; in Australia the Kauralgal go to 2+2+2+2; in New Guinea the Parb, Sisiami, and Anal

(a)

(b)

**Figure 2.10** (a) A specimen of the Babylonian system of number symbols, dating from *c*.1800–1600 BC. It displays horizontal groupings of vertical wedges made with a stylus, each signifying 'one', together with a horizontal wedge to denote 'ten'. (b) The ancient Persian system of number marks from an inscription of Darius the Great which counts from 1 to 10, dating from around 500 BC. It is similar to that adopted by the Babylonians, but with the wedges aligned in vertical pairings.

tribes all extend the system only up to the word for 2+2+1; whilst the African Bushmen go all the way to 2+2+2+2+2.

There are interesting examples which support the idea that 2-counting arose in higher centres of culture and was then modified and improved upon later as other methods of counting were found to be more efficient. The most interesting is evidence left by the ancient Persians from the time of Darius the Great, around 500 BC. An inscription from this period (Fig 2.10b) gives a list of the number symbols for the numbers from 1 to 10. There is no great mystery over the symbolism being used here. An extra stroke (these are impressions made in wet clay by a stylus) is added to the collection for each increase in number by one; then, when 10 is reached, a new symbol is created by altering the orientation of the writing stylus in order to denote one collection of ten strokes. Thus this is a base 10, or 'decimal', counting system like the one that we use today. The interesting feature of this system is highlighted only when we compare it with the earlier (1800–1600 BC) Babylonian system of number symbols shown in Fig. 2.10a. First, we see that the symbols being used for one and ten are actually identical, whilst the remaining numbers are formed by producing various arrangements of the simple 'stroke' used for the number one. But whereas the Persians stacked these strokes in vertical piles of two, the Babylonians arranged them in horizontal groupings as well. So why did the Persians alter the way of representing the Babylonian symbols when they encountered them?

A defining feature of the Persian scheme is the way in which the strokes are arranged in easily recognizable vertical *pairs* that can be read at a glance by an habitual 2-counter. The Babylonian scheme requires some concentration in order to be sure of the larger numbers—although the different shapes of the collections are no doubt designed to assist in this respect. Hence we might conclude that the Persians adapted the Babylonian notation in accord with their own previous habit of counting in pairs so that the symbolic representation of the numbers actually looks like combinations of symbols for 'one' and 'two'. If we look at the Babylonian arrangement again in the this light, the reason for its layout becomes clear. It reflects a residual counting system that counted in *threes*: the horizontal rows each grow to contain three members and then a new row is added below the old one.

Hence, it is very plausible that those 2-counting systems shown on the earlier map are a remnant of widespread 2-counting and this accords with their locations at the margins of world. Such a view has a number of interesting features. It does not take seriously the idea of independent invention of 2-counting systems as some sort of primary intuition in all the human minds; it looks instead to the gradual spread of useful ideas from higher cultures to lower ones. But we should be wary of categorizing cultures as advanced or retarded merely on the basis of their desire to elaborate a number-word system. We are not saying they are intellectually inferior for failing to graduate from the 2-counting system. Rather, we appeal to something particular about their culture that prevents or fails to necessitate the step being taken to a different system. Moreover, just because a culture lacks words for quantities beyond five or six does not mean that they do not possess that number concept. They may simply represent it by gestures or other non-verbal images.

Even if this grand scheme of birth, death, and transfiguration of the 2-counting systems leaving a few scattered remnants is not the way that it happened, the existence of the 2-counting systems shown in Fig. 2.9 teaches us an important lesson. They surely reveal that counting does not necessarily begin with finger counting. Finger counters count six as 5+1 not as 2+2+2. If 2-counting was a very widespread phenomenon on the Earth in very early times, predating finger counting, then its existence must be a major ingredient in any theory about what concepts are innate to the human mind. But the picture of a widespread 2-counting system is linked to the idea that it emanated from a single cultural source. It was not suggested that it sprang up naturally in human minds all over the Earth. And, given that the notion must have originated somewhere, how might that have happened?

Despite our natural inclination to regard counting up to two, or in pairs, as just about the simplest imaginable manifestation of the brain, we find no such ability amongst animals. Whilst they can distinguish between zero and one (for example in the difference between 'dinner' and 'no dinner'), and keep track of quantity up to four or five in some cases, they cannot count in pairs. There are primitive cultures, we have already seen, with words for 'one', 'two', and 'many', but no system of counting at all. Clearly, counting, whether in pairs or in any other way, is not a universal instinct. Whilst we can give examples of pairs and notions of twoness that would have impressed themselves upon early observers—things like our two hands, feet, eyes, ears, arms, and legs, or the mating pairs of animals and humans—it is hard to make the leap from those concrete examples to see why anyone would want to compare two snakes with two stones and extract a common factor. Would they not be more likely to compare like with like, snake with snake, stone with stone, as we do indeed witness much later in the way that separate number words are associated with collections of different things? A ubiquitous source of 'twoness' is in the use of the hands. If you are always working with your hands, then when you pick up things there is an immediate relationship between the number of things that you can pick up and your two hands or arms. You might easily come to measure how many logs you have to move by the number of journeys you had to make to transport the pile from the woods to your camp. Nonetheless, the sceptic might argue that any clear notion arising from such a simple correspondence would easily get clouded out by the simple fact that one can carry more than one object at a time.

## THE NEO-2 SYSTEM OF COUNTING

*Two's company, three's a crowd.* FOLKLORE

If the 2-counting system was a transient stage through which many, if not all, of the counting systems of the ancient world passed we should look to the neighbours of the 2-counters to discover the next stage in the evolution. The most obvious way in which the limitations of the pure 2-system, with its two number words, can be superseded is by introducing more number words whilst retaining the idea of pairing words to build up expressions for larger numbers. There are a variety of clues as to how this step might have taken place. If one used a pure 2-counting system supplemented by some system of tallying marks, there would be a natural way to represent slightly larger numbers than those which could be grasped at a glance. One might denote numbers like six, seven, and eight by the following system:

```
111  1111  1111
111  111   1111
```

If one reads these patterns vertically they appear as columns of 'twos and 'ones' and are the natural aggregates of the pure 2-counter. But they can also be read horizontally whereupon they appear as combinations of two larger quantities 3+3, 4+3, and 4+4. Expressions like these are the essence of the so-called neo-2-counting system. The simplest examples to be found amongst Australian aboriginals have words for the numbers 'one', 'two', and 'three' and then compose successors as 'two-two', 'two-three', and 'three-three'. Schemes like this are tremendously widespread, and there is good evidence that they may have been universal at one time. Accordingly, one finds some internal diversity in the detailed formulation of the aggregating. In some cases, like the example just given, the aggregating is additive; but in other cases a more sophisticated feature is found—that of multiplication. There are Paraguayan tribesmen whose sequence of number words follow the pattern: 1, 2, 3, 4, 2+3, 2×3, 1+(2×3), 2×4, (2×4)+1, 2+(2×4). Elsewhere, one finds the presence of a principle of subtraction, so the number word for 'seven' means 'one less than two-fours' or that for 'nine' is 'one less than two-fives'. But this is not very common and tends to follow from the appearance of another important development in this type of extension of the pure-2-system which sees the introduction of the hands as counting elements. One finds the pure 2-counter's expression 'two-two-one' replaced by the word for 'hand'. As a result the expression for 'four' might become 'one less than a hand'. This is occasionally tempered by the fact that three might have been the largest number word in use, so one might not find 'seven' represented as 'eight-less-one' or 'two-fours-less-one'. Instead, in such a case, one would find 'seven' as 'three-three-one' or 'two-threes and one'.

It seems therefore quite reasonable to view the neo-2-counting systems as an elaboration of the pure 2-counting system that was promoted by the systematic use of finger counting. This naturally leads to a counting system that is in effect based upon the number five—the fingers of one hand. Clear examples are found amongst South American Indians where the sequence of number words have the following meanings:

| | | | | | |
|---|---|---|---|---|---|
| 1 | one | 7 | hand-two | 15 | three-hands |
| 2 | two | 8 | hand-three | 16 | three-hands-one |
| 3 | three | 9 | hand-four | .... | |
| 4 | four | 10 | two-hands | 20 | four-hands |
| 5 | hand | 11 | two-hands-one | | |
| 6 | hand-one | ....... | | | |

Similar examples can be found which use both hands and feet to extend the system up to 20. Again, as with pure 2-counting, there is no reason why any number cannot be represented in this way and all quantities be measured in 'hands'. But like the pure 2-system, it soon becomes cumbersome by stringing together so many quantities that one needs a separate accounting scheme to keep track of them. The whole system becomes self-defeating as soon as the number of words being aggregated exceeds the quantity described by the base word (here it is five) which defines the system.

What one notices from a survey of systems of this type is that they are all found on the edges of the regions still characterized by pure 2-counting. Most likely, they subsequently evolved into the more sophisticated systems of counting based upon ten or twenty that one finds in more rapidly advancing cultures. This distribution is shown in Fig 2.11. The advance was mediated by the introduction of finger counting as the basis for 5- and then 10-counting. We might therefore credit the neo-2-counters with the invention of systematic finger counting. This opens the way for the replacement of the base-2 counting systems by others based upon 5, 10, or 20.

Just as we found vestigial remnants of the structure of pure 2-counting in later, more powerful, counting schemes; so there are examples of latent neo-2-counting in relatively recent counting practices. A curious one, dating from the nineteenth century, was reported by anthropologists studying the development of complicated forms of calendar near Madras in India. The investigators discovered that very large numbers were represented by the positioning of counters on the ground in groups in such a way that a residue of the neo-2-counting practice of pairing was conjoined with the modern way of representing numbers wherein the relative position of numbers determines their relative value. The number 1687 was represented by a groupings of shells or stones, like this:

```
.    ...   .....   ....
     ...   ...     ...
1    6     8       7
```

Elsewhere in India, amongst some traders in the Bombay region, there are still traces of an early base–5 method of counting which uses finger counting in a novel and powerful fashion, enabling much larger numbers to be dealt with without taxing the memory unduly. The left hand is used in the normal way counting off the fingers from 1 to five, starting with the thumb. But when five is reached this is recorded by raising the thumb of the right hand whereupon counting of the next five begins again with the left hand until ten is reached, then the next finger of the right hand is raised, and so on. This system enables the finger-counter to count to thirty very easily, so that even if he is interrupted or distracted he can determine at a glance where the count has reached.

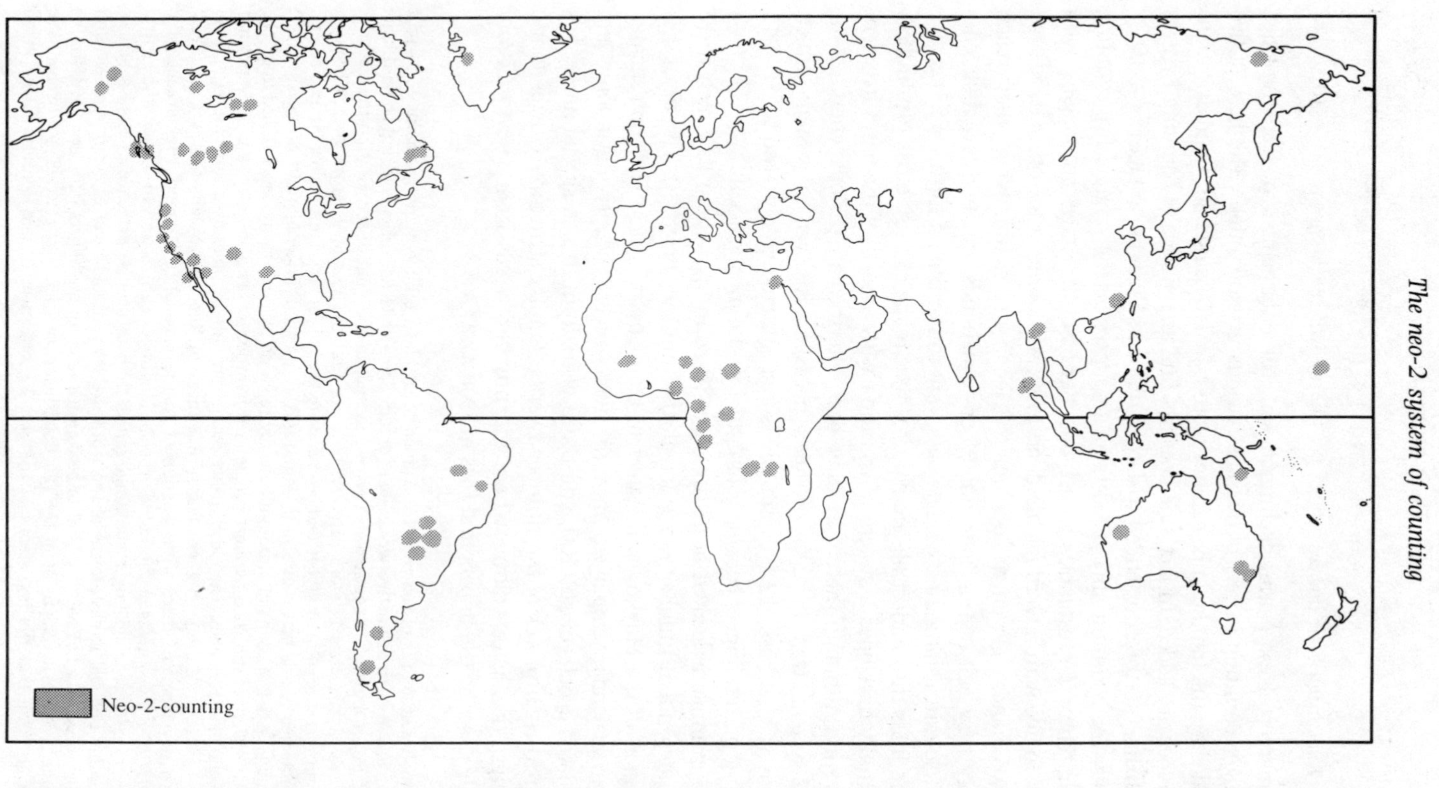


**Figure 2.11** The distribution of neo-2-counting systems that still exist. By comparison with Fig. 2.9 one sees that they lie close to areas which practise pure 2-counting.

## COUNTING IN FIVES

*Though thy beginning was small, yet thy end will be very great.* JOB

The merger of the pairing technique with the enlargement of scope provided by finger counting in sets of five has obvious generalizations. The base of five provided by the hand can be extended either to 10, using both hands, or to 20 if the toes are added. In fact, one does not find any system of counting by fives which has not been extended to larger numbers. The advantages of a 10 or 20 base for counting always led them to supersede the original five-counting system. Thus, for example, whilst one finds South American Indians who count to five with a word for hand, on to six with 'one on the other hand', ten as 'two hands', eleven as 'one on the foot', and so on, when they reach twenty they call it, not 'four hands' or 'two hands and two feet' but 'one-man', so establishing a new single basic unit rather than a further aggregate.

One finds that almost all the cultures that made extensive use of numbers eventually developed a system of counting with aspects of the bases five, ten, or twenty. Although there may have been intermediate periods when a 2- or neo-2-system was in use, or later occasionally even a system which employed the base 12,* eventually some system closely associated with tallying on ten fingers became established as the most efficient.

No example is known of a true base–20 counting system in which there are independent number words for all the numbers up to 20. The nearest is the habit of the Mayans and Aztecs to count by twenties. The sophisticated Mayan calendar employed 'months' which were twenty days long and calculated epochs of 20, 400, and 8000 years from this. A special name was given to each power of the number twenty. Similar structures can be found amongst the Aztec culture where the number words from one to twenty simply describe the fingers and toes in sequence.

* Such a system is of course not unknown to us and we have many remnants of its past influence, We have distinctive words, like 'dozen', for a collection of twelve and some things, like eggs, are habitually counted in twelves. Indeed, in the English-speaking business world, this base-12 structure is more extensive and we find the use of a 'gross' = 12×12 as a common measure of quantity. It derives from the French 'grosse douzaine' or 'strong dozen'. The influence of this base can also be found in the so-called 'Imperial' units of measurement (12 inches = 1 foot), which, although officially replaced by a metric system founded upon the base-10 system is still used by most people in Britain for all practical purposes. In America it is still the official system. The old British sterling currency partially used this base (1 shilling = 12 pence) and although there were 16 ounces in a pound weight this is a relatively modern development; there were 12 ounces in the original 'old pound'. Remarkably, the traces of base-12 systems in the Indo-European languages can be seen to support the central picture of reliance upon finger counting, rather than to provide evidence of rival schemes. The Gothic and Old English forms of the words for 'eleven' and 'twelve' mean 'one left' and 'two left'; that is, what was 'left' still to be counted after all the fingers had been used. In this way one sees the development of the Old English for 'two left' from 'twalif' to 'twelve'. We might also add that the persistence of the base-12 system in certain contexts is not at all

There are two ways of enlarging the simple finger-counting system up beyond five or ten. The first, called the *5–10 system*, builds up larger numbers as multiples of ten only; thus 70 would be described in words as 7×10. The other, called the *5–20 system*, builds up larger numbers by reference to multiples of 20: thus 60=3×20 and 70=3×20 + 10. The 5–10 system and its relics are to be found mainly in Africa and North America as shown in Fig. 2.12, and it is very often closely linked to the practice of finger counting. The 5–20 system is to be found amongst the Aztec, Mayan, and early Mexican and Yucatan cultures. From there it spread both north and south through the American continent, supplanting the primitive 2-counting systems to be found in South America together with the less extensively developed versions of the 5–10 system in North America. One even finds the 5–20 system amongst the Eskimo populations of the northern Polar regions although the linguistic and other pieces of evidence connecting those peoples to migrant groups from north-east Asia indicate that they derived their counting system from Asia as well.

The question of whether the 5–10 system preceded the 5–20 one, or vice versa, has no persuasive answer. The historical facts are too complicated and patchy to argue for any compelling trend. On the face of it one might have expected the 5–10 system to be the earlier. It is clearly easier and more natural to start counting on one's fingers up to ten and later extend this enterprise to twenty by incorporating the additional digits offered by the toes. But in these ancient cultures living in warm climates there was at first no form of footwear and, later, simply forms of undersole support which displayed the toes. Then, later still, complete foot-coverings developed which covered all the toes. Hence, one could imagine that early peoples with uncovered feet might find it natural to count with all fingers and toes, so establishing a 5–20 system. Only later when the feet were

surprising. It is extremely convenient. Twelve has factors 2,3,4, and 6, whereas ten has only 2 and 5. Twelve can be divided into halves, quarters, and thirds without the need for the invention of fractions. This is very attractive if one is involved in measuring and weighing things in the course of trading, perhaps changing the quantities offered to match the funds available. However, if one uses numbers solely for tallying quantities, then the attraction of all these dividing factors goes unnoticed. Yet, if we were starting from scratch today, there would be much to be said for the adoption of a base-12 (duodecimal) system rather than a decimal one. Indeed, during the eighteenth century the famous French naturalist Georges Buffon argued that the decimal system should be abandoned and replaced by the duodecimal, or base-12, system. He argued that the many factors of 12 were of overwhelming advantage and the existence of so many systems of measurement which made use of them bore witness to the utility of this system. However, others argued that the existence of divisors is not entirely a good thing. It results in different fractions having identical values. For example $^{28}/_{100}$, $^{14}/_{50}$, and $^{7}/_{25}$ are all equal in the decimal system. Even more ambiguity of this sort would arise in a duodecimal system, whereas if one chose a prime number, like 11 or 13, as the base, there would be no such duplications because prime numbers have no divisors at all.

covered would it become inconvenient to use the toes for counting and so there would be a tendency to use a truncated 5–10 version of the 5–20 system. At first sight the Eskimos appear an awkward case for this rather fanciful scenario to accommodate. Living in Arctic conditions they would never have had uncovered feet and yet we find them employing the 5–20 system. However, since their system derived from one developed in the more temperate climes of Asia this cannot be considered as a counter-example.

One of the most intriguing features of the incorporation of elements of a base–20 system into ancient modes of counting is the manner in which so many traces of it can be found inhabiting the nooks and crannies of modern European languages. The reason is that it once held sway over a large portion of western Europe, including Spain, Britain, Ireland, France, and some of the Nordic countries. It was gradually replaced by the base–10 system we currently employ, which spread with the Celtic and Germanic languages. But traces remain. In French we recognize the formation of words for numbers like 80 as *quatre-vingts* = four twenties: in English we have the 'score' as an ancient word for twenty with which there is an associated form of counting, so 70 years is 'three-score years and ten' in the famous description of the expected human lifespan in the King James Version of the Bible. The 'score' is derived from the old Saxon word *sceran*, which means to shear or to cut. Subsequently, 'to score' meant either to cut a mark or to keep count when used as a verb but meant 20 when used as a noun. The connection is simple: the score was the mark made on the tally stick whenever 20 had been counted. We still use it in both senses today. Shakespearean English displays both the score and the 5–20 system and medieval French manuscripts use Roman numerals in a related way, denoting 80 by $\mathrm{IIII}^{xx}$ or 133 by $\mathrm{VI}^{xx}\mathrm{XIII}$; that is, $80 = 4\times20$ and $133 = 6\times20 + 13$.

In French and Latin the words for 20 (*vingt* and *viginti*, respectively) have no connection at all with those for 2 (*deux* and *duo*) or 10 ( *dix* and *decem*) which implies that they were both remnants of a system with a base of 20 rather than one which composed the number 20 from smaller primary numbers like 5 or 10. Interestingly, there is a very old hospital originally built in Paris to house 300 war veterans which still retains its original thirteenth-century name: '*L'Hôpital des Quinze-Vingts*', that is, 'The Hospital of the Fifteen-Twenties'.

Both the 5–10 and the 5–20 systems seem to have been swept into oblivion by the spread of the more efficient base–10 system and today they are to be found only amongst a very small number of peripheral and underdeveloped tribal cultures.

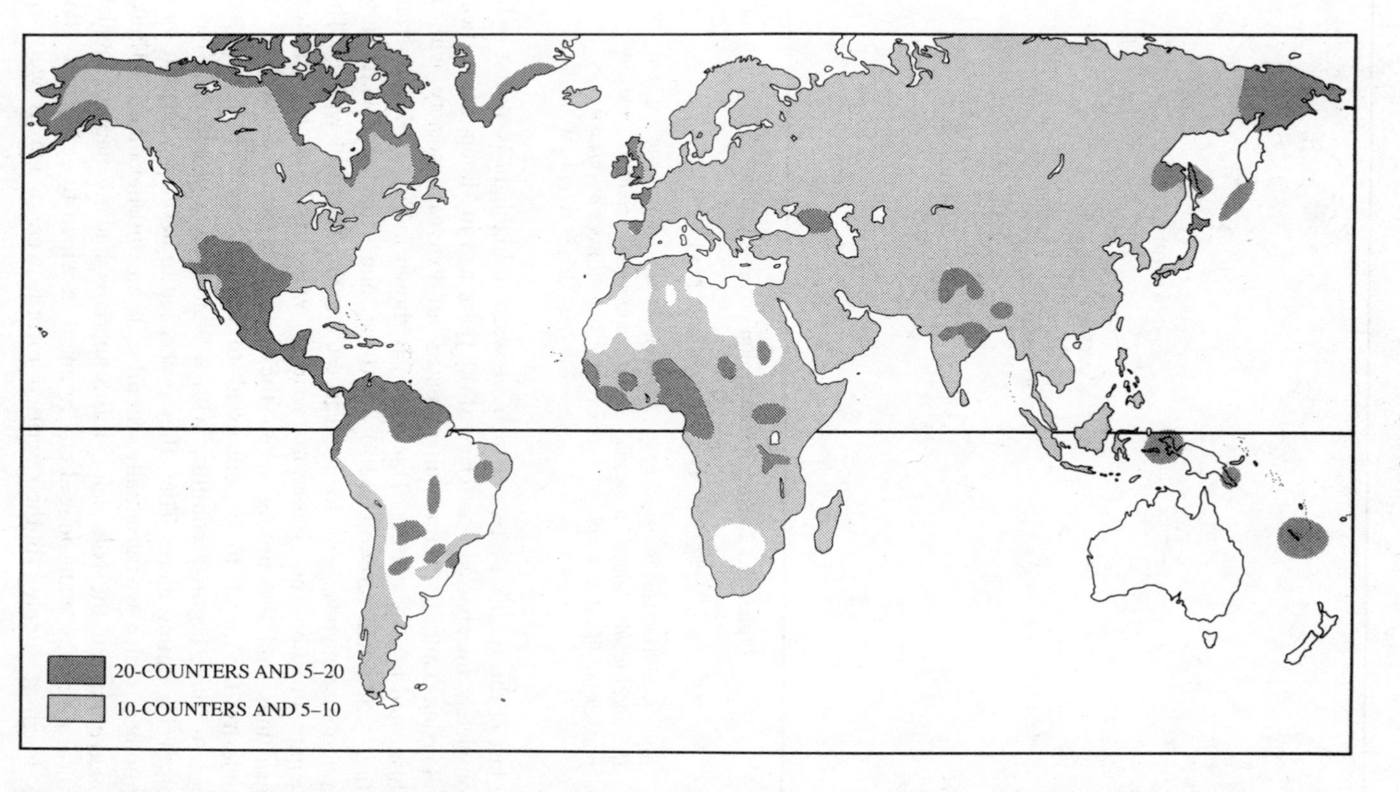


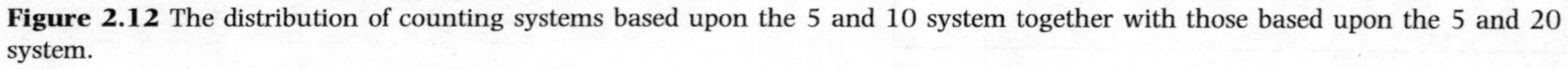
**Figure 2.12** The distribution of counting systems based upon the 5 and 10 system together with those based upon the 5 and 20 system.

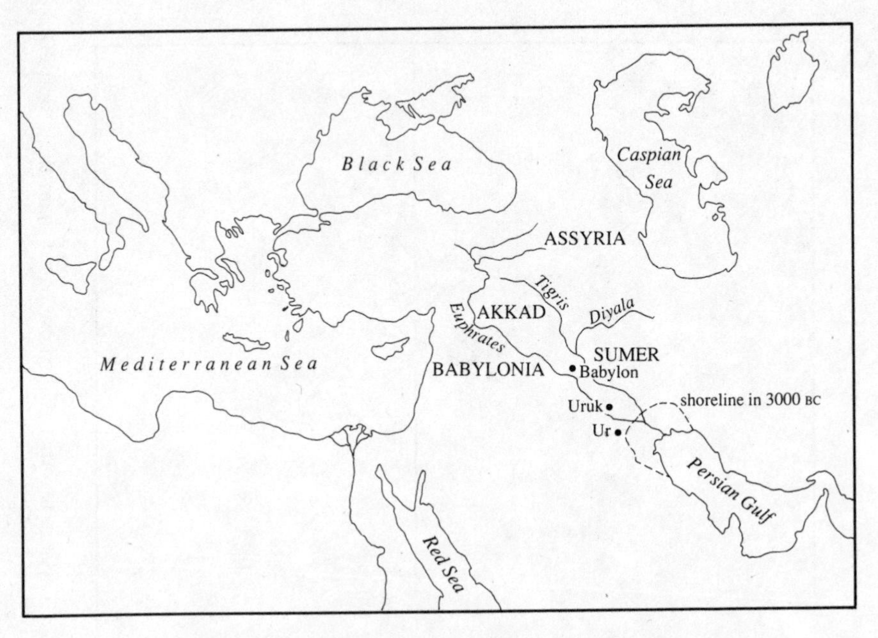


**Figure 2.13** The ancient region of Mesopotamia.

## WHAT'S SO SPECIAL ABOUT SIXTY?

*My second fixed idea is the uselessness of men above sixty years of age, and the incalculable benefit it would be in commercial, political, and in professional life, if as a matter of course, men stopped work at this age.*

WILLIAM OSLER

The last of the specific systems which we want to highlight is at first sight a completely unexpected one. Over 3000 years ago in Mesopotamia, near the Persian Gulf, the Sumerians created an extensive society with a sophisticated level of culture. This region is shown in Fig. 2.13. Sumerian writing originated around 3200–3100 BC and, together with early Egyptian hieroglyphs, provides the earliest known system of writing. The principal function of Sumerian writing was to facilitate business transactions and accounting when these dealings reached a level of complexity too great to be entrusted to human memory alone. They appear to have begun accounting with a 5–20 system of counting based upon oral counting alone. Then they proceeded to develop a written form of numbers which was gradually extended in an unusual manner. Unlike the Egyptians on the Nile delta, the Sumerians had no papyrus, which requires a marshy water-logged environment in which to grow, and stone was in short supply, so their written records were made in wet clay with a wooden wedge-shaped stylus. By changing the orientation of the wedge, different shapes could be impressed, and this diversity could be exploited to associate distinct meanings to different-shaped marks. The clay was then baked hard or left in the hot sun to create a semi-permanent inscription. These tablets are called cuneiform texts (the Latin *cuneus* means 'wedge') and are roughly contemporaneous with Egyptian records. When the Sumerian civilization was finally conquered, its intellectual heritage was absorbed into the Babylonian Empire, which adopted the Sumerian systems of writing and enumeration. Thousands of their cuneiform texts have been found which contain extensive arithmetical calculations.

The Sumerians did something that was both strange and unique amongst known human counting systems. First, they followed their oral tradition of counting, with steps at 5, 10, and 20, counting up to 50 using words which denoted the combinations

$$1,2,3,4,5,5+1,5+2,5+3,5+4,10,\ldots,20,\ldots,10\times3,\ldots20\times2,\ldots,20\times2+10.$$

The numbers in-between are filled-in in the usual way. A name is given to each of the multiples of 10 up to sixty, some of them being compound words, as indicated by the way the symbols are listed. This looks unremarkable. But after 59 we do not find 60 denoted by 20×3; instead, 60 is taken as a new unit and termed *gesh,** which is the same as the word for 'one' which started the whole system off (if there was ever a danger of confusion between 1 and 60, *geshta* was used instead of *gesh*). Larger numbers were denoted by both multiples and products of sixties and ran through the benchmarks 600 (*gesh-u* = 60×10), 3600 (*shàr* = 60×60), 36 000 (*shàr-u* = 60×60×10), 216 000 (*shàr-gal* = 60×60×60), 2 160 000 (*shàr-gal-u* = 60×60×60×10), and 12 960 000 (*shàr-gal-shu-nu-tag* = 60×60×60×60). This structure reveals that 600, although described in terms of the words for 60 (*gesh*) and 10 (*u*), becomes a new unit for expressing the numbers between 600 and 3600. The number 3600 has a new name and also acts as a new unit for expressing all the numbers from 3600 up to 36 000. New levels arise with the numbers 36 000, 216 000 2 160 000, and 12 960 000. The most symmetrical representation of this hierarchy is obtained by regarding it as constructed from two complementary bases of 10 and 6 in alternating fashion as shown below.

* Sometimes called *ash* or *dish*. Their first ten number words were 1=*gesh*, 2=*min*, 3=*esh*, 4=*limmu*, 5=*iá*, 6=*àsh*, 7=*imin*, 8=*ussu*, 9=*ilimmu*, 10=*u*. This perhaps betrays an earlier base-5 system with 7 = *imin* = *i* + *min* = *i*(*á*) + *min* = 5+2, and so forth.

| *Number* | *Decomposition* |
|---|---|
| 1 | 1 |
| 10 | 10 |
| 60 | 10×6 |
| 600 | 10×6×10 |
| 3600 | 10×6×10×6 |
| 36000 | 10×6×10×6×10 |
| 216000 | 10×6×10×6×10×6 |
| 2160000 | 10×6×10×6×10×6×10 |
| 129600000 | 10×6×10×6×10×6×10×6 |

So, we see that for some reason sixty was regarded as a special number and is used to restart the system all over again. It was a device that the Babylonians and the Assyrians took over from the Sumerians, and so its influence was extensively propagated through space and time. The Sumerian usage of base 60 was commonly employed by the Babylonians in 1800 BC and tablets survive from that period which display Sumerian number symbols being used to count herds of animals.

Why was 60 selected to play this pivotal role in the counting system and labelled in the same way as the number 'one?' No one knows for certain. The evidence for any viewpoint is purely circumstantial. It appears that the early Sumerian systems of weights and measures contained standard reference quantities which were in the ratio of 60 to 1. Later, this structure seems to have been taken over into the more general number system that was used for all forms of counting. This may well be so, but it tells us nothing about why the first measures incorporated the factor 60. Perhaps it is simply a reflection of the relative sizes of the very first standards of weight that were used and they happened to be in this ratio? But one is really looking for some natural system of weight or measure that gives rise to the 60 to 1 factor. Lacking this,* historians

* The best that one can do in this respect seems to be an appeal to the observations of Neugebauer that, if one studies the majority of cuneiform texts, then

> In economic texts units of weight, measuring silver, were of primary importance. These units seem to have been arranged from early times in a ratio of 60 to 1 for the main units 'mana' and 'shekel'. Though the details of this process cannot be described accurately, it is not surprising to see this same ratio applied to other units and then to numbers in general. In other words, any sixtieth could have been called a shekel because of the familiar meaning of the concept in all financial transactions. Thus the 'sexagesimal' order eventually became the main numerical system ... The decimal substratum, however, always remained visible for all numbers up to 60.'

There are later examples of this migration of terminology from weights and measures to the description of any quantity. For example, the Romans first used *as* as a weight equal to $^1/_{12}$ of an ounce and then as a quantity of time equal to $^1/_{12}$ of an hour. One should also stress that just as today we employ a mixture of different counting habits and bases for different situations so the Babylonians are found to mix base-10 with base-60 descriptions of the same quantity in documents which are not used solely for mathematical or astronomical purposes.

have considered possible astronomical motivations. An early belief that the year was made up of 360 days, dividing into twelve lunar months each of thirty days' duration, might have played a role. Alternatively, it has been suggested that the earliest Mesopotamian calendars were based upon a zodiac of only six constellations which would then divide the year into six periods of sixty days' duration. The importance of 60 would be underlined by the fact that it represented the interval of time which the Sun spent in each of the constellations of the zodiac. But the division of a circle into 360 'degrees' only originated in Babylonia in the last centuries BC, long after the establishment of a base-60 system. Another proposal has been that the significance of 60 emerged from geometrical knowledge about dividing a circular area into six equal parts by drawing lines through the centre. But it seems more likely that such a procedure was derived *from* the perceived importance of the number 60. A further possibility is that the 10–60 system was the outcome of the merger of two earlier systems of counting—one using base 6, the other base 10. But there is no independent evidence for this transient base-6 system. Finally, there are claims, which were first voiced as early as the fourth century, that 60 was chosen because it has so many divisors, including all of the first six whole numbers. This type of numerological consideration either points to the practical usage of the number system for measurement or to the priestly emphasis upon numbers with special properties having some particular significance. From what we know of early Sumerian culture the latter tendency is surprisingly absent, especially when contrasted with trends in later cultures, and the early employment of numbers and number words is largely confined to practical measurement.

Despite the peculiarity of the 60 unit, its presence is still evident in our own systems of measurement. Angular measure uses 60 seconds of arc to 1 minute, 60 minutes to 1 degree, and 360 degrees around the circle. Navigational positions are fixed by this system through the specification of latitude and longitude. Our measures of time in terms of minutes and seconds are based upon the very same circular system. We have inherited these systems from the Greeks but they derived them from the ancient Babylonian tradition. In French the different structure of the words for numbers above sixty* from those for sixty and below implies that at one time sixty may have had a special status in the hierarchy of numbers.

* In French we see compound structure above sixty, but independent number words before that: 20 = *vingt*, 30 = *trente*, 40 = *quarante*, 50 = *cinquante*, 60 = *soixante*, 70 = *soixante-dix*, 80 = *quatre-vingts*, 90 = *quatre-vingt-dix*.

## THE SPREAD OF THE DECIMAL SYSTEM

*It is a profoundly erroneous truism, repeated by all copy-books and by eminent people when they are making speeches, that we should cultivate the habit of thinking of what we are doing. The precise opposite is the case. Civilization advances by extending the number of important operations which we can perform without thinking about them. Operations of thought are like cavalry charges in a battle—they are strictly limited in number, they require fresh horses, and must only be made at decisive moments.*

ALFRED NORTH WHITEHEAD

The base-10, or 'decimal', system which is the most widespread in the world today is a happy compromise. The use of the base 10 derives from the systemless finger-tallying traditions of the past, but the use of ten, rather than, say, five or twenty, is really just a consequence of its convenient size: neither too big nor too small. Choose a base that is too large and one requires a very large number of tally marks or number words. Choose one that is too small and the counting system becomes very inefficient with many different collective units employed to represent quite small and frequently used numbers.

The pure decimal system that we have inherited is also pleasantly symmetrical. After counting up from one, the number 'ten' is taken to be a new unit; multiples of ten are then kept count of by exactly the same system until ten of them are counted. Ten tens define a new unit—a hundred; ten hundreds 'one thousand'; and so on, in the familiar way. The reason for the ubiquity of this method of counting is undoubtedly in part due to its ease of use and efficiency in calculation.

The successful propagation of the decimal system throughout so much of the Earth owes much to the vagaries of human language and its concomitant evolution. The decimal system of counting can be found in essentially the same form as part of all the so-called 'Romance' languages of Europe, like French, Spanish, and Italian, which derive from Latin roots. It also appears in the Germanic languages, like English and old forms of Gothic, as well as Greek back to 1200 BC, and Hittite, which was spoken in the Persian Gulf region as early as 1800 BC. The common factor behind all these decimal counting cultures is that their languages belong to the family of related tongues that linguists call 'Indo-European', which were spoken over a wide swath of the world from India to Europe. Their present distribution is shown in Fig. 2.14. They are assumed to derive from a single 'mother tongue' which emerged around 2500–3000 BC. The similarity of number words in all the Indo-European languages (some of which we saw in the table on p.44) which issued from the mother tongue means that it probably also had a decimal

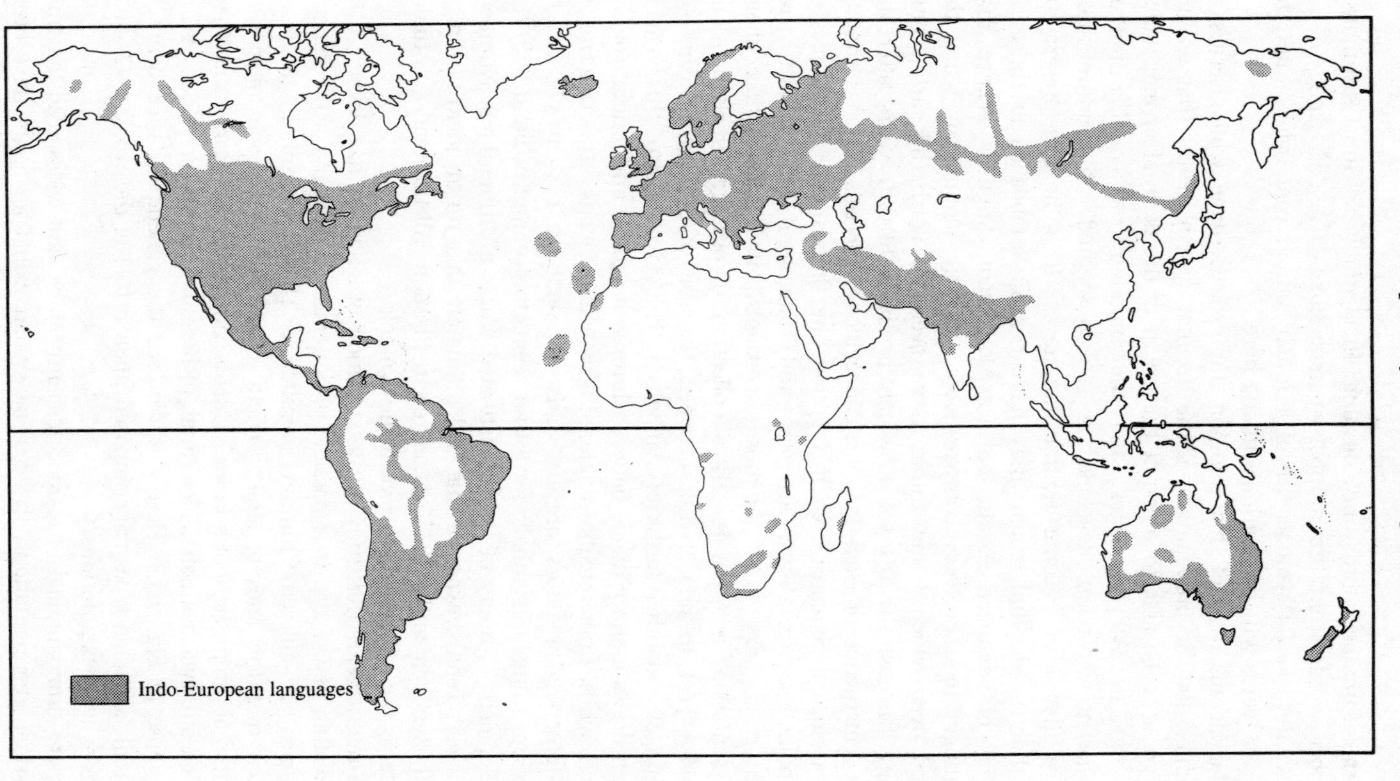


**Figure 2.14.** The distribution of the Indo-European languages today.

system of number words and counting. Incidentally, we must distinguish number words from the number symbols (like '1','2','3', etc.) we use today; the specific symbols are of a much more recent origin than the number words which we are discussing here.

The identification of this common linguistic heritage which serves as a vehicle for the propagation of the decimal system means that we are not going to be able to pinpoint the origin of the decimal system which we now use. All languages, and the Indo-European mother tongue will be no exception, developed as spoken methods of communication before they ever become written languages. It is in this pre-graphic era that the decimal system may have become established in the early centres of advanced Asiatic culture. At present, linguists believe that language has evolved comparatively recently compared with the period over which living things have been on Earth, appearing only during the last 40 000 years, which is about 60 000 years after the first emergence of *Homo sapiens*. The evidence for this belief is that in archaeological remains from this period one finds the first examples of symbolic representations, artwork, and images. If one believes that language is first and foremost a sophisticated form of symbolic representation, and not merely a collection of oral gestures, then one would expect to find a link between the use of oral symbolism (i.e. language) and other forms of symbolism. The advantage of this view is that it draws a sharp divide between human language and animal sounds which, whilst undoubtedly functional in warning friends, attracting mates, or frightening predators, are just signs, not symbols. It is this capacity to represent things in symbolic form that seems to go hand in hand with the development of many of the sophisticated works of humankind. Numbers are a very refined example that will ultimately take on an identity that is totally symbolic and devoid of concrete realization. All that will eventually matter is the relationship between the symbols.

In recent years the study of the evolution of language has developed in interesting new ways. It is possible to determine a measure of the genetic closeness of different human populations. These determinations have shown that the lines of sharp genetic distinction in Europe follow the traditional linguistic boundaries. This is not entirely surprising. When carried out over a much wider range of human cultures, one finds results like those in Fig. 2.15. Here one can see the genetic distance between different language types superimposed upon a tree of the world's 42 most representative populations.

By the time languages became sufficiently stable and widespread to admit of written representations, the decimal system would no doubt have been

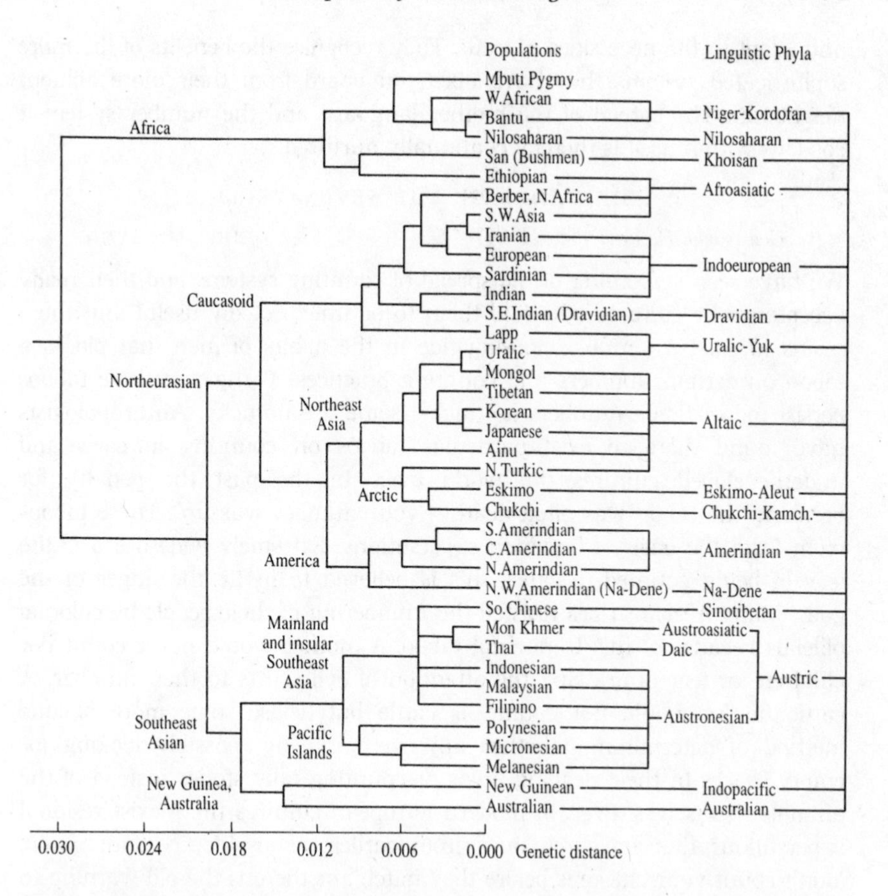


**Figure 2.15** A detailed comparison of genetic and linguistic divisions compiled by Luigi Cavalli-Sforza, Alberto Piazza, Paolo Menozzi, and Joanna Mountain. It groups together populations on the basis of a classification of their languages and then genetic distances are worked out by analysis of their specific gene frequencies. The relative genetic distances are shown along the base of the diagram and one can determine the genetic closeness of related linguistic types.

fairly well established in a useful form. The subsequent spread is easier to explain. Archaeological evidence gives credence to the idea that forms of writing began in order that property could be identified. The usefulness of distinguishing things has a natural connection with sharing and trading. If things change ownership, then marks are altered, records are revised, and a form of memory about the historical sequence of events is set in motion. This is merely reasonable speculation, but what is more certain is the existence of a natural one-way diffusion of counting systems from the advanced centres of culture to the more primitive ones as the latter adopt and adapt to the necessities of trade. They recognize the benefits of the more sophisticated systems they have taken on board from their more affluent neighbours. The spread of the mother language and the number system it cocoons within itself is thereby continually nurtured.

## THE DANCE OF THE SEVEN VEILS

*God is odd. He loves the odd.* MUSLIM SAYING

We have seen something of the spread of counting systems and their ready acceptance by cultures who find them to be unexpectedly useful. But there seems to exist a curious opposing tide in the affairs of men that places a taboo on certain numbers and counting practices. Vestiges of these taboos persist today. Some numbers are lucky; some are unlucky. Anthropologists have found there to exist particular taboos on counting in early and underdeveloped cultures the world over. In the past the penalty for breaking the taboo was often death—'your number was up'. These taboos often form the basis of lingering superstitions. Extremely widespread is the fear of being counted, because this is believed to invite the anger of the gods. Many African tribes resisted the numbering of their people by colonial officials because of this traditional taboo. A mother would never count her children for fear of drawing the attention of evil spirits to their number. A cattle herder would not count his cattle but devise some more oblique method of determining whether any were missing, possibly looking for empty spaces in their sleeping areas or counting tally stones instead of the animals themselves. Even in modern European cultures there exist residual superstitions that have remained from earlier taboos: don't count sheep; 'don't count your chickens before they hatch'; or there is the old warning to the miser that each time you count your money it decreases. One suspects that many of these taboos led in the distant past to the development of sets of different number words for the same quantities of different things that we find in all languages. By using new descriptive terms for numbers, one could avoid 'counting' in the usual sense by referring to quantities by new words.

This widespread belief is an interesting cultural feature because it appears to be a baseless psychological prejudice. There would appear to be no possible reason why such prejudices should appear independently in different cultures. Maybe they have spread from a single source? Only if there was some real reason why counting your sheep made them die off would we expect this to have been learned by experience in independent cultures. Thus, a commonality in irrational beliefs is a very suggestive and important trend. In this case it adds credibility to the general notion that counting was not an easy thing that sprung up as a by-product of sentience. Rather, it flowed outwards from the world of its peculiar inventors.

We are also familiar with the tradition that a census of the people is wrong. In the Bible there is a story about King David being moved to number his people. This was regarded by God as a sin and brought a plague upon the Hebrew people. It appears that the motive for this census was to discover how many people there were and this distinguishes it from a later census which David conducted with the intention of conferring a blessing upon the soldiers as they paraded before him. The implication here is that the first census was somehow performed for the wrong reason or in an inappropriate way. The 'right' way was associated with some aspect of religious worship or ritual, hence the link with the blessing being conferred. The 'wrong' way is for pride or knowledge. In other early societies, like that of ancient Egypt, numbering of people and animals is also found, but again the numbering is part of an elaborate ritual celebration. This pattern is found amongst the Incas also. It has been suggested that these traditional taboos are telling us something important about the origins of counting.

## RITUAL GEOMETRY

*It is the spirit of the age to believe that any fact, however suspect, is superior to any imaginative exercise, no matter how true.* GORE VIDAL

The most elaborate examples of the ritual origin and use of mathematical concepts are to be seen in the development of geometry with its focus upon symmetry and pattern. The most striking are those to be found in a collection of ancient Hindu manuals called the *'Sulba-Sûtras* which gave the detailed geometrical instructions for the construction of altars. The literal meaning of the word *'Sulba-Sûtras* is 'Manual of the cord' because the surveying for the altar constructions was laid out by pegging cords into the ground just as bricklayers are seen still to do today.* These guides were written between 500 and 200 BC, although they draw upon much earlier practices and there are written records of similar geometrical constructions in Indian writings back to 1000 BC. The *'Sulba-Sûtras* reveal the deep knowledge of geometry that the authors required in order to perform their ritualistic requirements and special sacrifices, the *saumikí vedi*, the *paitriké vedi*, and so on. The famous theorem of Pythagoras was clearly already known to them together with procedures for constructing right angles and squares.

Their interest in mathematics was driven by some rather elaborate ritual requirements. There was a tradition that the only way to overcome a plague that threatened the lives of the people was to appease the gods by doubling the size of the altar. The altar was a sacred object imbued with its own

* The manuals do not make use of the word *Sulba*, for cord, but use *rajju*; they form a supplementary portion of the more extensive *Kalpasûtras*.

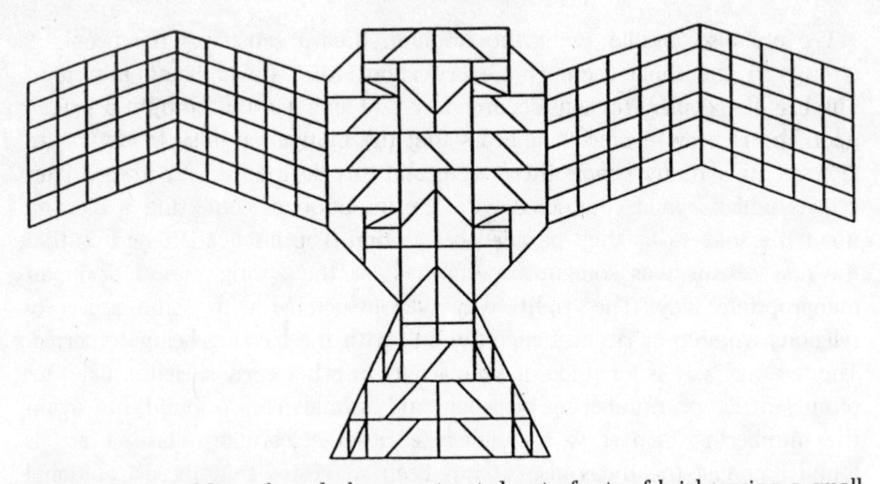

**Figure 2.16** A falcon-shaped altar constructed out of sets of bricks using a small number of basic shapes. There would have been five layers of bricks on the altar shown here. Each layer would have contained two hundred bricks so that the entire *agni* contained a thousand. The first, third, and fifth layers possessed the same pattern of bricks, but a different arrangement was used for levels two and four, so that no brick ever rested upon another of the same size and shape. In the Vedic religion households would maintain altars of specific design for different types of worship. Their construction required great care in order to conform to detailed specifications regarding shapes and areas. Many were square or semicircular but the larger ceremonial altars, like the one shown here, offer the most challenging mathematical problems in their construction and ritual magnification into structures of similar shape but with twice the area. The falcon-shape shown here was chosen for symbolic reasons, because, we read in the *Taittiríya Samhitá*, 'He who desires heaven, may construct the falcon-shaped altar; for the falcon is the best flyer among the birds; thus he [the sacrificer] having become a falcon himself flies up to the heavenly world.'

powers, and so it was viewed almost like a champion to fight against the forces of adversity. If plague or disease appears, then it indicates that the forces of evil have begun to get the upper hand and the forces opposing them must be reinforced. However, in order to invoke the powers of the gods, it was not sufficient simply to increase the area of the altar; it was essential that it be done in the correct way by performing the ritual of traditional and intricate geometrical calculations because as one of their ancient texts warns,

> Those who deprive the altar of its true proportions will suffer the worse for sacrificing.

These ceremonial duties were made absurdly demanding by the unusual shapes of the altars. A common example was an altar in the shape of a falcon built up to about knee height, in five layers, from bricks of simple rectangular and triangular shapes (see Fig. 2.16). The *'Sulba-sûtras* pose and solve the problem of constructing a new altar with the same falcon-like shape as the old one but with its area doubled. The solution involves being able to construct a square with the same area as a given rectangle by first regarding the rectangle as the difference between two square areas before using what we now call Pythagoras' theorem to turn this area difference into a square of equal area. These are the same constructions that we find in works of Euclid two hundred years later. This fact persuades some to argue that both derived them from an earlier common source. Certainly the result was known to the Babylonians and might have been discovered in the same context of ritualistic altar worship, thereby ensuring that, as the geometrical knowledge spread, it carried with it the ritual association. The Sumerian predecessors of the Babylonians laid out elaborate temple plans and believed that they were the works of the gods themselves, conveyed to their architects in dreams. For detailed later examples we have only to look at the elaborate rules of construction and ritual which the Hebrews were required to follow in their temple-building and worship. Even with the early Greeks one still finds this ritual association between geometry and sacrifice mediated by the presence of an altar constructed in a precise way. Abraham Seidenberg has drawn attention to a puzzling reference to Pythagoras which, seen in this light, takes on a new significance. In his commentary on Euclid's geometrical work Proclus proves Pythagoras' theorem for triangles and then comments that

> If we listen to those who wish to recount ancient history, we may find some who refer this theorem to Pythagoras, and say that he sacrificed an ox in honour of the discovery.

Other early Greek histories tell the same story. But it is a very strange one. The casual reader might assume that Pythagoras was doing the equivalent of opening a bottle of champagne to celebrate his great discovery. But this seems most unlikely. Pythagoras was the leader of a fraternity that was fanatically opposed to the sacrifice of animals. It is more likely that this account is partly legendary and has drawn together the traditional elements of geometry and ritual sacrifice. In all probability Pythagoras was not an independent inventor of the theorem that bears his name, but learned of it during his travels around the Mediterranean region and with the passage of time became strongly associated with it in the way that legends have a tendency to crystallize around famous people. And like all legends it contains a grain of truth: it is the ancient link between geometrical theorems and altar sacrifices.

Besides the formidable geometrical problems solved in the *'Sulba-sûtras*, there exists another extraordinary set of geometrical ideas which grew out

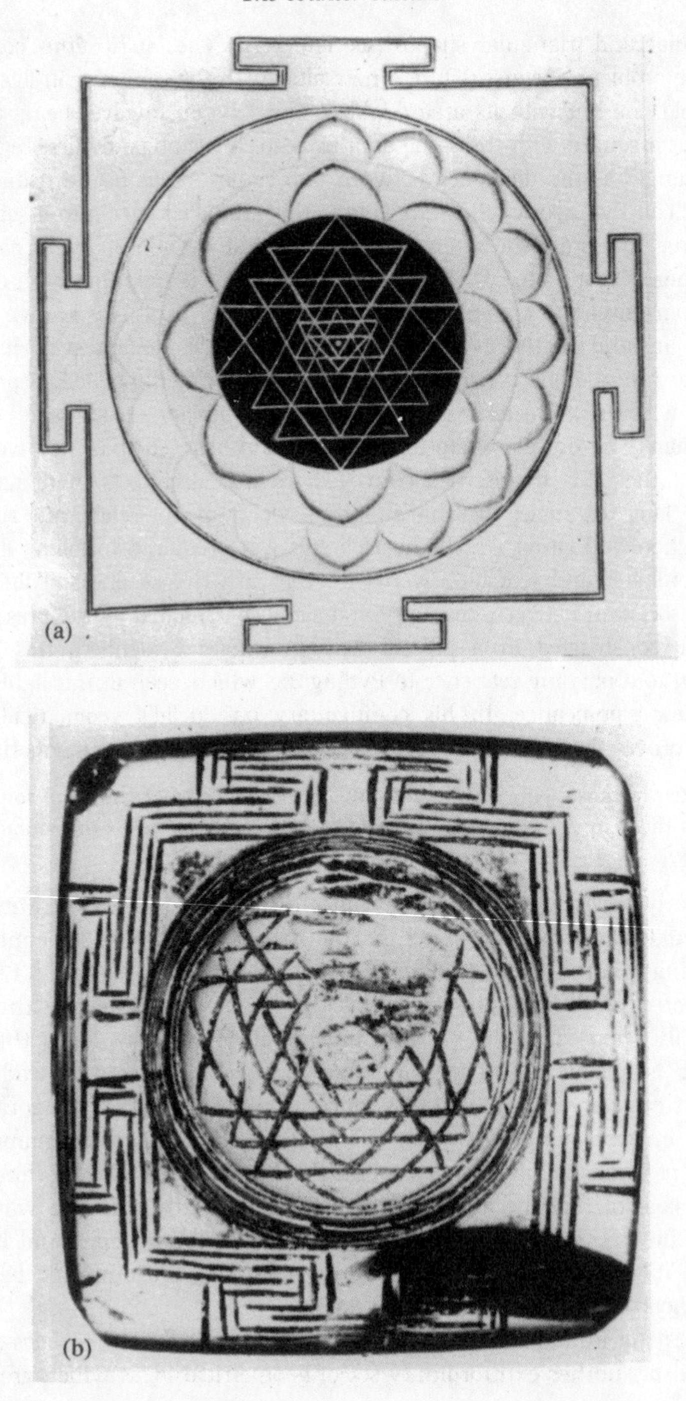
(a)

(b)

**Figure 2.17** (a) The *Sriyantra* is a geometrical construction used for meditation in various parts of the Indian tantric tradition. The earliest known examples and descriptions date from the seventh century AD, but there are Vedic writings as old as the twelfth century BC which are dedicated to the contemplation of geometrical *yantras* which are amalgamations of nine triangles. The *Sriyantra* consists of an intricate scheme of polygons, triangles, circles, and lines enclosing the central *bindu* point. The most difficult mathematical problem is the construction of the system of triangles within the innermost circle. This is called the 'seal'. Almost all the known seals have the structure shown here, with only small modifications. There are a few rare examples of a more exotic structure still, shown here in (b). It uses arcs from ovals, rather than straight lines, to construct three-sided figures with curved sides. These pictures are fascinating to the modern mathematician because they look like 'triangles' that appear in non-Euclidean geometries. In fact, it is known that some of the original drawings of this sort were constructed on a curved solid surface. The earliest examples of this type that survive date from the seventeenth century, but they are quite sophisticated and clearly do not mark the start of this form of seal. These *yantras* were made originally with coloured pastes or powders on the floor or the ground; more permanent copies were made on many materials: paper, metal, cloth, and, most symbolically, rock crystal which was seen to represent an all-inclusive reality because it could be carved to focus incoming light at a single point on its boundary. It is still not clear that early Indian mathematics was sophisticated enough to construct these figures systematically; perhaps they were entwined with highly developed geometrical imagination rather like that displayed in modern times by the Dutch draughtsman Maurits Escher.

of the Indian tantric religious tradition. The earliest known examples date from the seventeenth century. Meditation and worship focused upon the contemplation of intricate geometrical objects, the most elaborate of which is the *Sriyantra*, or 'supreme object', which is a member of a large class of *yantras* ('objects') used for meditational and ritual purposes. Two of these are pictured in Fig. 2.17.

The *Sriyantra* is extraordinary; it consists of a fourteen-sided polygonal construction at the centre surrounded by eight- and sixteen-petalled lotus flowers surrounded by three circles before an outer boundary containing four 'doors' leads to the outside world. The central pattern consists of forty-three small triangles, where the gods reside, produced by the intersection of nine large triangles. Very precise draughting is requiring in order to achieve the intersections of more than two lines at single points as required in a perfect rendering. This structure was used for meditation in two ways. Either beginning from *bhupura*, the outside realm of disorder, proceeding through the doors, into the realm of the gods, past the three circles to the central point, or *bindu*; or outwards from the central point to the *bhupura*. The outward path is taken to represent the path by which the Universe evolved from nothingness and static harmony into greater

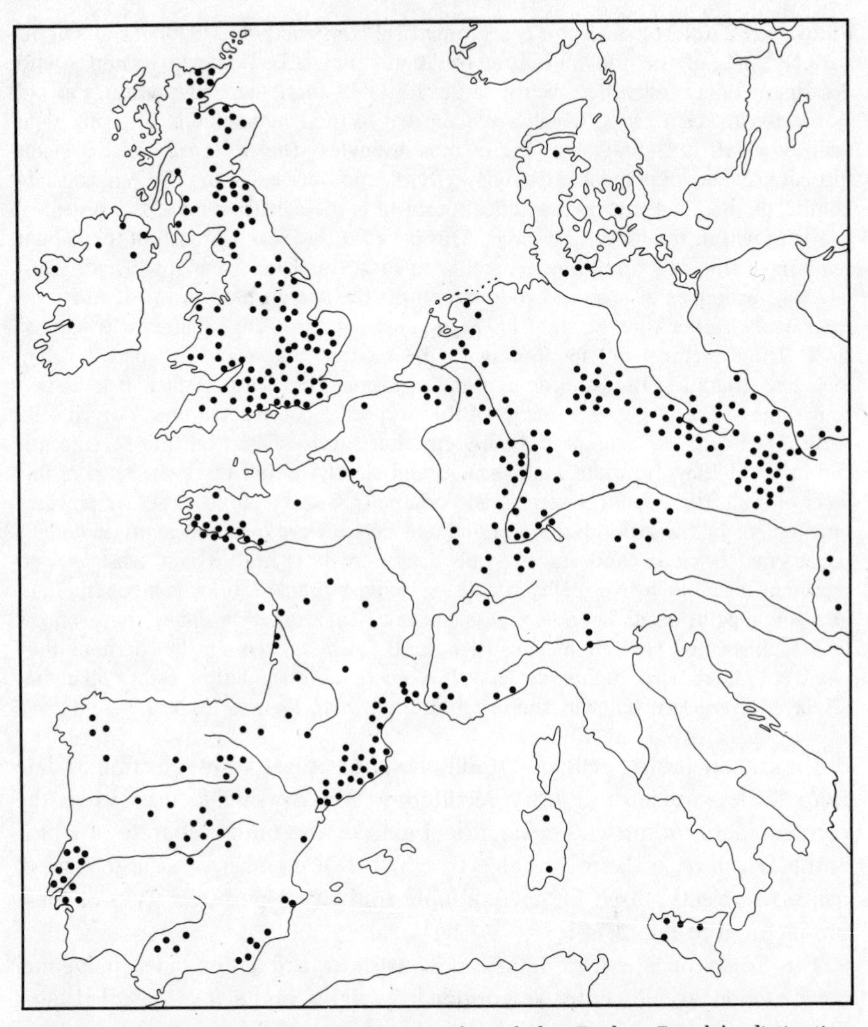

**Figure 2.18** The region in which examples of the Beaker People's distinctive lipped and patterned pottery has been found. Their sites are particularly common around early English standing-stone monuments like Stonehenge, which they erected in the late neolithic period around 2100 BC. The Beaker People migrated to England from continental Europe in about 2500 BC, and played an important role in the spread of Indo-European languages and the decimal counting system.

diversity and complexity and is characteristic of one tantra sect. The other, whose path is from the outside inwards, represents the gradual destruction of the Universe is emphasized by other strains of Tantrism.

It has long been known that in Continental Europe, Britain, and Ireland there are many extraordinary megalithic monuments, of which Stonehenge is the most spectacular. There have been extensive investigations of their possible purposes and the way in which they encode quite advanced geometrical and astronomical knowledge. The earliest examples witnessing to such knowledge by patterns of stones which allow the rays of the sun to shine through special openings on the days of the summer or winter solstice appear in Ireland around 3300 BC.

Around 3000 BC the first stone circles were constructed; the first part of Stonehenge appeared in 2800 BC during the latter part of the neolithic age. Very little is known about the people who had this astronomical knowledge of the motions of the Sun and the Moon. What excavations there have been of the sites where they lived near Stonehenge indicate that they were an unusual and rather élite group whose food supplies were brought from elsewhere by others who were hunters and gatherers. Perhaps they received provisions as a form of tribute from the awestruck people round about? After 2500 BC, a new population of physiologically distinctive immigrants from southern Europe appears in southern England. They have become known as the 'Beaker People' because of the distinctive lipped pottery and beakers covered with intricate designs that are found in all the places where they settled. It was the 'Beaker People' who built the spectacular trilithons at Stonehenge in two periods of building in 2100 BC and then in the Bronze Age after 2000 BC. The spread of the Beaker People seems to be intimately connected with the propagation of Indo-European languages and the cultural aspects embedded in them. They emerged from a region where an Indo-European language was spoken at a very early date. The extent of their spread and influence can be seen in Fig. 2.18 which is based upon a compilation by B. L. van der Waerden.

The remarkable common features of the ritual origins of geometry and the idea that punishment might be averted by performing certain exact calculations in Greece and India undoubtedly has a linguistic aspect. Greek and Sanskrit are both Indo-European languages deriving from a common mother language. The spread of this common root language throughout the centres of civilization during the period from 3000 to 2000 BC undoubtedly carried with it many religious and mathematical ideas. Pythagoras' theorem seems to have been known as early as 2000 BC when the ancestors of both the Greek and the Hindus shared a common geographical region near the Danube. It is reasonable to suppose that as the Greeks spread to the south and the Hindus to the east through Iran and northern India they took this common knowledge with them to their new centres of civilization.

All these things—the taboos, the rituals, and the origin of counting—may be connected in a profound way. Some historians, notably Seidenberg, have placed great emphasis upon the detailed form and function of ancient pagan rituals as the source of many mathematical intuitions to do with counting and geometry. They argue that counting originated in the detailed form of tribal rituals which revolved around the celebration of the most primitive recognitions of fertility rites and the pairings of male and female. Such rituals are the dramatic rendering of many mythopoeic notions which we find dominating primitive views of the origin and consistency of the world and the peoples in it. Certainly in some ancient cultures—Egyptian, Babylonian, or Mayan—we find numbers associated with particular participants and deities involved in elaborate rituals. These rituals are already rather sophisticated but they must have developed from more primitive forms in which the numerological elements were rather more basic. The simplest fertility rites celebrate the creation of life in the male and female form and offer natural formats in which people then appear in pairs. Some of the primitive 2-counters we have looked at already in this chapter certainly possessed ritual usages of even and odd numbers in naming places or in the divining activities of the high priest, or 'medicine man', who played a central role in all their ritual activities. We have seen examples of 2-counters who develop larger numbers in ways that cannot have evolved directly from finger counting. The pattern that suggests itself is that perhaps finger counting was just one ingredient of body counting, but that all these practices of gesturing with parts of the body to indicate numbers were part of primitive ritual. They represented a division of the body, and by replacing the process of counting by the mechanics of gesturing one enables the taboo to be respected. The particular ritual that has been picked as a potential candidate for initiating the practice of counting is the widespread 'creation' or 'fertility' ceremony that is very significant in the ancient world. The appearance of male and female in pairs offers a natural setting for 2-counting. One can then ask whether the strange tradition of attributing sexual significance to inanimate objects and numbers, for example regarding, as Pythagoras did, 'even' numbers to be female and 'odd' numbers to be male, is a remnant of such a beginning. Perhaps men and women were numbered during a ritual, starting with the men and proceeding alternatively male–female so that all the odd-numbered were male and all the females were even-numbered. We know there certainly did exist counting rites in a very large number of ancient cultures. The Mayans and the Babylonians both had number gods and there are many numbering rituals associated with Orthodox Jewish worship. The most famous echo of such things in the Old Testament writings is in the story of Noah, where we read of the ceremonial boarding of the Ark:

There went in two and two unto Noah into the ark, the male and the female, as God had commanded Noah.

The natural pairing associated with the male and female fertility rites is used by Seidenberg as an explanation for the prevalence of 2-counting around the Earth and for its priority over systems of finger counting as well as the ritual and superstitious distinction between odd and even numbers. The extension of the system to higher numbers is seen as being provoked by the growth in the number of participants in the ritual. There are further baroque embroideries that can be added to this basic hypothesis. Seidenberg finds all this, and much more, partially suggestive evidence completely compelling, feeling that it all lends support to one and only one possible explanation for the origin of counting and the strange superstitions that seem to have issued irrationally from it:

Counting was invented in a civilized center, in elaboration of the Creation ritual, as a means of calling participants in ritual onto the ritual scene, once and only once, and thence diffused.

## THE PLACE-VALUE SYSTEM AND THE INVENTION OF ZERO

*Now thou art an O without a figure. I am better than thou art now. I am a fool, thou art nothing.* WILLIAM SHAKESPEARE (King Lear, I: iv)

Our exploration of the evolution of number words and number systems in the ancient world suggests that intuition for number was far from spontaneous and inevitable. Most systems of counting arose from practical application and were strongly wedded to finger counting. There was no abstract sense of number. Moreover, the systems we have examined are all somewhat cumbersome and difficult to use for anything more sophisticated than simply counting. To employ numbers in a more powerful and adventurous fashion than merely as mnemonics—to multiply and divide numbers by reference *only* to the number symbols rather than to the collections of things being multiplied—requires another innovation that was made in very few centres of advanced civilization. To see what that innovation was we shall take a look at its emergence in the Sumerian and Babylonian cultures.

Written languages are able to express a vast array of ideas simply by the permutation of a small number of symbols because both the identity of the symbol and its position relative to others have meanings. The same three symbols 'g', 'd', and 'o' can produce the two strings 'god' and 'dog', which convey completely different ideas. The Babylonians were the first to exploit this simple but powerful principle in the writing of number

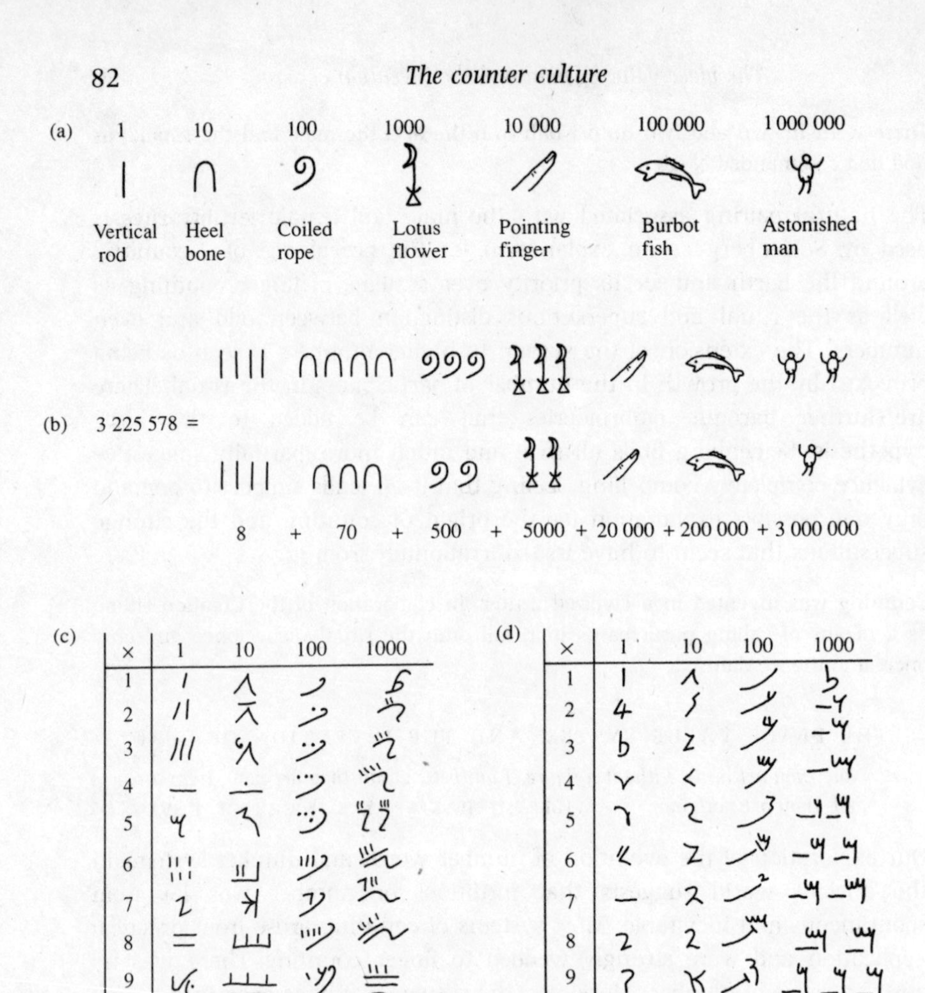


**Figure 2.19** Egyptian 'hieroglyphic' (meaning 'sacred picture writing') symbols for numbers developed earlier than 3000 BC which were used for making permanent records on stone or metal. When writing on papyrus the priests used a simpler script called 'hieratic' (meaning 'priestly'); for mundane purposes the Egyptians used the 'demotic' (meaning 'popular') script. The hieroglyphic number symbols are shown in (a) along with a translation of the pictures used. Since the Egyptians usually wrote from right to left (occasionally they wrote vertically downwards or even from left to right, as I do), they would have represented the number 3 225 578 by the cluster of symbols shown in (b). The order of the symbols has no numerical significance and the arrangement was made for aesthetic reasons alone. After the introduction of the hieratic script, hierogylphic number symbols tended to be used in special circumstances where a formal antique style was appropriate, rather as we still use Roman numerals. The hieratic system was much more economical in the use of symbols, which are shown in a multiplication format in (c). Five or six times fewer penmarks were required to

write some large numbers in hieratic rather than hieroglyphics. Later, some abbreviations were introduced for numbers which required large numbers of marks. These abbreviated forms of the hieratic and demotic number symbols are shown again in a multiplication table, in (d), based on an original representation by Carl Boyer.

symbols. It was an innovation that seems to have appeared around 2000 BC, conceived by the palace mathematicians and astronomers of Babylon and it made their system superior to all others used in the ancient world. It is a very familiar device to us today. If we see the number 123, we must interpret it to mean $1\times10\times10 + 2\times10 + 3$. The positions of the digits determine the values that are associated with them. Besides economizing on the number of symbols that are required to express numbers of unlimited size we recognize that this *place-value* system leads to very great ease and convenience when carrying out arithmetic. We can list numbers to be added one below the other and add them column by column carrying over totals beyond ten to the next column on the left. Learning to do this under columns headed 'hundreds', 'tens', and 'units' is probably our earliest memory of doing sums at school.* To appreciate how difficult it is to develop any systematic study of mathematics without a place-value system one has only to imagine trying to do arithmetic with Roman numerals, where there is no place-value system. If one adds 365 (=CCCLXV) to 651 (+DCLI), then we have

| | |
|---:|---:|
| CCCLXV | 365 |
| DCLI | 651 |
| MXVI | 1016 |

There is simply no system in the Roman symbols which permits the calculation on the left to be performed systematically. You end up doing the sum on the right and then converting the answer into Roman numerals. The columns mean nothing in the Roman sum. You have to count by some other more primitive method of tallying on an abacus before describing the answer using the Roman symbols. The notation has no intrinsic algorithmic power. And these problems arise just at the stage of doing addition; systematic division and multiplication are unthinkable with such symbols. We see immediately how a good place-value notation like our own actually does some of the thinking for you.

The place-value scheme is very economical in the use of symbols. What happens when you do not employ it is well illustrated by the situation

* This principle of 'place-value' does not only work in the decimal system of course; it can be applied to arithmetic systems with any base.

| Number | Symbol | Number | Symbol |
|---|---|---|---|
| 1 | | 11 | |
| 2 | | 16 | |
| 3 | | 25 | |
| 4 | | 27 | |
| 5 | | 32 | |
| 6 | | 39 | |
| 7 | | 41 | |
| 8 | | 46 | |
| 9 | or or | 52 | |
| 10 | | 55 | |
| 20 | | 59 | |
| 30 | | | |
| 40 | or | | |
| 50 | or | | |

**Figure 2.20** The first fifty-nine numbers in the Babylonian system.

**Figure 2.21** The number 69 is not written as in (a) but as in (b) with a new symbol representing 60 and a place-value system, so this number is read as 1×60 + 9. This distinguishes the Babylonian system from that of the Sumerians, from which it evolved.

with early Egyptian mathematics which was summarized in Fig. 2.19. A real ambiguity arises because one can write down the same collection of number symbols in a variety of different patterns and the sum of their values will still be interpreted to be the same. The situation was made even more cumbersome by the existence of two forms of writing: the hieroglyphic and demotic, the former for sacred writing, the latter for popular consumption. The two forms of representing numbers are described in more detail in the caption to Fig. 2.19.

Since the Babylonian place-value system was built around the base 60, when it records symbols represented by the stylus marks corresponding to the combination 123 this denotes 1×60×60 + 2×60 + 3, which we would call 'three-thousand seven-hundred and twenty-three'. An easy way to get used to the Babylonian system is to recall that we still use it for measuring time and we could think of 123 as one hour two minutes and three seconds, which equals 3723 seconds.

Although our own place-value system uses nine symbols to mark the digits from 1 to 9, the Babylonian one did not use 59 symbols to record the numbers from 1 to 59. They managed to represent all these basic numbers using combinations of just two shapes which the wedge-shaped stylus could make in the clay tablets: a vertical wedge shape for the number 1 and a crescent shape for 10 (see Fig. 2.20). So we see that all the numbers below the base 60 were written according to the base-10 system, using addition of symbols without the use of the place-value idea. But numbers above 60 are written using the place-value system, as illustrated in Fig. 2.21. There were two difficulties with this system which could cause error and ambiguity when it was read by people who did not already know precisely what any given combination of symbols was supposed to represent. The number 10;10 = 10×60 + 10×1 (601) was denoted by the double wedge <<, but it could easily be confused with the symbol used for twenty. Attempts were made to overcome this by leaving spaces so that the 10;10 representation looked like < <. But when one needs more than one adjacent space it becomes difficult to tell whether a scribe has just created one slightly wider than average space or whether there are meant to be two gaps. Idiosyncrasies in writing become sources of serious error. Eventually the strategy of the blank space was replaced by inserting a new symbol to indicate the space.

It is clear what the real problem is here. The ambiguity in the system arises because there was no symbol for *zero*. Only with such a symbol would it be possible to indicate that particular multiples of 60 were absent in an expression for a number. Without a zero, the symbol for 2 could mean 2, 2×60, 2×60×60, 2×60×60×60, and so on. The simple device of leaving spaces to clarify the meaning of an expression was introduced by the Babylonians somewhere around 2000–1800 BC. But it was not until about 200–300 BC that the Babylonians began to employ a special symbol to denote the absence of a unit (see Fig. 2.22). It was still limited in concept, being used to mark gaps between symbols, but never used at the end of a number so there was still the ambiguity that any given symbol

$2 \times 60 \times 60 + 0 \times 60 + 15$

**Figure 2.22** An example discovered on an astronomical tablet at Uruk dating from the period between the late third and early second century BC displaying the Babylonians' use of a 'zero' symbol to denote a blank space in their place-value system. It shows the adjacent number symbols 2;0;15 which, in their base-60 system, signify the number 2×60×60 + 0×60 + 15. We would write this as 7215. The Uruk tablet is now in the Louvre Museum, Paris. This is one of the earliest examples of the use of a zero symbol. The symbol string we have shown here occurs on line 24 of the original.

could mean its interpreted value or that value multiplied by any power of 60.* Curiously, it appears that the employment of the symbol zero at the end of a string of numbers began to be used by the astronomers but not by the mathematicians. Indeed, the astronomers developed the notation to such an extent that they were able to represent fractions in sixtieths without difficulty. In so doing they also employed the zero at the start of a number string as well as at the end, so that they could represent, say, one-sixtieth unambiguously as zero (whole units) plus one-sixtieth.

Here we see the development of a system of recording numbers that was poised for dramatic and abstract extension. The Babylonians had developed the place-value system and the essential associated notion of the zero symbol. When employed in conjunction they empowered their astronomers to begin representing fractions in symbolic form. When Greek astronomers later inherited the Babylonian system they introduced a small circle to denote the zero symbol and this is the origin of the 'nought' that we use today to denote zero. They, like Jewish and Arab astronomers, replaced the cuneiform notation by their own system of using adaptations to their alphabet of letters for denoting numbers.

Yet, for the Babylonians the zero symbol did not carry with it all the meanings that we might readily associate it with today. It had the definite technical meaning of being just a blank entry in a particular representation of things. It was not thought of as being a symbol for 'nothing' in the more general or abstract sense. Indeed, it was never used to represent the result of, say, subtracting ten from ten. The result of

* However, the cuneiform examples show that usually the Babylonian scribes were very careful in the spacing of symbols to avoid such ambiguities. As a general point, it might be remarked that errors of calculation and transcription in ancient texts can often be more revealing of the writer's thought processes than correctly performed calculations.

that sum was 'nothing', a quite different concept and one which they clearly had very great difficulty in conceiving a representation for. It certainly was not conceived of as a number. One reason for treating zero differently is that one cannot use it like other numbers when making fractions—you cannot divide by zero.* It is almost as if there appeared to be some latent contradiction in the idea that one should use a 'something' to denote the absence of anything. The difficulty of assimilating this idea will be familiar to those involved in teaching counting to very young children.

The place-value system with its benefits and stimulus to represent the zero concept was invented first by the Babylonians, but it was also introduced independently later on by a small number of other advanced civilizations. The ancient Mayan civilization, which began in the fifth century BC and reached its peak between 300 and 900 AD in Central America, employed a place-value system in conjunction with its base-20 counting system. They used a dot to represent 'one' and a horizontal line for 'five'. For example, they would represent our numbers 3, 8, and 10 as follows:

| | . . . | ― |
|---|---|---|
| . . . | ― | ― |
| 3 | 8 | 10 |

Larger numbers (and they were interested in very large numbers for religious reasons, since they assiduously maintained calendar dates from the assumed mythological beginning of their civilization thousands of years in the past[†]) were represented by stacking these symbols in columns with the units in the bottom row, 20's in the next row up, 360's in the next, 7200's in the next, and so on. For example our number 37 373 would be represented as $5\times7200 + 3\times360 + 14\times20 + 13\times1$ in the

* This was the reason why Aristotle had difficulty in accepting zero as a number. The early Indian mathematicians seem to have been the most sophisticated in their treatment of zero in this respect. In the sixth century AD Brâhmagupta addressed the problem of division by zero and in the twelfth century Bhâskara actually used division by zero as a means of defining infinity.

† The sophistication of their system of time-keeping was quite staggering. For instance, the duration of the solar year as determined by modern astronomy is 365.242 198 days. The Gregorian calendar which we use for everyday purposes gives the year a duration of 365.242 500 days, and so this means we have an error of 3.02 ten-thousandths of a day in our measure of the true duration of the astronomically defined year. The Mayans, by contrast, created a calendrical system which gave their year a duration of 365.242 000 days, an error of only 1.98 ten-thousandths of a day when compared with the true astronomical duration. They would think us primitive indeed! All this and more they did in building a vast culture enhanced by architecture and art of great quality, and yet they never invented the wheel nor employed beasts of burden. How nervous this should make us about our speculations as to what ancient and distant cultures 'must' have been like.

following a vertical array with the units at the bottom of the stack:

(= 5 × 7200)
(= 3 × 360)
(= 14 × 20)
(= 13 × 1)

A shell-like symbol was used to mean 'zero' and has a suggestive resemblance to a closed eye. Like the Babylonians, they used it to indicate the absence of any unit of quantity. Curiously, they introduced it in ways that seem at first superfluous. We find periods of time recorded as 'zero hours, zero minutes, and five seconds'. Why not merely say 'five seconds'?

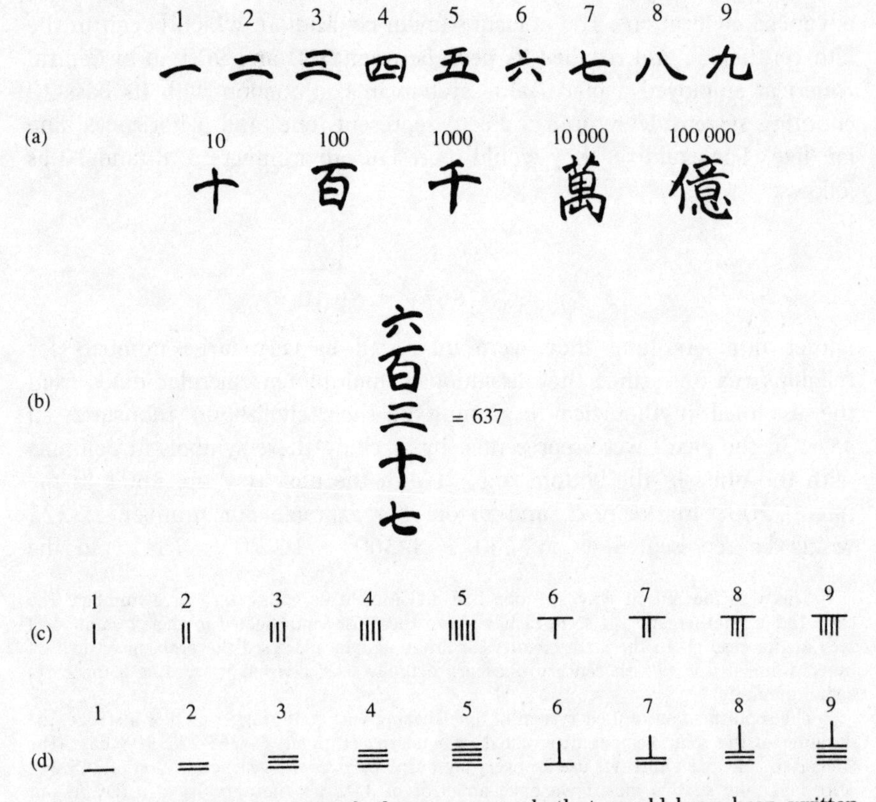


**Figure 2.23** (a) The traditional Chinese numerals that would have been written with an ink-brush on paper, silk, or wood, and used for fiscal accounting; (b) our decimal number 637 written in traditional symbols; (c) The scientific Chinese numerals from 1 to 9. They are images of bone and bamboo calculating rods. When these numbers are used in the tens, hundreds, or thousands position, they are written the other way around, as shown in (d). Our decimal number 6666 would have been written as ⊤⊥⊤⊥.

The answer, and with it a clue to their motivation in independently inventing the zero symbol, comes when we realize that the Mayans depicted periods of time and other quantities as *pictures* in which each part of the design symbolized a part of the total sum. Thus the picture for 'one hour, one minute, and five seconds' would have three ingredients to fill the frame allowed for it. Unless one had a cartoon for zero, the overall picture for 'five seconds' would have two empty spaces. The zero sign was therefore vital to retain the balance and imagery of the representations of the numbers and all the aesthetic and religious associations that went with them.

The early Chinese also invented a decimal system with 'ten' represented by a single horizontal mark '—' in the tens position, with further multiples of ten up to fifty denoted by more horizontal lines stacked on top of this, one line for each ten; so = means twenty, and so on. Beyond the number fifty, new symbols were introduced and stacked with themselves or other symbols repeated in an analogous fashion. This representation derives from the custom of using bones and then bamboo counting-rods for representing numbers. These numerals were used for scientific purposes and should be distinguished from the traditional pictorial symbols employed by Chinese accountants and administrators. They are shown in Fig. 2.23. Whereas the place-value system can be found in use on Chinese coins in the sixth to fifth centuries BC, the zero symbol, again denoted by a circle, was only added very late in the development of the system, in the eighth century AD, and was imported from India.

The Indian development of these ideas is unusual in many ways. It was a flowering of the great Indus culture that arose near Mohenjo-daro and Harappa at about the same period (3000 BC) as the early Egyptian and Sumerian civilizations. These Hindu cultures developed a true place-value system that has been much studied because by the sixth century AD it evolved into the system of numerals that we employ in the West today. But whereas the mathematicians and astronomers of Sumer and Babylon laboured for nearly 1500 years before they introduced the notion of a 'zero' symbol, in India it was introduced much earlier. Moreover, unlike in other examples, the notion of 'zero' or the 'null number', as it was known, was originally associated with the notion of 'nothing' in the abstract sense. The literal meaning of the number word was 'void' and it represented both an empty slot in a counting system and the answer to a sum like 'ten minus ten'. Its subsequent evolution as a symbol was not dissimilar to that in Mesopotamia, but the development of the terminology during the gradual synthesis of Indian and Arab cultures is interesting. In Sanskrit, the Hindu name for zero is *sunya*, whilst the Arabic is *as-ṣifr*; both mean 'the empty one'. When it was written in medieval Latin, the

Arab word transcribed as *zefirum* or *cefirum*. In Italian it gradually evolved from *zefiro* to *zefro* and *zevero*. When the latter was expressed in the Venetian dialect, it became our 'zero'. The other Latin word, *cifra*, evolved differently and acquired a less specific meaning, being used to denote any of the numbers from 0 to 9. From it we obtain our English word 'cipher' or 'cypher', meaning a string of number symbols. The *Oxford English Dictionary* records the verb 'to cipher' being used as a synonym for 'to count' in the sixteenth century. But we also find recorded there the largely defunct usage of the term as meaning 'nothing' or a term of abuse for a person whom one wanted to call a nonentity. In the nineteenth century Thackeray writes of an unbalanced domestic situation of a hen-pecked Lord and his Lady in which he finds

> his lordship being little more than a cypher in the house.

At the beginning of the nineteenth century we read in Edgworth of a predictable dispute in English Academe:

> It was said that all Cambridge scholars call the cipher aught and all Oxford scholars call it nought.*

The epigram which appears on p. 81 gives an example from Shakespeare's *King Lear* in which the symbol 'O' appears in the text and a comic point is made which turns upon a distinction between zero and nothing.

The use of a symbol for, and concept of, zero might have been expected to create the notion of negative numbers rather readily. This seems not to have been the case. Negative numbers do not appear to have become generally recognized as 'numbers' until the sixteenth century. Thus, Diophantus described as 'absurd' equations with negative answers. The early Chinese used counting rods coloured black and red to represent negative and positive numbers. Sixth-century Indian mathematicians mention negative numbers and Hindu mathematicians would signify a negative number by placing the number symbol inside a circle (a practice that is still retained on the scoresheets of some games like the 48-card game of pinochle). Two hundred years later one finds Arab mathematicians adopting a similar strategy by placing a dot over a number symbol to denote what we would call a minus sign. Much later, in thirteenth-century Europe, one finds a ready appreciation of negative quantities by mathematicians like Leonardo Fibonacci, to signify losses in financial problems; but formal equations like three minus seven were still not regarded as having any meaning. By the sixteenth century negative

* With the passage of time, it should come as no surprise that the wisdom of Oxford's choice in this matter has evidently prevailed.

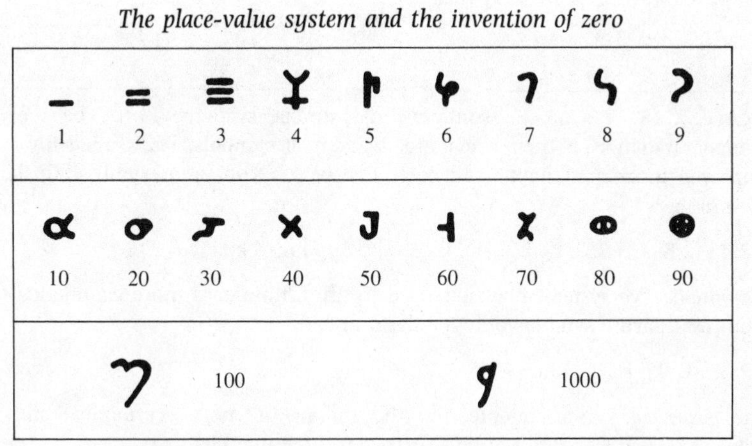


**Figure 2.24** An early form of Indian numerals in Brâhmî script. Our own number symbols are descended from them.

numbers seem to be readily accepted and are referred to as 'false' numbers in contrast to 'true' positive numbers.*

The real power of the place-value system of Indian origin was that it combined four advantageous features. There were unique symbols for the numbers 1 to 9 (see Fig. 2.24) which were truly abstract and did not need to convey pictorial information about their meaning; it was a base-10 system throughout; it employed a place-value notation; and it used a zero. The Mayan system was irregular in the way in which it stepped up to new units. The positions of a basic unit did not mean 1, 20, 20×20, 20×20×20, etc.; but 1, 20, 18×20=360, 18×20×20=7200, etc. The Babylonians had signs only for 1 and 10 and just repeated these in combinations to get the numbers up to 60; the Chinese were similarly restricted, with special signs only for 1 and 5. Thus none of these systems could deploy them in an operational way like the Indians could. Their system was a fully regular decimal one—essentially that which we use today with levels at 10, 10×10, 10×10×10, etc. (see Fig. 2.25). Moreover, the Indians also developed the quick and easy means of referring to large numbers that we habitually use today. A number like 3456 would correspond in the decimal system to

$$3\times10\times10\times10 + 4\times10\times10 + 5\times10 + 6$$

* The subtraction sign '–' arose as an abbreviation of 'm' and 'm̅' which were both used to denote 'minus'. The plus sign '+' arose as a contraction of the Latin word *et* meaning 'and' when written quickly. The equality sign '=' was introduced in 1557 by Robert Recorde in a book entitled *The Whetstone of Witte*. It signified a pair of parallel lines because in his words 'noe 2 thynges can be moare equalle'. It took a long time for this to become accepted usage, however, and eighty years later we find Descartes using a proportional sign (α) for equality. It looks to be a hastily written form of 'æ' an abbreviation of *æquales*, the Latin for 'equal'.

**Figure 2.25** A symbolic summary of number systems. If the base of the number system is B then a ciphered system of symbols, like the *additive* ones employed in ancient Egypt and early Greece, uses different symbols to denote the numbers

$1,2,3,\ldots\ldots,B-1,B,2B,3B,\ldots B(B-1); B^2, 2B^2, 3B^2,\ldots B^2(B-1);\ldots$etc

A *multiplicative* system, like that used by the Chinese, is more economical and requires separate symbols only for the numbers

$1,2,3\ldots(B-1), B, B^2, B^3,\ldots$

The *positional* system adopted by the Indians is more economical still and requires only B separate symbols (with zero now included)

$0,1,2,3,\ldots(B-1)$

Any number, N, is expressed as a formula

$N = a_nB^n + a_{n-1}B^{n-1} +\ldots+ a_2B^2 + a_1B + a_0$

and written in place-value notation simply as the string of symbols

$a_na_{n-1} \ldots a_2a_1a_0$

and would be spoken as three-thousand, four-hundred, five-tens, and six in other decimal number systems. The Indians used the abridged oral place-value method of description that we use: reading it as 'three, four, five, six'. Simple as this strategy is, it was never developed anywhere else.

Another curiosity of the Indian system is that it used symbols to denote numbers, in the way that we have adapted from them, *and* it used words with symbolic meanings to represent numbers, so one finds mathematics effectively written in verse. This seems to have played an important role in preserving records accurately over the passage of time as they passed through the hands of many copyists.

The Indian system of counting has been the most successful intellectual innovation ever made on our planet. It has spread and been adopted almost universally, far more extensively even than the letters of the Phoenician alphabet which we now employ. It constitutes the nearest thing we have to a universal language. Invariably, the result of any contact between the Indian scheme and any other system of counting was the adoption of the Indian system with perhaps a different set of names for the symbols employed. When the Chinese encountered the Indian system through the influence of Buddhist monks in the eighth century they adopted the Indian circular zero symbol and moved towards a full place-value system with a simplified system of just nine number signs. The

Hebrew tradition imported the Indian ideas through the travelling scholar Rabbi Ben Ezra (1092–1167), who travelled extensively in the Orient during the twelfth century. He described the Indian system of numbers in his influential *Book of Number* and used the first nine letters of the Hebrew alphabet to represent the Indian numbers from 1 to 9 but kept the Indian circle to symbolize zero, naming it after the Hebrew word for 'wheel'. Arab cultures had their own sophisticated arithmetical traditions and, upon encountering the Indian system, they gradually altered the appearance of its number symbols until they took the forms which, with the exception of some later changes to the form of 'five' and 'zero', we use today. Although Arabic writing reads from right to left, the Indian convention of reading numbers from left to right has been retained. The evolution of number systems and number symbols is summarized in Figs 2.26 and 2.27.

The spread of the Indo-Arab system of numbers into Europe is traditionally credited to the influence of the French scholar Gerbert of Aurillac (945–1003). He spent significant periods of his early life in Spain, where he became acquainted with the science and mathematics of the Arabs. Later, he directed the education of generations of theological scholars in Rheims and his intellectual influence spread widely throughout Europe. He held several other influential positions in the Church before finally being elected Pope Sylvester II in 999. He is generally credited with the introduction and spread of the Indian-Arab numerals throughout Europe. They spread, not through written documents, but through the practice of teaching people how to count using a particular type of abacus called the Roman counting-board. The written tradition overtook all others at the end of the twelfth century when Gerbert's counting-board started to fall into disuse. Gradually, the advantages of the Indo-Arab system became apparent and in the thirteenth century we find its widespread use in commerce and trade. But it was not entirely without opposition. Florentine merchants were forbidden to use them and were instructed to employ Roman numerals or number words. They were not permitted in contracts and other official documents. In 1299 a law was passed in Florence prohibiting their use. Apparently, the opposition was not simply the result of a Luddite conservatism. Rather, in these days before printing, it was a fear of fraud. Roman numerals were written in a fashion that was designed to prevent them being surreptitiously altered. Thus 'two' was written as IJ rather than II to prevent some unscrupulous person adding another I on the end. It is rather like our practice of adding the word 'only' to the written amount on a cheque. The Indo-Arab system of numerals had far less defence against this type of fraud by interpolation. The pattern of the Roman system stopped most additions simply because

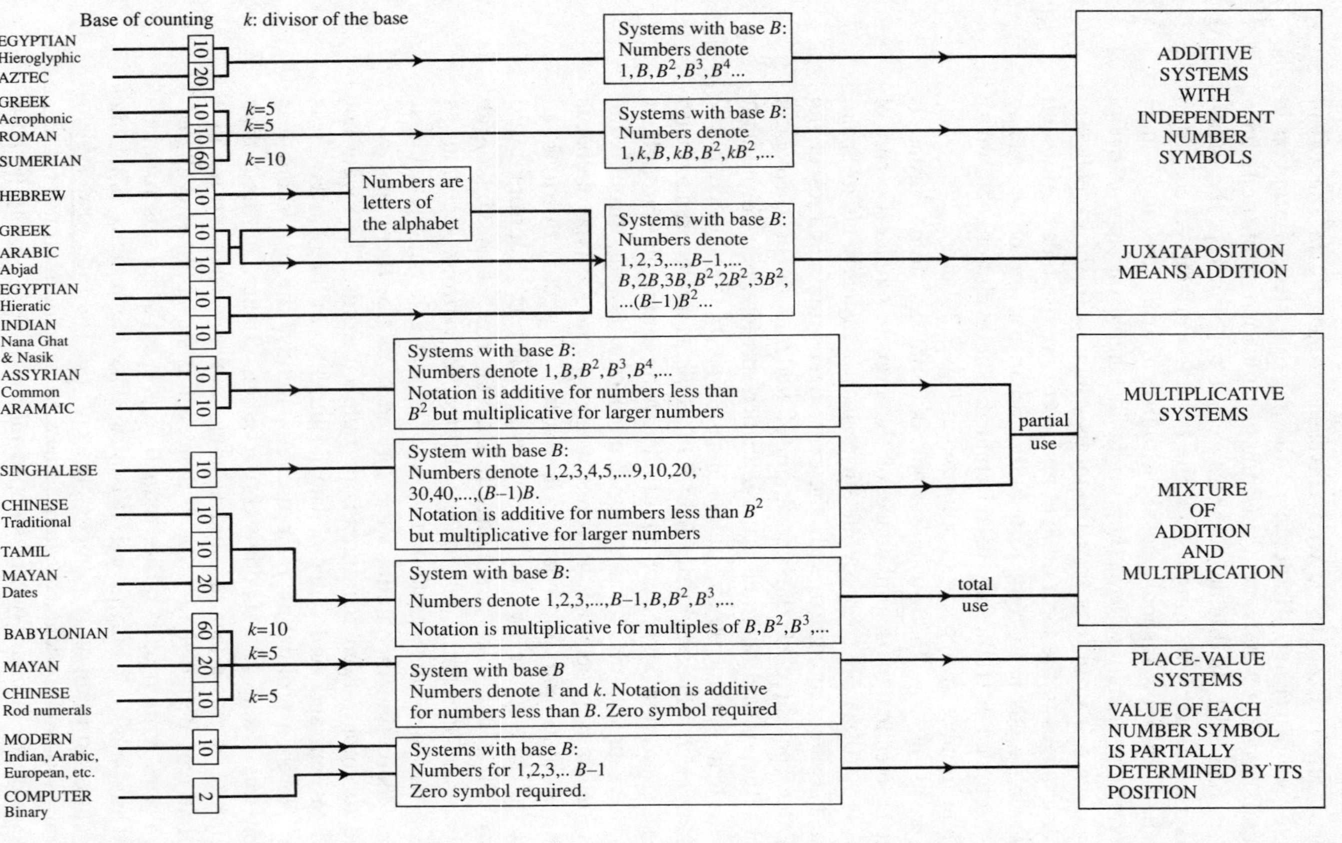


**Figure 2.26.** The development and interrelations of written counting systems showing their choice of base and systematization. They divide into the three types of system described in Figure 2.25: *additive, multiplicative,* or *positional.* This classification is a modified and updated version of that originally devised by the French historian G. Guitel.

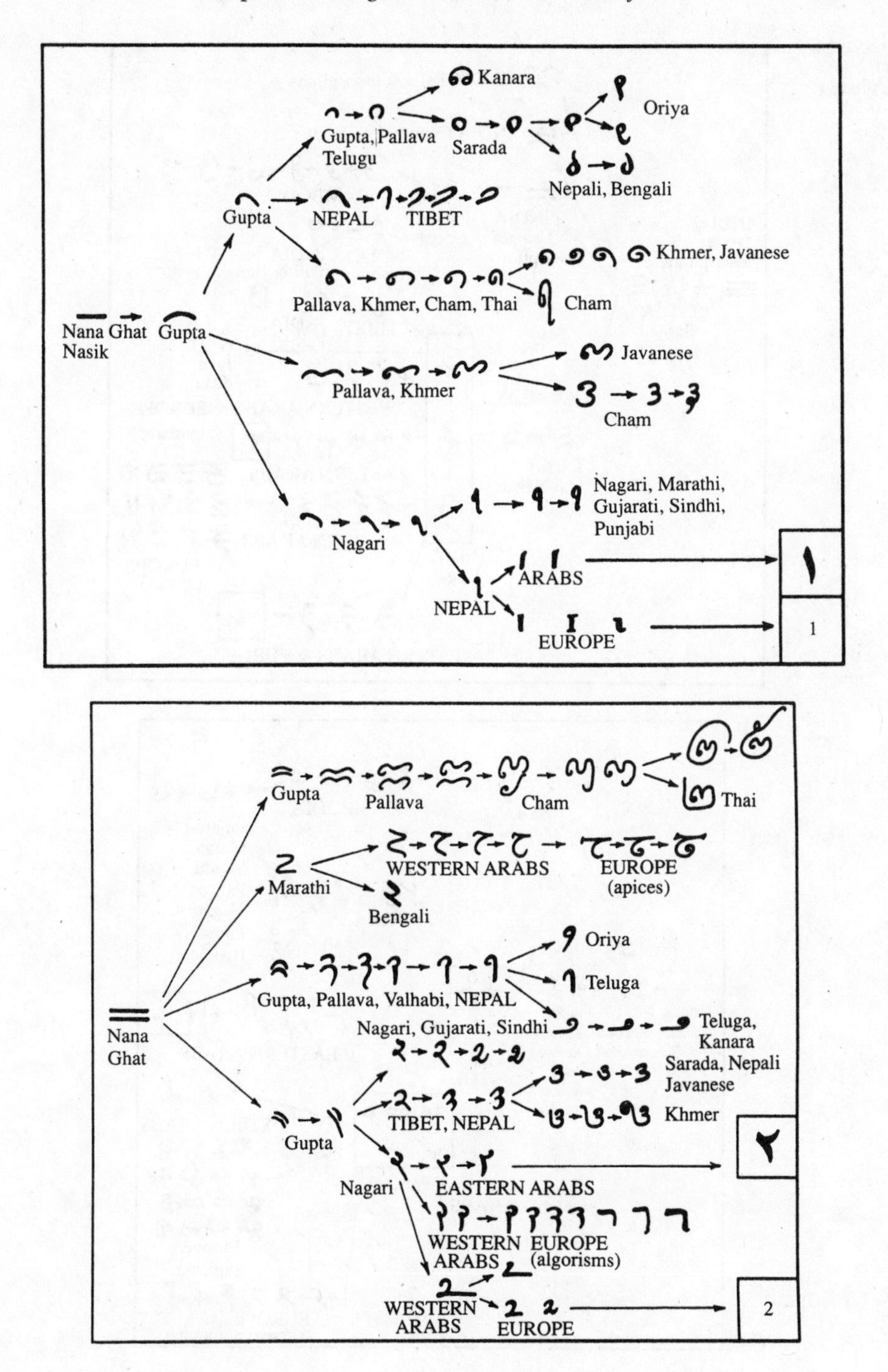
Kanara
Oriya
Gupta, Pallava
Telugu
Sarada
Nepali, Bengali
Gupta
NEPAL
TIBET
Khmer, Javanese
Pallava, Khmer, Cham, Thai
Cham
Nana Ghat
Nasik
Gupta
Javanese
Pallava, Khmer
Cham
Nagari, Marathi,
Gujarati, Sindhi,
Punjabi
Nagari
ARABS
NEPAL
EUROPE
1
Gupta
Pallava
Cham
Thai
Marathi
WESTERN ARABS
EUROPE
(apices)
Bengali
Oriya
Teluga
Gupta, Pallava, Valhabi, NEPAL
Nana
Ghat
Nagari, Gujarati, Sindhi
Teluga,
Kanara
Sarada, Nepali
Javanese
TIBET, NEPAL
Khmer
Gupta
Nagari
EASTERN ARABS
WESTERN
ARABS
EUROPE
(algorisms)
WESTERN
ARABS
EUROPE
2

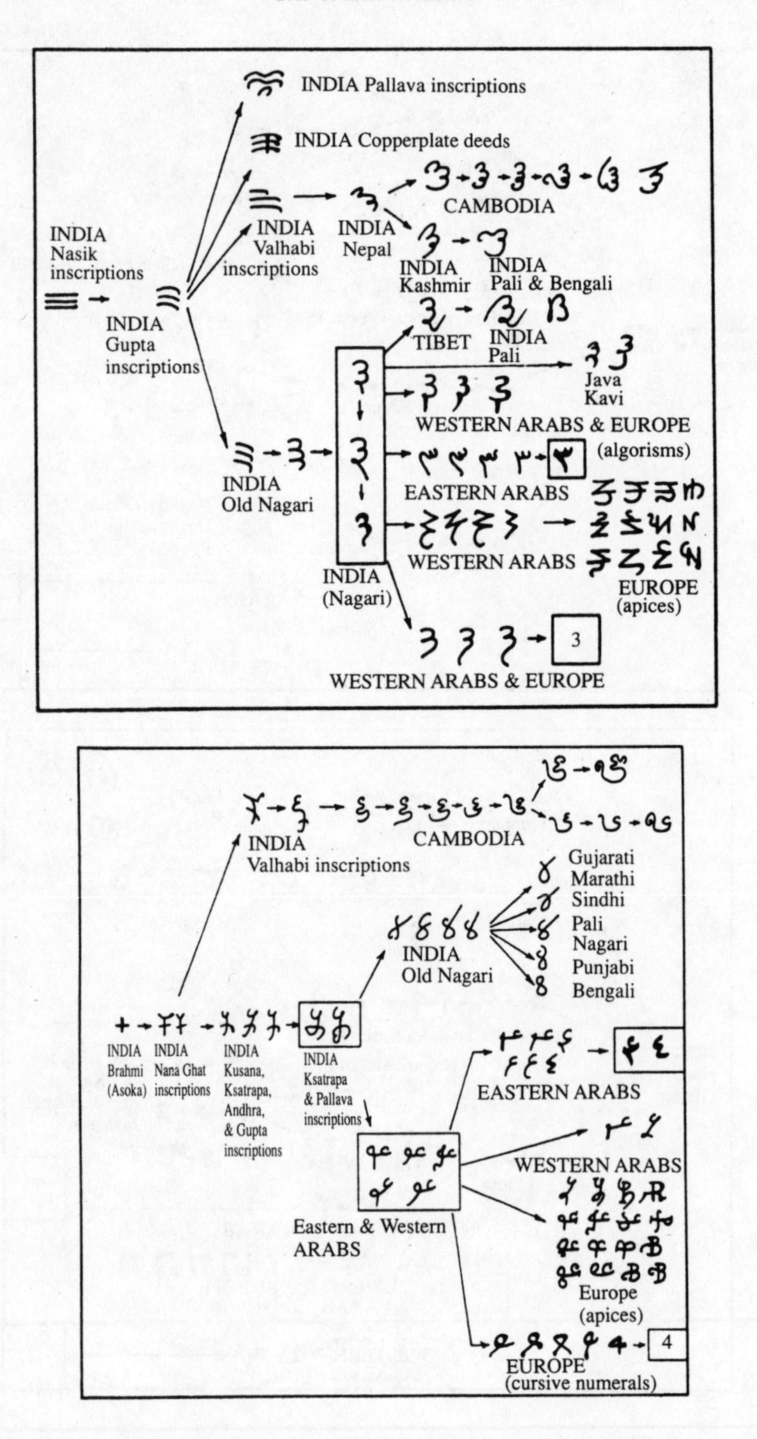
INDIA Pallava inscriptions
INDIA Copperplate deeds
CAMBODIA
INDIA Nasik inscriptions
INDIA Gupta inscriptions
INDIA Valhabi inscriptions
INDIA Nepal
INDIA Kashmir
INDIA Pali & Bengali
TIBET
INDIA Pali
Java Kavi
WESTERN ARABS & EUROPE
(algorisms)
INDIA Old Nagari
EASTERN ARABS
WESTERN ARABS
EUROPE (apices)
INDIA (Nagari)
3
WESTERN ARABS & EUROPE
INDIA Valhabi inscriptions
CAMBODIA
Gujarati
Marathi
Sindhi
Pali
Nagari
Punjabi
Bengali
INDIA Old Nagari
INDIA Brahmi (Asoka)
INDIA Nana Ghat inscriptions
INDIA Kusana, Ksatrapa, Andhra, & Gupta inscriptions
INDIA Ksatrapa & Pallava inscriptions
EASTERN ARABS
WESTERN ARABS
Eastern & Western ARABS
Europe (apices)
4
EUROPE (cursive numerals)

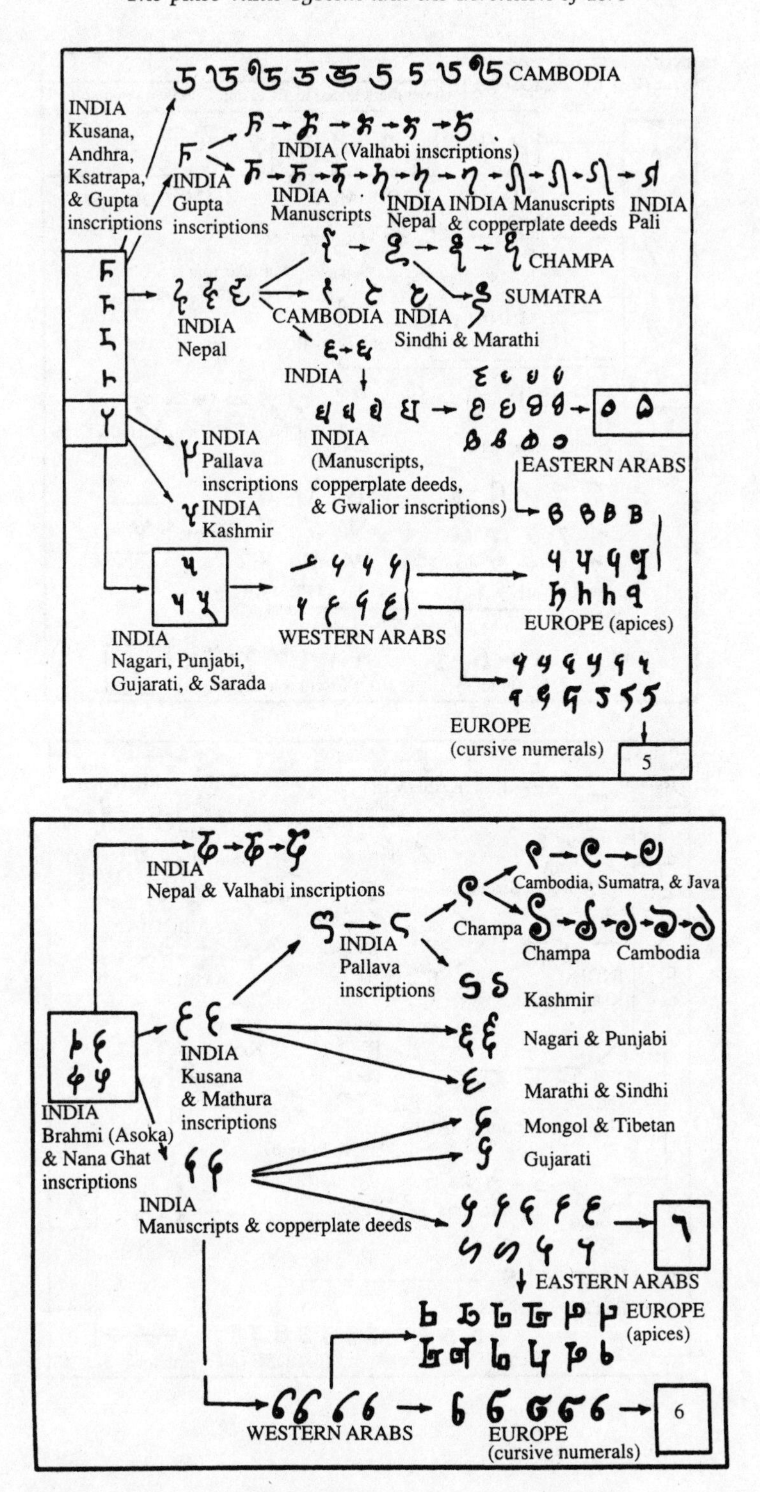
CAMBODIA
INDIA
Kusana,
Andhra,
Ksatrapa,
& Gupta
inscriptions
INDIA
Gupta
inscriptions
INDIA (Valhabi inscriptions)
INDIA
Manuscripts
INDIA
Nepal
INDIA Manuscripts
& copperplate deeds
INDIA
Pali
CHAMPA
SUMATRA
INDIA
Nepal
CAMBODIA
INDIA
Sindhi & Marathi
INDIA
INDIA
Pallava
inscriptions
INDIA
(Manuscripts,
copperplate deeds,
& Gwalior inscriptions)
EASTERN ARABS
INDIA
Kashmir
INDIA
Nagari, Punjabi,
Gujarati, & Sarada
WESTERN ARABS
EUROPE (apices)
EUROPE
(cursive numerals)
5
INDIA
Nepal & Valhabi inscriptions
Cambodia, Sumatra, & Java
Champa
INDIA
Pallava
inscriptions
Champa
Cambodia
Kashmir
Nagari & Punjabi
INDIA
Kusana
& Mathura
inscriptions
Marathi & Sindhi
INDIA
Brahmi (Asoka)
& Nana Ghat
inscriptions
Mongol & Tibetan
Gujarati
INDIA
Manuscripts & copperplate deeds
EASTERN ARABS
EUROPE
(apices)
WESTERN ARABS
EUROPE
(cursive numerals)
6

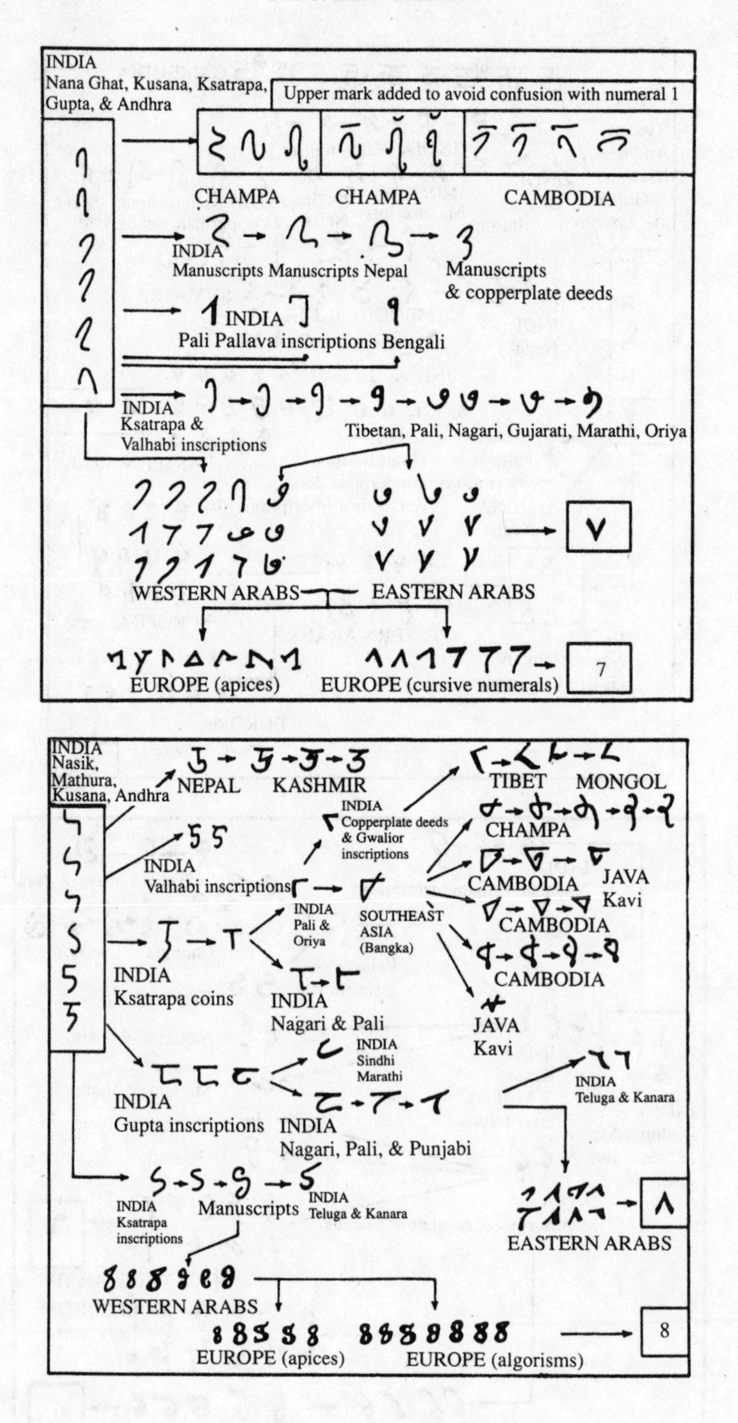
INDIA
Nana Ghat, Kusana, Ksatrapa, Gupta, & Andhra
Upper mark added to avoid confusion with numeral 1
CHAMPA
CHAMPA
CAMBODIA
INDIA
Manuscripts Manuscripts Nepal
Manuscripts & copperplate deeds
INDIA
Pali Pallava inscriptions Bengali
INDIA
Ksatrapa & Valhabi inscriptions
Tibetan, Pali, Nagari, Gujarati, Marathi, Oriya
WESTERN ARABS
EASTERN ARABS
EUROPE (apices)
EUROPE (cursive numerals)
7
INDIA
Nasik, Mathura, Kusana, Andhra
NEPAL
KASHMIR
TIBET
MONGOL
INDIA
Copperplate deeds & Gwalior inscriptions
CHAMPA
INDIA
Valhabi inscriptions
CAMBODIA
JAVA
Kavi
INDIA
Pali & Oriya
SOUTHEAST ASIA (Bangka)
CAMBODIA
CAMBODIA
INDIA
Ksatrapa coins
INDIA
Nagari & Pali
JAVA
Kavi
INDIA
Sindhi
Marathi
INDIA
Teluga & Kanara
INDIA
Gupta inscriptions
INDIA
Nagari, Pali, & Punjabi
INDIA
Ksatrapa inscriptions
Manuscripts
INDIA
Teluga & Kanara
EASTERN ARABS
WESTERN ARABS
EUROPE (apices)
EUROPE (algorisms)
8

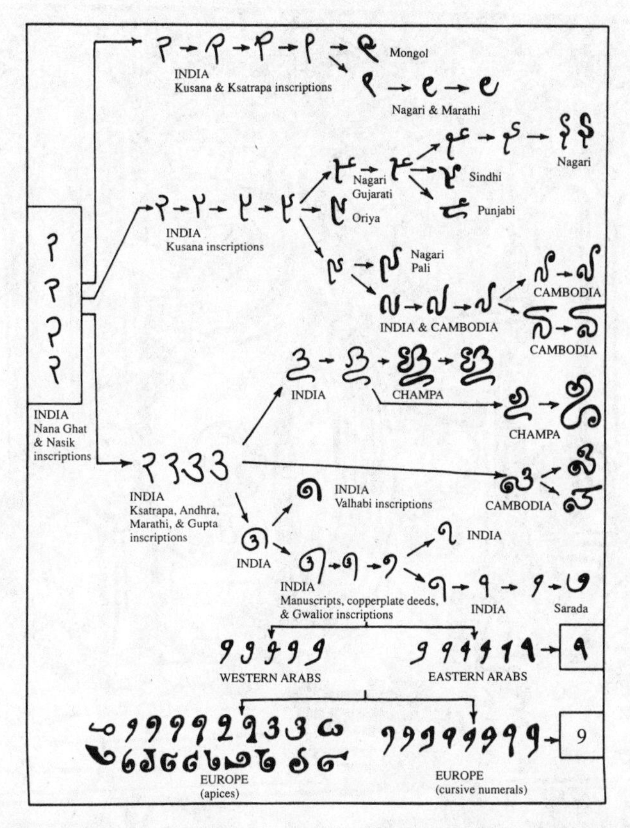


**Figure 2.27** Nine charts, drawn by Georges Ifrah, showing the key steps in the evolution of the Hindu–Arab number symbols which we have inherited in the West today.

they did not result in meaningful combinations of symbols, but in the Arab system you can add any number you like on the end of the one you have and you will invariably have another (bigger!) number. Moreover, the presence of the zero symbol '0' in the Indo-Arab scheme tempts alterations to '6' or '9'. It was not until the latter part of the sixteenth century that we find the Arab numerals being adopted by the majority of merchants in England and northern Europe. Until then the antiquated Roman numerals held sway in this domain.

This rivalry between the Roman and Indo-Arab number systems had another origin. When the Indo-Arabic number symbols appeared in Europe, their first systematic use in the twelfth century was for the Latin translation of the treatise on arithmetic by the great Arab scholar Al-Khowârîzmî. As a result the Indo-Arabic system became associated with a

**Figure 2.28** Gregor Reisch's early sixteenth-century woodcut *Margarita philosophica,* which means 'The pearl of philosophy'. It shows a miserable-looking Pythagoras calculating with the abacus that tradition maintained he invented. He has lined up the number 1241 on his left hand (one sees one counter in the top 'thousands' row, two in the 'hundreds' row, four in the 'tens', and one at the bottom) whilst on his right he has just set up the number 82, with the counter midway between the 'tens' and 'hundreds' column representing fifty, together with three 'tens' and two 'units'. By contrast, on the left sits a smugger-looking Boethius who is working on an algorithmic calculation using Indo–Arab numerals which he is adding in columns; he displays the use of a zero symbol as well as his facility with fractions. Presumably they are each racing to prove the superiority of his method, whilst the Latin inscription identifies Dame Arithmetica standing above them in judgement. Were I Pythagoras, I would be a little concerned about the fact that the supposedly impartial judge has her gown embroidered with Indo–Arab number symbols.

particular form of 'algorithmic' calculation (*algorithmi* is the Latin form of Al-Khowârîzmî's name*). It was characterized by the manipulation of written symbols on slates or a blackboard. The Indians had developed their calculating practices in the same way, writing with sticks on sand or by tracing out symbols with flour or seeds. Such a technique allows one to erase symbols very easily later. The rival method to the algorists' technique was that of the abacists. The proponents of Roman numerals were in the habit of counting by moving counters or beads on a counting table or abacus. They could not erase and had far less scope for storing information during and after calculations. By the fourteenth century the increasing demands of science for sophisticated mathematical calculations and the growing availability of paper inspired by the invention of the printing press proved that the pen was at least mightier than the abacus. An amusing representation of the rivalry is given in the picture by Gregor Reisch shown in Fig. 2.28. Drawn in 1504 it contrasts the efficiency of the algorithmists personified by Boethius on the left, with the pedestrianism of the abacists, who are represented by a rather unhappy looking Pythagoras to his right.

## A FINAL ACCOUNTING

*If we knew history better, we would find a great intelligence at the origin of every innovation.*
EMILE MALÉ

We have endeavoured to draw some lessons from the early history of human counting. The situation that we find is a diverse and peculiar one. Nothing is quite what one would expect and simple conclusions are remarkably elusive. One issue that we had before us is that of innate human propensity for mathematical concepts:

- Does the human brain contain such a natural structure 'hard-wired' into its make-up in some way?
- Does the human mind therefore possess a natural intuition for simple mathematical concepts?
- If so, does this mean that our mathematical picture of the physical world is primarily a mind-imposed projection of our own mental structure upon the outside world?

* In early English this gave rise to two words: *algorism*, which was used in scholarly writings until being fairly recently corrupted to *algorithm*, and *augorime* or simple *augrime* which one finds in popular literary usage. One finds the latter usage in Chaucer's 'Miller's Tale' (c1386) that describes the Miller's astrolabe and his counting stones in the words

'His astrelabie, longinge for his art,
His augrim-stones layen faire a-part.'

- Did we discover counting or invent it?
- Could we have just missed it and developed in a literate but innumerate fashion?
- Did an intuition for counting arise all over the world wherever there was language and society of any sort, or was it come upon rather infrequently and then only by those in the midst of sophisticated cultural developments?
- And what of counting itself; it appears to be necessary for mathematics but is it sufficient to guarantee the sophisticated abstract notion of number and structure that forms the basis of the modern scientific picture of the world?

At the outset we must recognize that we are examining historical evidence. It is in the nature of this subject that the best it can offer is the story of what *did* happen. Even if that story demonstrated that certain events or facets of the human mind were responsible for the development of mathematics, we can never say that those things were *necessary* for its development; that, had things been otherwise, mathematics would not have developed. All we know is that they were *sufficient* to allow its development.

First, what of our human aptitude for counting? Does it really exist? The ancient historical evidence makes such a view difficult to sustain. Whilst we have seen that rudimentary counting systems were almost universal in the ancient world, they were not completely so. Certainly there exist primitive tribes with no systematic idea of number at all and a great many who could only really distinguish one from two or many.

Whilst not every society could count, they could all speak. Language predates the origin of counting and numeracy. Thus, any natural propensity the brain might possess for particular patterns of thought or analysis are likely to have evolved with greater bias towards effective general linguistic or gesticular communication rather than those features which focus upon counting practices. Counting appears to have evolved out of the general desire for symbolic representation which language first meets. It is a specific form of this mental symbolism and there is much evidence to suggest that it may have developed from entirely symbolic ritual practices in which notions of pairing, patterns, succession, and harmony were paramount. This very elementary intuition probably has its roots in something more basic than number. It derives from conscious notions of distinction, like those of male and female, hot and cold, high and low, without which it would not be possible for us to exist in the world. Our brain must possess the ability to distinguish states of the world.

Some distinguished biologists and linguists, like Stephen Jay Gould and Noam Chomsky, have proposed that the evolution of language may be merely a by-product of a process of natural selection which was primarily selecting for other quite different abilities, although this view has not found much favour amongst other students of those subjects. One might approach the evolution of a numerical sense in the same fashion. It could itself be a by-product of selection for some more primary attribute, like language, or it could be an offshoot from another neutral evolutionary development.

The introduction of an evolutionary viewpoint muddies the waters in a number of ways. For, we recognize that the existence of some order in Nature is a necessary prerequisite for our own existence. If what we mean by 'mathematics' is the variety of ways we have developed for recording aspects of that order, then there emerges a new possibility. When there exists some intrinsic order in Nature, either in the form of laws of Nature or the run of outcomes of those laws, then the evolutionary process which gave rise to our minds must bear some faithful representation of that order in so far as its domain intersects with those things necessary for our survival. It is necessary for us to be able to distinguish different states of the world at some level of accuracy and to have a means of noticing if quantities have altered. Thus, if the world *is* intrinsically mathematical in some sense, we should expect to find the signature of that feature imprinted upon our minds because of the selective advantage that is bestowed by a true, rather than false, perception of reality. Just as the physiology of our eyes and ears can be relied upon to tell us something about the true nature of light and sound, so the mental processing within our brains will give us information about the ambient world in which it has developed.

Nevertheless, having a notion of quantity is a long way from the intricate abstract reasoning that today goes by the name of mathematics. Thousands of years passed in the ancient world with comparatively little progress in mathematics. What progress there was occurred primarily in a small number of significant population centres with advanced general culture. In retrospect, we can identify the critical steps that enabled the elementary notions of counting to graduate into the sophisticated and efficient schemes which we have inherited. It is not enough to possess the notion of quantity. One must develop an efficient method for recording numbers and we have seen how crucial was the choice of a good notation, a sensible base number for arithmetic with a regular ladder of steps up to higher multiples of the base number (in our own system, tens, hundreds, thousands, ...); but more crucially still, the adoption of a place-value system in conjunction with a symbol for zero was a watershed. Again, we find that these discoveries are deep and difficult; almost no one made them. But those cultures that did found that they now had a notation that could do so much more than merely *record* numbers. The notation possessed a logic all of its own. Its very use ensured that certain simple logical steps were automatically carried out. Only with such systems of notation is it possible to do more than simply add and subtract numbers. A good notation permits an efficient extension to the ideas of fractions and the operations of multiplication and division.

Besides highlighting how easy it would have been for all human cultures to have missed these discoveries, this optimal structure for the creation of an effective and efficient arithmetic offers us two quite different conclusions. On one hand, there may really exist a 'natural' structure to mathematics which we finally stumbled upon after much primitive trial and error, and when found, its evident benefits ensured that it would supersede all rival schemes. On the other, the situation might be far less idealistic. Perhaps it is merely the case that some inventions work better than others? And, after all, we can think of some ways of improving our own method of counting for special purposes; indeed, we program our computers to use a base-2 place-value system that we call 'binary arithmetic'.

The existence of a natural human propensity for counting would lead one to expect that counting would spring up independently all over the globe. But we have seen that the rival picture, in which distinctive counting practices develop once in an advanced cultural centre and then spread to lesser societies, has much evidence to support it. The idea of independent discovery simply does not square with the subsequent record of creativity of the inhabitants of the places where it was supposedly invented from scratch.

Finally, we should highlight one key feature of our search for the origins of counting. Whilst we have traced all manner of curious practical developments of numbers and number words in the ancient world, we have found not one example of what we might call the abstract notion of number. All numerate cultures seem to have developed counting practices from their practice of numbering particular collections of things. We see this emphasized most strongly where we find different words in use to describe the same quantities of different things. Modern *counting* is still like this; but modern *mathematics* is not. The mathematics done by mathematicians today, and that part of it that is effective in understanding the workings of Nature, differs in principle from the activities of the early counters in this one important way. Mathematicians have abstracted the mathematical process away from the specific examples that were used to motivate their introduction and they study the concept of 'number', or 'shape', or 'distance', in the abstract. This is done by focusing attention upon the operation by which numbers are changed rather than upon the numbers themselves. Thus, a simple counting process like 1,2,3, ... is seen not as a list of particular numbers but as the result of carrying out a particular operation of change upon a number, thereby generating its successor. In this way we have come to appreciate a notion of say 'threeness' that need not be tied only to specific collections of three things. Once an abstract notion of number is present in the mind, and the essence of mathematics is seen to be not the numbers themselves but the collection of relationships that exist between them, then one has entered a new world. That world is one we shall now begin to explore a little further.

## CHAPTER THREE

# *With form but void*

*You can only find truth with logic if you have already found truth without it.*

G.K. CHESTERTON

### NUMEROLOGY

*I love her whose number is 545.*

GRAFFITO FROM POMPEII

Abstract ideas and concrete realities were once interwoven and interdependent to such an extent that no significant wedge could be driven between them. For the ancients and the medievals symbolic meanings of things assumed a natural significance that rests upon associations of ideas that we no longer possess. We do not feel at home with it, but its strangeness is less unnerving if we view it merely as we would the use of metaphor or simile in modern writing. And, as in this modern literary usage, ancient symbolism is not just for show. It plays the role of the simple analogy or apposite metaphor in making the abstract plain by bringing it into the realm of the concrete. Numerology was a widespread and deeply felt manifestation of this desire to symbolize. For the numerologist, numbers and their symbols were not merely lifeless marks on parchment; they were fundamental realities, pregnant with meaning and alive with memories of the past. For the medievals numbers drew symbolic meaning from several traditional sources. There were the several parts of the human anatomy—two arms, four limbs, ten fingers and so forth—but there were also the numbers in the stars. If a number could be divined from the motions or structure of the heavens then it was ascribed a special reverence commensurate with it having been divinely ordained. The root of these astrological beliefs was the search for the mysterious connection between the workings of the heavens and human affairs. The way in which numbers could act as the medium through which such connections between local and cosmic connections could be channelled can be seen in many situations. We find the seven days of the week are named after the seven then-known planets. The deep meaning of such examples was underwritten for the medievals by the recognition of number after number in the pages of the Bible. Interwoven with these biblical references was the numerological thread provided by the Pythagorean tradition of number symbolism. It was a thread that bound tight any who sought to look at the Universe in new ways. Mathematics was not *just* mathematics. Equating numbers meant strange things to many people. We feel still a vestigial remnant of this powerful tradition.

Astrology created the view that fixed quantities, described by the same number, were related in some way. Thus there were connections between the seven in the heavens and the sevens to be found in Holy Scripture and the calendar. The twist offered by the Pythagorean heritage was that there was new knowledge to be had by manipulating these numbers—adding them, subtracting them, organizing them in geometrical patterns or sequences. All these activities required interpreters skilled in extracting the true meanings of things from the mundane world of appearances. The Church was accommodating to this nexus of numerical mysteries and astrological numbers, seeking to exploit numerical symbols as a way of demonstrating the inspiration and deep harmony of the Scriptures. It was by no means alone in this regard. The other great monotheistic faiths all had their numerological aspects with similar origins and aims. No mention of a number in their holy writings was ever thought to be superfluous. Its true meaning in the cosmic scheme of things had to be searched out by the study of intricate common factors. And the hierarchy of religious offices and ceremonies had its basis in this numerical scheme.

In this way numbers came to possess one aspect that was within the reach of human computation, whilst always possessing others which could be fathomed only by divine revelation. We see that this endows numerology with what we might anachronistically call an anti-Copernican perspective. Every user of numbers adds his or her own subjective ingredient to the question of their true *meaning* and its link to the meanings of other aspects of reality. Nearly as bad was the prejudice that everything had to have a numerological aspect. As a result subjects like medicine, which have little or no need of numbers for diagnosis, introduced them as a display of their philosophical significance.* An amusing example of the power of this approach when wielded by someone who knows it to be meaningless is the famous occasion on which Leonhard Euler, the great Swiss mathematician who was sometime tutor to Catherine the Great of Russia during the eighteenth century, decided to bamboozle the Voltairean philosophers at Court in an argument about the existence of God. Calling for a blackboard, he wrote

* This state of affairs is not entirely unknown today and its manifestations in some subjects are tellingly documented by the sociologist Stanislav Andreski in his book *Social Sciences as Sorcery*.

$$(x + y)^2 = x^2 + 2xy + y^2$$
therefore God exists.

Unwilling to confess their ignorance of the formula or unable to question its relevance to the question at hand, his opponents accepted his argument with a nod of profound approval.

The lesson we draw from this thumbnail sketch of a part of our past is that once upon a time all numbers and their symbols possessed deep and many-faceted meanings. The combination of numbers in equations or by other logical operations produced new meanings that resounded in the heavens. It is important to bear this aspect of the past in mind when we come to consider one of the more recent attempts to explain what mathematics is in an attempt to place it very firmly under the control of the human mind.

## THE VERY OPPOSITE

1. *Resolved, by this Council, that we build a new Jail.*
2. *Resolved, that the new Jail be built out of the materials of the old Jail.*
3. *Resolved, that the old Jail be used until the new Jail is finished.*

*RESOLUTION OF THE COUNCIL BOARD OF CANTON, MISSISSIPPI*

In the early years of the twentieth century logicians and mathematicians found themselves the unwilling inheritors of a rather embarrassing collection of logical paradoxes which threatened to undermine the very foundations of human thought and the edifice of mathematics that had been erected upon it. Whilst, as we saw in Chapter 1, there had been an erosion of belief in mathematics as a source of absolute truth and in the uniqueness of logic, there was still every confidence in the logical soundness of the deductions it drew from whatever axioms it freely chose. There may be geometries other than Euclid's but they arose by sequences of unimpeachable logical deductions from sets of axioms nonetheless.

In the years just before and after 1900 we find a real fascination with logical paradox in the works of European philosophers and mathematicians. Some of these paradoxes are not purely logical, but involve a variety of semantic elements which create contradiction when made to refer to themselves. An ancient example is the paradox of the liar who states that 'I am lying'.*

This statement is clearly self-contradictory and cannot be said to be true or false. A more interesting example is that proposed by the German

* St Paul cites this paradox, quoting Epimenides a Cretan philosopher of the sixth century BC, in his letter to Titus where he writes that 'All Cretans are liars...one of their own poets has said so'.

logician Kurt Grelling in 1908. It is known as the paradox of 'heterologicality' and is set up as follows:

> Some adjectives such as *short* and *English* apply to themselves, others such as *long* and *German* do not. Let us call those of the first group *autological*, and those of the second group *heterological*. Is the adjective heterological itself *heterological*? If it is, then according to the definition it does not apply to itself, and so it cannot be heterological.

Although it is tempting to regard these semantic contradictions as mere curiosities devoid of real interest it would later be discovered that there were deep lessons to be learnt from them. The fact that entirely correct use of simple English leads to a blatant contradiction reveals that a language like English cannot express its own semantics completely. Alfred Tarski would eventually show that any language that is logically consistent necessarily fails to be semantically complete in this way. What this reveals is that one needs to conceive of a hierarchy of languages. Each is consistent but insufficiently rich to contain all of its semantics; completeness requires a richer overarching language which in turn requires another to express all of its semantics and so on *ad infinitum*. These overarching languages are called 'metalanguages' for the ones below. This enables us to distinguish statements *of* logic from statements *about* logic and thereby avoid serious problems with statements of the 'this statement is false' variety.* To say 'this statement is false' is to introduce a confusion between the language of the sentence and its metalanguage because we are talking *about* statements in some language not merely making them. Without this distinction the linguistic paradoxes are the harbingers of disaster. I can use one to prove the truth of any claim at all. Take any statement you care to choose, like 'pigs can fly' or 'the moon is made of green cheese', and label it by *S*. Now consider the new statement

Either this entire sentence is false or the statement *S* is true.

This whole sentence must be either true or false. If false, we see that S must be true. If, on the other hand, the entire sentence is true then one or other of the clauses 'this entire sentence is false' or 'the statement *S* is true' must be true. Since the sentence has been assumed true, the first of these clauses cannot be true in this case; hence the second must be true. In conclusion, we have deduced that the statement

*S* is true

is always true whatever the statement *S* is chosen to be!

* This is Eubulides' paradox and dates from the fourth century BC.

Alfred Tarski's concept of a metalanguage averts this crisis. The statement used to generate the final conclusion is not admissible because it mixes statements in a particular language with statements about that language and these belong to its metalanguage.

Although this approach works with these linguistic paradoxes one should appreciate that many human languages like English are ill-defined and intrinsically illogical (ask anyone engaged in learning a foreign language!) and they cannot really be dealt with in a totally logical manner. When Tarski analysed the hierarchies of languages and their metalanguages, together with the definition of truth that each introduces, he did so using the completely rigorous and formal languages of logic and mathematics. What this establishes is that we cannot have a criterion of truth in a language or logical system without stepping outside the system to make statements *about* it in its metalanguage. If we recall our earlier discussion of how the concept of absolute truth was undermined first by the discovery of geometries other than Euclid's and then by logics other than Aristotle's classical laws of thought, then we can see this relativization of the formal notion of truth dealt another blow to any such grandiose associations between human thinking about mathematics and absolute truth.

But there are other paradoxes that are truly logical and show up in any symbolic or mathematical representation. They are highlighted by a graphic example of Bertrand Russell's that is now known as the Barber paradox:

A man of Seville is shaved by the Barber of Seville if and only if the man does not shave himself. Does the Barber shave himself?

If he does then he doesn't; yet if he doesn't then he does. Russell's deepest fears about the self-consistency of logic were aroused by a more general formulation of this type of problem. Suppose one considers the idea of a 'set' as being a collection of objects. Then some sets, like a collection of stamps, do not themselves constitute another possible member of the collection being considered: a stamp collection is not a stamp. But the collection of all the things that are not stamps is a thing that is itself not a stamp and so is a member of itself. Russell then invites us to consider the set of all the sets that are not members of themselves. Is this a member of itself or not? Again we find that it is if and only if it isn't.

This discovery shook Russell's confidence as he sought to finish *Principia Mathematica*, his project of founding mathematics upon symbolic logic. In his autobiography we find the story of his struggle to come to terms with it:

At first I supposed that I should be able to overcome the contradiction quite easily, and that probably there was some trivial error in the reasoning. Gradually however, it became clear that this was not the case ... Throughout the latter half of 1901 I supposed the solution would be easy, but by the end of that time I had concluded that it was a big job ... The summers of 1903 and 1904 we spent at Churt and Tilford. I made a practice of wandering about the common every night from eleven until one, by which means I came to know the three different noises made by nightjars. (Most people only know one.) I was trying hard to solve the contradiction mentioned above. Every morning I would sit down before a blank sheet of paper. Throughout the day, with a brief interval for lunch, I would stare at the blank sheet. Often when evening came it was still empty. We spent our winters in London, and during the winters I did not attempt to work, but the two summers of 1903 and 1904 remain in my mind as a period of complete intellectual deadlock. It was clear to me that I could not get on without solving the contradictions, and I was determined that no difficulty should turn me aside from the completion of *Principia Mathematica*, but it seemed quite likely that the whole of the rest of my life might be consumed in looking at that blank sheet of paper. What made it the more annoying was that the contradictions were trivial, and that my time was spent in considering matters that seemed unworthy of serious attention.

After discovering this paradox, Russell had communicated it in a letter to the great German logician Gottlob Frege who was completing the second volume of his classic treatise on the logical foundations of arithmetic. The news was devastating to his work. He responded as few other scholars would have done when faced with such a disaster at the culmination of twelve years of meticulous work, adding to his book one of the most deflating acknowledgements one could ever wish to read:

A scientist can hardly meet with anything more undesirable than to have the foundation give way just as the work is finished. In this position I was put by a letter from Mr. Bertrand Russell as the work was nearly through the press.

The existence of these and other paradoxes served to widen the gulf between mathematics and reality which had been established by the realization that one could generate consistent logical systems willy-nilly. Now one saw that statements could be made in the language of mathematics which could not correspond to any situation in the real world. At first one might think that such examples are artificially contrived and easy to set on one side as irrelevant to the mainstream of logical deduction as it is used in mathematics and science. But this is not so easy. It is not as if we have a list of all the possible paradoxes. Russell managed to classify the known ones into seven sorts, but who knows how many other varieties there might be? Moreover, any logical system that contains an inconsistency is totally worthless, since it is possible to use that inconsistency to prove the truth of *any* statement whatsoever, including its opposite. Any definition of what mathematics really is and why it works—if indeed it does work—as a description of the physical world needs to find its way through this minefield of paradoxes.

One attempt to exorcize paradoxes of Russell's sort was made by the French mathematician Henri Poincaré. He recognized that the paradoxes of Eubulides, Epimenides, and Russell all involve some collection and a member of it whose definition depends upon the collection as a whole. Such definitions, which he called 'impredicative' are in some sense circular reasonings—thus the definition of the Barber involves the people of Seville and the Barber himself is one of the people of Seville. Poincaré thought that impredicative definitions were the sources of all these paradoxes and Russell himself was attracted to this opinion, expressing it in his 'vicious circle principle'

> No set *S* is allowed to contain members definable only in terms of *S*, or members involving or presupposing *S*.

This was a new restriction upon the concept of a collection, or a set. It seems a good solution to the problem of paradox. However, things were not quite so simple. It transpires that there are all sorts of parts of the established body of mathematics that are founded upon impredicative definitions. In 1918 Hermann Weyl attempted to establish all of known mathematics without the use of these semi-circular definitions. He failed.*

## HILBERT'S SCHEME

> *The present state of affairs is intolerable. Just think, the definitions and deductive methods which everyone learns, teaches and uses in mathematics, the paragon of truth and certitude, lead to absurdities! If mathematical thinking is defective, where are we to find truth and certitude?*
>
> DAVID HILBERT

Russell and his followers sought to reduce mathematics to logic. This they hoped to achieve by defining pure mathematics to be all those prescriptions where one true statement implies the truth of another together with some appropriate criterion of truth. This amounts to treating mathematics as the collection of all possible tautological statements, that is, all those statements that are necessarily true. We are all familiar with tautologies and regard them as rather unsatisfactory and disappointing

* For instance, he was unable to derive the crucial result that every non-empty set of real numbers having an upper bound has a least upper bound.

conclusions to investigations which we expected to reveal something novel. A tautology is a statement that is necessarily and hence invariably true, like for instance:

Either it will rain today or it won't.

However, not all tautologies are ridiculously obvious statements like this. All the theorems of mathematics—like that of Pythagoras—may be viewed merely as more complicated examples. A nice example of the life-saving power of tautology can be appreciated by the following problem.

Suppose that you are incarcerated along with a demon and an angel inside a room which has two doors. One of these doors leads to safety; the other leads to destruction. The demon always tells lies and the angel always tells the truth. You have the opportunity to ask only one question in order to discover the door to safety and you will receive only one answer. Unfortunately, the two spirits are both invisible, and so you have no way of knowing which of them has answered your question. What question should you ask in order to make good your escape?

What you require is a question that is tautological; that is, it has the same answer in all circumstances and hence regardless of the respondent. Your strategy is to ask what door the *other* spirit would have recommended you exit through to escape to safety. Then leave through the *other* door. If you were answered by the demon, he would have lied and pointed towards the door to destruction, the opposite of what the angel would have done. On the other hand, if the angel had replied, she would also have pointed to the door of destruction, to which she knows the demon would have directed you. Thus both answer your question by indicating the wrong door irrespective of their identity. If you take the other door, then it always leads to safety.

Although Russell and Whitehead's *Principia* sought to found mathematics upon logic alone in a paradox-free form, the concept of a metalanguage necessary to carry this out is not altogether clear in their work. Moreover, Russell was primarily a philosopher rather than a working mathematician. These two weaknesses were repaired by the contribution of David Hilbert to the quest to represent mathematics as a taut web of logical connections. Hilbert was the greatest mathematician of his day. He was born in 1862 in Königsberg, the home town of Immanuel Kant, and became professor of mathematics in Göttingen in 1895, where he remained until his death in 1943. Hilbert's breadth of expertise was extraordinary. There is virtually no part of mathematics to which he did not make profound original contributions. In 1900 he gave a famous lecture to the International Congress of Mathematicians in Paris in which he welcomed the new century by listing twenty-three problems which he challenged mathematicians to solve. By so doing, they would

lift the veil under which the future lies concealed, in order to cast a glance at the advances that await our discipline and into the secrets of its development in the centuries that lie ahead.

Hilbert's choice of problems displayed an almost clairvoyant prescience for the deepest and most interesting aspects of the unknown whose probing would reveal new and fascinating structures. Some of his problems remain unsolved to this day. These and others are the focal points for whole areas of mathematical research. The second in his roster of problems was to

demonstrate the consistency of the axioms of arithmetic.

This challenge reveals the additional factor which motivated Hilbert's approach to the foundation of mathematics, which took it a step beyond the quest of Russell and Whitehead. Whereas Russell wished to banish logical paradoxes, Hilbert wished for something much stronger—a proof of the self-consistency of the axioms of arithmetic and all the deductions that could be drawn from it. This was Hilbert's answer to our question: *What is mathematics?* He sought to trammel up all of mathematics within a web of logical interconnections following from perfectly consistent axioms by precisely prescribed rules of reasoning. In other words, he wanted to define mathematics in such a tightly controlled way that it had to be free of logical problems. As such it need have no *meaning*. Indeed, Hilbert believed that the logical paradoxes that so concerned Russell were merely due to the semantic content of language and so could be banished by making them meaningless. Mathematical statements need not have any particular relationship to things and events around us in the outside world or to the natural intuitions of our minds—although they might well do. The sets of consistent axioms one might choose are required only to be consistent; this consideration overrides all others. Mathematics need have no further content: it possesses consistency but no *meaning*.* In this process of logical

* Although this cornerstone of the formalist doctrine is generally credited to Hilbert, one finds it spelt out very clearly in the writings of L. Couturat as early as 1896. In his book *De l'infini mathématique* he writes at some length that 'A mathematician never defines magnitudes in themselves, as a philosopher would be tempted to do; he defines their equality, their sum, and their product, and these definitions determine, or rather constitute, all the mathematical properties of magnitudes. In a yet more abstract and more formal manner he lays down symbols and at the same time prescribes the rules according to which they must be combined; these rules suffice to characterize these symbols and to give them a mathematical value. Briefly, he creates mathematical entities by means of arbitrary conventions, in the same way that the several chessmen are defined by the conventions which govern their moves and the relations between them.'

deduction Hilbert thought it self-evident that all possible mathematical problems could be resolved, and indeed this was one of the reasons he placed such emphasis upon the solving of key problems in his famous Paris lecture to the world's mathematicians. He believed all mathematical questions to be resolvable by a sequence of deductions from preordained axioms. In this vein he writes:*

> The conviction of the solvability of every mathematical problem is a powerful spur to us in our work. We hear within us the steady call: There is the problem, seek the solution. You can find it by pure thought; for in mathematics there is no Ignorabimus! We must know. We will know.

Hilbert began the search for the Holy Grail of mathematical consistency with a well-thought-out plan. He would start with sets of axioms which defined smaller and simpler logical schemes than those spanning the whole of arithmetic, prove them to be consistent, and then move outwards from there, adding extra axioms to enlarge the scope of the system, showing that consistency is retained despite each extension, until the whole of arithmetic is encompassed. He began by proving the consistency of Euclidean geometry showing that Euclid's choice of axioms had been an act of intuitive brilliance because they are indeed independent of each other—none is implied by any, or all, of the others.† Next he moved on to demonstrate the consistency of a simple system of logic which made use of only the concept of 'negation' together with the operation of and/or.

The general strategy for establishing the consistency of different axiomatic systems that he followed exploits a theorem about mathematics—that is, a piece of metamathematics—*that a system being consistent is completely equivalent to some statement in the system being underivable.* It is not difficult to see that this odd statement is true. For if the system is consistent then it must not be possible to derive the statement which says that something is true and also its negation is true. So this is an underivable statement. Conversely, if the system is inconsistent, then, by definition, some statement and its opposite can both be proved true. If this is so, then any statement whatsoever can be shown to be true. So there would be no underivable statements at all in an inconsistent system.

* The last words of this exhortation, which he used on many occasions, are engraved on Hilbert's tombstone:

Wir müssen wissen.
Wir werden wissen.

† Euclid made use of many 'intuitive' but unstated axioms and rules of reasoning in his original proofs. In particular, he often relied upon diagrams being drawn in particular ways. As a result some nineteenth-century pedants had shown that by drawing alternative diagrams one could use his axioms to prove theorems about triangles which were clearly false—for example, that every triangle has two equal sides. Hilbert's work swept all this ambiguity away.

By this method of ensuring there exist underivable statements, Hilbert proceeded to demonstrate that a simplified version of arithmetic was free of contradictions. However, he would like to have been able to prove something rather stronger: that any formula one could state in the symbols of the arithmetic could be shown to be either true or false using the rules of deduction. If such mathematical omnipotence is possible, then the logical system is said to be *complete*.

This programme of work lasted until about 1930. Hilbert had high hopes that his 'formalist programme', as it was called, could inject complete mathematical rigour into every part of the subject whether it be geometry, algebra, or arithmetic. Indeed, he had even more grandiose plans in mind. One of his challenge problems at the Paris Congress was to develop a full mathematical treatment of the axioms of physics, thereby extending the web of logical self-consistency over the sciences as well, maintaining that

> whenever mathematical concepts arise, whether from the side of epistemology, or in geometry, or out of the theories of natural science, mathematics is faced with the problem of probing into the principles behind the concepts and isolating these principles by means of a simple and complete system of axioms in such a way that the precision of the new concepts and their applicability in deductive argument shall be in no respect inferior to what obtains in arithmetic.

The ultimate belief in the solvability of all problems that could be posed was a powerful psychological spur to Hilbert. By stripping mathematics of any form of special philosophical meaning and divorcing its essence from descriptions of the real world, Hilbert was free to imagine that all questions could be answered in the mathematical world that he had created, a world where the only criterion of meaningfulness was logical self-consistency. Mathematics is simply what mathematicians do.

Hilbert's programme was a greater novelty than it might at first sound. Before him, the way in which the self-consistency of some mathematical edifice would have been established was by producing a specific interpretation, or 'model' as it was called, of the axioms in question. If this model actually existed in the real world, then the mathematical system was taken to be consistent under the assumption that physical reality is free from contradictions. If one demanded a demonstration that the model was consistent, then one was faced with the production of another model, and so ultimately either an infinite regress or a circular argument if one returned to use the same model at a later stage of the regress. Hilbert abandoned this doomed procedure to seek proofs of consistency that did not make use of some other model of the meaning of the axioms, whether it be physical or mathematical. In order to carry such a proof of consistency from within, it was necessary for the symbols and deductive steps of the mathematical scheme to be put into an appropriate form. They needed to be *formalized*. That is, they had to be arranged as separate axioms and a series of definite rules of deduction from which 'theorems' followed by application of the latter to the former. The key point about this set-up is that it should be possible to manipulate the mathematical symbols according to the rules without associating any physical 'meaning' to the symbols: one should not, for example, have to draw diagrams on pieces of paper or make models in order to establish a step in a proof about some property of triangles. If that is not possible, it is a signal to the formalist that there exist some more mathematical rules governing the symbols which have not yet been identified. If you appeal to the fact that 'these symbols and their combination correspond to adding apples together', in order to draw upon your experience of what happens to apples, you are trying to add further information about the rules of combination being applied to the symbols. One can do this by adding a further rule to the formalism. When the meaning of the symbols has been distilled off, it is then a formal system, free of interpretations in terms of concrete objects.

## KURT GÖDEL

> *I wanted to visit Gödel again, but he told me that he was too ill. In the middle of January 1978, I dreamed I was at his bedside. There was a chessboard on the covers in front of him. Gödel reached his hand out and knocked the board over, tipping the men onto the floor. The chessboard expanded to an infinite mathematical plane. And then that, too, vanished. There was a brief play of symbols, and then emptiness—an emptiness flooded with even white light. The next day I learned that Kurt Gödel was dead.*
>
> RUDY RUCKER

All the noonday brightness of this confident picture of the formalists' little mathematical world was suddenly extinguished by the appearance of a twenty-five page paper in a technical mathematics journal in 1931. The ramifications of this paper have been immense and have reverberated through subjects far removed from mathematical logic. Its author, Kurt Gödel, was a young Austrian working in Vienna, who was destined to become the most famous logician of all time and spend the better part of his life as a member of the élite Institute for Advanced Study that was created around Einstein in Princeton. Gödel was a strange man and seems to have become quite mentally deranged near the end of his life. In effect, he gradually starved himself to death out of paranoia, suspecting that everyone was seeking to poison him. There are numerous strange Gödel stories which exhibit how uncomfortable this totally logical man found ordinary life to be. Colleagues tell how, if anyone telephoned to arrange to see him, he would readily make an appointment but would never be there at the appointed place and time. When asked on one occasion why he made such definite arrangements if he did not want to meet the people concerned, he replied that his procedure was the only one that guaranteed that he would not meet his visitor.

After many years of residence in the United States, the time came for him to take on American citizenship. This required him to answer a number of simple questions about the American Constitution in order to demonstrate his general knowledge and appreciation of it. Moreover, he needed two nominees to vouch for his character and accompany him to this oral examination before a local judge. Gödel's sponsors were impressive—Albert Einstein, who needs no introduction, and Oskar Morgenstern, a famous mathematical economist and co-inventor with John von Neumann of 'game theory'. Einstein tells the story of how he and Morgenstern became increasingly worried about Gödel's instability and lack of common sense in the run-up to this simple citizenship interview. Apparently, Gödel called Morgenstern on the eve of the interview to tell him that he had discovered a logical loophole in the framing of the Constitution which would enable a dictatorship to be created. Morgenstern told him that this was absurdly unlikely and under no circumstances should he even mention the possibility at his interview the following day. When the day of the interview came, Einstein and Morgenstern tried to distract Gödel from thinking too much about what was in store by generating a steady stream of jokes and stories, hoping that he would be content to turn up, mouth a few rote answers and pleasant platitudes, and depart with his citizenship. John Casti's account of what actually transpired at the interview confirms the distinguished witnesses' worst fears:

> At the interview itself the judge was suitably impressed by the sterling character and public personas of Gödel's witnesses, and broke with tradition by inviting them to sit in during the exam. The judge began by saying to Gödel, 'Up to now you have held German citizenship.' Gödel corrected this slight affront, noting that he was Austrian. Unfazed, the judge continued, 'Anyhow, it was under an evil dictatorship ... but fortunately that's not possible in America.' With the magic word dictatorship out of the bag, Gödel was not to be denied, crying out, 'On the contrary, I know how that can happen. And I can prove it!' By all accounts it took the efforts of not only Einstein and Morgenstern but also the judge to calm Gödel down and prevent him from going into a detailed and lengthy discourse about his 'discovery'.

Fortunately, things were smoothed over. Kurt Gödel did become an American citizen.

Einstein was Gödel's closest friend during their Princeton years and he clearly had to deal with some difficult situations. As a complement to the last story we have just told, one might recall the difficulty Einstein had when Gödel saw a picture of General MacArthur on the front page of the *New York Times* the day after his Madison Avenue ticker-tape welcome home from Korea. Gödel told Einstein that he believed that the man in the photograph was an impostor masquerading as MacArthur. He owned an older picture of MacArthur and claimed that if he measured the length of MacArthur's nose and divided it by the distance from the tip of his nose to the point of his chin, then the ratio of one divided by the other was different in the two pictures. So the man in the parade must have been an imposter. It is not recorded how Einstein sorted this one out.

Using Hilbert's metamathematical technique in a wholly new and powerful way, Gödel established two theorems about mathematics which showed Hilbert's goals to be unattainable. First, he established that any logical system large enough to contain our ordinary arithmetic was necessarily incomplete. There will always exist statements that are expressible in the language of such a mathematical or logical system whose truth or falsity can never be demonstrated. Then, to make matters worse for Hilbert and his followers, he showed that it is never possible to prove that the system is logically self-consistent.

These results were totally unexpected. So much so that we know that his preliminary presentation of them in lectures and at a gathering of mathematicians in Hilbert's home town created almost no general interest at all amongst the logicians present.

Gödel's doctoral thesis had made an important advance in Hilbert's programme. It established the logical completeness of a simple system of logic called the 'predicate calculus'. Still virtually unknown outside Vienna, he came to present this work in a twenty-minute talk at the Königsberg conference on 7 September 1930. Although important and difficult to achieve, this result was just what the formalists had expected. Indeed, it was a problem which Hilbert himself had posed in 1928. But the following day there were a series of presentations on the principal interpretations of mathematics. In the ensuing discussion Gödel was very critical of the formalist philosophy that logical self-consistency was the only criterion required for the existence of mathematics. He went on to announce for the first time that

> one can even give examples of propositions ... which are contentually true but are unprovable in the formal system of classical mathematics.

In retrospect, this was earth-shattering to the formalist programme but appears to have just passed the participants by, since the report of the conference that was subsequently published makes no mention of it at all. The only mathematician who seemed to seize immediately upon the significance of Gödel's remarks when he heard them presented was John von Neumann. After the meeting he took Gödel aside and pressed him for more details. Indeed, two months later he wrote to tell Gödel that he had been able to show, using Gödel's methods, that the self-consistency of axiomatic systems could never be proved. However, in the time since their first discussion Gödel had established this for himself and added it to the text of his paper which was already in the hands of the publishers. Hilbert himself never really came to terms with Gödel's discovery but there is a detailed new version of its proof in the book he wrote with Paul Bernays, published in 1939, which is a mark of the acceptance and widespread assimilation of Gödel's insights by the world of mathematicians. There were exceptions though—some very striking. Judging by his later writings about this period, Bertrand Russell did not seem to appreciate fully what Gödel had done and laboured under the misconception that he had proved arithmetic to be *inconsistent* rather than incomplete. After his monumental efforts in producing the *Principia Mathematica* with Whitehead, Russell made no further significant contributions to logic, turning his attentions to philosophy and politics. He once remarked that the mental effort involved in writing the *Principia* irretrievably drained his mental abilities.*

Ironically, today this great work is mainly of historical interest and is rarely studied, even by mathematicians. Perhaps Russell's worst fears about its fate might yet be fulfilled. Hardy told of a fascinating exchange he had with Russell about its ultimate significance:

> I can remember Bertrand Russell telling me of a horrible dream. He was on the top floor of the University Library, about AD 2100. A library assistant was going round the shelves carrying an enormous bucket, taking down book after book, glancing at them, restoring them to the shelves or dumping them into the bucket. At last he came to three large volumes which Russell could recognize as the last surviving copy of *Principia Mathematica*. He took down one of the volumes, turned over a few pages, seemed puzzled for a moment by the curious symbolism, closed the volume, balanced it in his hand and hesitated...'

The long-term impact of Gödel's two theorems reminds one of the legendary crisis that the discovery of the irrational numbers created for

* *Principia* consisted of three volumes. A fourth, on geometry, written only by Whitehead was originally planned but never appeared. Of the joint writing project as a whole, Russell commented: 'Neither of us alone could have written the book; even together, and with the alleviation brought by mutual discussion, the effort was so severe that at the end we both turned aside from mathematical logic with a kind of nausea.'

Pythagoras and his followers. Hilbert and his school thought their demonstrations of the consistency of logical systems that were less complex than arithmetic were merely stepping stones on the road to extending their results to the whole of arithmetic. They believed these extensions to be simply a matter of plugging away at the logical methods they had developed. They did not suspect that any qualitatively new element would enter the game when one tried to come to grips with something like arithmetic—something that we have a good intuitive understanding of, more so probably than we have of the simpler logical systems (like arithmetic devoid of the operation of subtraction or Euclidean geometry) which proved to be complete. Some mathematicians expressed a general scepticism as to the significance of Gödel's results by speculating that there existed methods of proof other than the finite deductive steps used by Gödel.* Even Wittgenstein advanced rather ill-conceived criticisms which Gödel dismissed as 'completely trivial and uninteresting'. Thus, although the earlier disciples of formalism seem to have readily appreciated what Gödel had done,† others were far more sceptical and antagonistic, often trying to limit the significance of his conclusions. Gödel

* Hilbert's programme, and Gödel's proof of the incompleteness of arithmetic, permitted only what were called 'finitistic' logical steps. This meant that the the number and the length of the axioms employed together with the rules of reasoning employed should all be composed of a finite number of individual steps. This proviso amounts to a restriction upon the type of metamathematical argument that can be employed in attempts to prove completeness. Within a few years of Gödel's demonstration, Gerhard Gentzen showed that if we relax these finitistic restrictions so as to include an operation called 'transfinite induction', which allows one to draw a conclusion from an infinite set of premisses, then both the consistency and completeness of arithmetic can be proved. This method of proof generalizes the method of mathematical induction in which one shows that the $n$th step implies the truth of the $(n+1)$th after explicitly showing that the result of the 1st step $(n=1)$ is true; hence it is true for all steps. Just as mathematical induction can be used to show that all the natural numbers (1, 2, 3 ... and so on) possess some property, so transfinite induction is employed to prove that a given property is possessed by all members of certain ordered sets. These sets may be infinitely bigger than the collection of natural numbers. Thus mathematical induction is a special case of transfinite induction. However, although this method allows the deduction of all the truths of arithmetic, it does not allow the deduction of all the truths of mathematics as a whole, because the latter contains truths about infinities whose truth or falsity remain undemonstrable using the logical machinery of the system.

† It is an interesting but little-known fact that there had been earlier attempts at showing that logical systems were incomplete. Paul Finsler published some ideas along these lines in 1926 which Gödel showed to be incorrect and confused in their conception when Finsler drew his attention to them in 1933 in a brief dispute about priority. But potentially more significant had been the early premonitions of Emil Post who, unlike Finsler, was under no misapprehension that he had beaten Gödel to the discovery of incompleteness. He realized that the methods used in Russell and Whitehead's *Principia* could yield undecidable statements that were true in higher metamathematical systems. Yet, he seems to have recoiled from following this up because he thought it was leading him too far into uncharted waters of human psychology, and concluded that 'mathematical proof was [an] essentially creative [activity]', the proper elucidation of which would require an understanding of 'all finite processes of the human mind'.

himself always shunned the limelight and controversy, only considering it important to have convinced a small collection of mathematicians whom he respected of the truth of his claims. One surprising fact about him is his admission in later life that he never met Hilbert, nor had any correspondence with him, although this may tell us more about the rigid nature of the German academic system at that time than anything else. When Hilbert first returned to his studies of logic and the foundations of mathematics in 1934, after pursuing other mathematical interests, he persisted in his old beliefs about the desire for completeness and demonstrable consistency until his death in 1943, and refused to see Gödel's work as undermining his desire to prove the consistency of arithmetic, mistakenly basing his doubts perhaps upon the sinking sands of Gentzen's 'infinite' induction proofs which we mentioned above. He reiterated (wrongly) that it was

> the final goal of knowing that our customary methods in mathematics are totally consistent. Concerning the goal, I would like to stress that the view temporarily widespread—that certain recent results of Gödel imply that my proof theory is not feasible—has turned out to be erroneous. In fact, those results show only that, in order to obtain an adequate proof of consistency, one must use the finitary standpoint in a sharper way than is necessary in treating the elementary formalism.

It is surprising to learn from some of Hilbert's contemporaries that, despite the way in which he is associated with the formalist programme that he initiated, he did not much care for the type of mathematical work that it involved and much preferred working in other areas of mathematics. Two other logicians, Fraenkel and Bar-Hillel, say that 'this kind of research for him was not too pleasant a duty that he felt obliged to perform but one which distracted him from other more attractive occupations'.

In view of this state of affairs it is interesting to see what Gödel had to say about what led him to think in the opposite direction to the milieu. The most interesting information about this is to be found in Gödel's collection of private papers. He received many letters from students and other interested parties asking him about the development of his ideas. Gödel seems to have been meticulous in drafting and redrafting replies to these letters but, more often than not, never actually mailed the reply. Whether this was because of a perfectionism which left him dissatisfied with his responses, or because he really only wanted to convince himself that he had an answer to all questions, or just one more aspect of his very strange personality, one can only guess. In 1970 a student wrote to ask him how he had arrived at the idea that some statements might be undecidable. In one of the drafts of his unsent reply he writes:

In consequence of the philosophical prejudices of our times 1. nobody was looking for a relative consistency proof because [it] was considered axiomatic that a consistency proof must be finitary in order to make sense 2. a concept of objective mathematical truth as opposed to demonstrability was viewed with greatest suspicion and widely rejected as meaningless.

Unlike the formalists, Gödel believed that mathematical truth was objective truth about something that really existed, not merely a facet of a creation of the human mind. But such an idea would have been scorned in 1930, so this philosophical view was not mentioned explicitly in the presentation of his incompleteness theorems. Here, as in all other aspects of his life, we find Gödel being paranoiacally cautious. The papers he left after his death reveal vast quantities of incisive and deep thought about mathematics and vast areas of philosophy (written in an obscure form of Germanic shorthand), none of which he revealed during his lifetime. One of his rare visitors spoke of his hypnotic presence 'able to follow any of my chains of reasoning to its end almost as soon as I had begun it. What with his strangely informative laughter and his practically instantaneous grasp of what I was saying, a conversation with Gödel felt very much like direct telepathic communication.'

The most intriguing discovery amongst his unpublished work was his 'ontological proof' for the existence of God. Such a proof of God's existence from pure reason, rather than from any evidence in the physical world, was first attempted by Archbishop Anselm back in the eleventh century. Gödel's version was based upon his use of Leibniz's concept of 'positive' and 'negative' properties. He reasons as follows:

*Axiom 1:* A property is positive if and only if its negation is negative.

*Axiom 2:* A property is positive if it necessarily contains a positive property.

*Theorem 1:* A positive property is logically consistent (that is, possibly it has some instance).

*Definition:* Something is God-like if and only if it possesses all positive properties.

*Axiom 3:* Being God-like is a positive property.

*Axiom 4:* Being a positive property is logical and hence necessary.

*Definition:* A property $P$ is the essence of $x$ if and only if $x$ has the property $P$ and $P$ is necessarily minimal.

*Theorem 2:* If $x$ is God-like, then being God-like is the essence of $x$.

*Definition:* $x$ necessarily exists if it has an essential property.

*Axiom 5:* Being necessarily existent is God-like.

*Theorem 3:* Necessarily there is some $x$ such that $x$ is God-like.

When questioned about such flights of logical fancy he remarked with a smile that 'the axiomatic method is very powerful'.

Gödel was clearly, in many ways, a rather mystical thinker who believed that we participated in a Cosmic Mind separate from all matter, in which we found the truths of mathematics when we searched in the right way. Because of this belief about the reality of mathematical concepts, the formalist notion that mathematics had no 'meaning' was anathema to him. This focus upon the ubiquity of meaning we find associated with his religious views in letters to his mother where he tells her:

> We are of course far from being able to confirm scientifically the theological world picture ... What I call the theological worldview is the idea, that the world and everything in it has meaning and reason, and in particular a good and indubitable meaning. It follows immediately that our worldly existence, since it has in itself at most a very dubious meaning, can only be the means to the end of another existence. The idea that everything in the world has a meaning is an exact analogue of the principle that everything has a cause, on which rests all of science.

He recognizes that there are many speculations about the world which seem to have no basis in current scientific thinking, but which may one day be set on a firm scientific basis just as currently accepted ideas about the atomic constitution of matter first arose thousands of years ago for religious and theological reasons when there was no shred of observation evidence for their truth.

## MORE SURPRISES

> *Mathematics is not a book confined within a cover and bound between brazen clasps, whose contents it needs only patience to ransack; it is not a mine, whose treasures may take long to reduce into possession, but which fill only a limited number of veins and lodes; it is not a soil, whose fertility can be exhausted by the yield of successive harvests; it is not a continent or an ocean, whose area can be mapped out and its contour defined; it is as limitless as that space which it finds too narrow for its aspirations; its possibilities are as infinite as the worlds which are forever crowding in and multiplying upon the astronomer's gaze; it is as incapable of being restricted within assigned boundaries or being reduced to definitions of permanent validity, as the consciousness of life.*
>
> JAMES SYLVESTER

Gödel's discoveries about the limitation of formal reasoning led others to seek restrictive results of the same general ilk. Gödel had established his theorems by setting up a correspondence between arithmetic and statements about arithmetic. He invented a procedure for uniquely associating a number, now called a 'Gödel number', with any statement about arithmetic. So, given any whole number we can write down the mathematical statement it corresponds to and vice versa. Gödel then considered the statement

The theorem possessing Gödel number *G* is undecidable.

He found the Gödel number of this statement and then used it as the value for G in the statement. The resulting statement is then a theorem which establishes its own unprovability. It is intriguing that Gödel makes use of what is in effect one of the infamous linguistic paradoxes to generate a perfectly rigorous proof of undecidability.

Alfred Tarski took Gödel's arguments a little further to show that logical systems are also semantically incomplete as well. He showed that if a mathematical system is consistent then the notion of truth—that is, the collection of all true theorems of the system—is not definable in the system itself. This shows that there are concepts which just cannot be defined within the compass of some formal systems. The upshot of this discovery is that logical and mathematical systems rich enough to contain arithmetic are not only formally incomplete, in the sense that some of their truths are unprovable using the paraphernalia of the system, but they are also semantically incomplete, in the sense that some of their concepts cannot be defined using the language and concepts within the system. One can always define them using a bigger system, but only at the expense of creating further undefinable concepts within the larger system. This means that there is no formal system in which the truth of all mathematical statements could be decided, or in which all mathematical concepts could be defined.

These developments scuppered many of Hilbert's original aims. But one remained whose fate was more peculiar. Besides consistency and completeness, Hilbert's programme was interested in establishing another property of mathematical systems: he wanted to show that they were 'categorical'. What this means is that if we take some part of mathematics and formalize it by noting its axioms and rules of deduction then that piece of mathematics must be the only interpretation of that system of axioms with which it can be made to correspond by direct one-on-one association of the axioms. This type of one-on-one relationship is called an 'isomorphism'. In 1934 Skolem showed that if a finite number of axioms is employed to characterize an arithmetical system then more than one interpretation of the axioms is possible and they are not necessarily isomorphic. In fact, nearly thirty years later, it was shown that if the system was categorical then it would necessarily have been complete and hence at variance with Gödel's theorem.

These investigations brought to light another awkward and unexpected problem for the formalist definition of mathematics. When Hilbert began work on his programme, most mathematicians believed that all mathematical systems could be axiomatized by reducing them to a finite list of axioms. And, if one could not achieve such a thing, it was merely a reflection of one's lack of ingenuity. But it transpired that the situation was more problematic. There exist mathematical systems which cannot be axiomatized. In fact, Skolem had discovered some systems without formally specified axioms that were both consistent and complete arithmetics. It followed from Gödel's theorem, therefore, that they could not be finitely axiomatizable or we would have established that all statements of these arithmetics were decidable—in contradiction to Gödel's theorem. Skolem thereby established that no finite set of axioms can characterize the natural numbers uniquely; there will always exist some structure which is not isomorphic to the natural numbers but which satisfies the same finite set of axioms. However, further investigations of this situation revealed more interesting things. If, instead of restricting our mathematical theories to those which are generated from a finite list of specific axioms, we allow there to exist *axiom schemes*, which are equivalent to infinite lists of specific axioms, then the system of natural numbers can be characterized uniquely. An axiom scheme is a prescription that delivers a specific axiom each time it is applied to a different statement of the system. As in the case of the natural numbers, there could be an infinite number of them to apply the axiom scheme to; this is like having an infinite set of the usual unadaptive axioms. A simple example of an axiom scheme is the procedure called mathematical induction. Suppose the number 0 satisfies some proposition $P$, and also suppose that if the number $N$ satisfies $P$ then so does the next number, $N+1$; this ensures that every number 0,1,2,3,... satisfies the proposition $P$. This is mathematical induction. It is an axiom scheme because there are an infinite number of different propositions that could be substituted for the label $P$ and the rule would apply to each of them. Mathematical induction cannot be replaced by a finite number of specific axioms. Only an infinite number of specific axioms would do.

Gödel showed that rather elementary mathematics necessarily contains unprovable statements and rests partially upon axioms of faith concerning its own consistency. If one constructed a statement whose truth or falsity was undemonstrable, then one is at liberty to add either this statement or its negation to the set of axioms which define the system. Each strategy would create a larger logical system, but each would necessarily contain new unprovable statements. Thus, mathematics always seems to be just out of reach and the formalists' hope that it could simply be defined as the sum total of all the deductions that could be made from consistent axioms by finite methods foundered. In effect, one can make arithmetical statements that have greater information content than the defining axioms of the language in which they are stated.

## THINKING BY NUMBERS

*Music is the pleasure the human soul experiences from counting without being aware that it is counting.*
GOTTFRIED LEIBNIZ

Hilbert's attempt to trammel up the consequences of mathematics in a network of axioms and rules was not the first attempt to codify human knowledge in a formalistic scheme. The French philosopher René Descartes had been among the first to embark upon a programme to establish a firm foundation for human knowledge of the world and had singled out mathematics as the only reliable route to such unimpeachable knowledge. Newton's great contemporary and long-time rival, Gottfried Leibniz, was the first to embark upon the implementation of such a grandiose codification of knowledge by means other than mere words. Leibniz sought to construct a universal language which would be adequate to represent all human statements and arguments. He foresaw a multitude of benefits flowing from this. Although the procedure would be used primarily to determine indisputable mathematical and scientific truths, its scope was wide enough to include the study of all logical arguments. Human disputes would be amicably resolved without ambiguity and the demonstrable truth would produce harmonious agreement in every case. For example, in 1686, he proposed that it be used in theology to produce a core of doctrine that all variants of Christianity could agree upon. Needless to say this particular application never materialized. However, the mathematical part of the project was carried forward by Leibniz in a novel way. He sought to split up logical parts of argument into components, each of which was to have a prime number associated with it. This is what Gödel would do hundreds of years later. Subsequently, he even explored the use of Chinese pictograms to create a special symbolic language. Complicated arguments were to be represented by unique strings of symbols. The laws of reasoning are listed as specific rules for manipulating strings and all logical deduction is reduced to the performance of algebra. Although this scheme was too great a step of the imagination for Leibniz to realize fully, it did make him the founder of the symbolic logic that would one day become the bread and butter of Russell and Whitehead's great project to reduce mathematics to logic. He was also the inventor of the first calculating machine in his quest to mechanize the use of his universal language of calculation and reasoning.

Leibniz's ambitious scheme to formalize human thought and mathematics into a universal language—a *lingua characteristica*—was begun in the 1670s. Here are some descriptions of it in his own words, which convey the flavour of his thinking:

All our reasoning is nothing but the joining and substituting of characters, whether these characters be words or symbols or pictures ... if we could find characters or signs appropriate for expressing all our thoughts as definitely and as exactly as arithmetic expresses numbers or geometric analysis expresses lines, we could in all subjects in so far as they are amenable to reasoning accomplish what is done in Arithmetic and Geometry. For all inquiries which depend on reasoning would be performed by the transposition of characters and by a kind of calculus ... And if someone would doubt my results, I should say to him: 'let us calculate, Sir,' and thus by taking to pen and ink, we should soon settle the question.

And of the procedure for creating a correspondence between mathematics and statements about mathematics—the analogue of Gödel numbering*—he prescribes the following

rule for constructing the characters: to any given term (that is, the subject or predicate of a statement), let there be assigned a number, but with this one reservation, that a term consisting of a combination of other terms shall have as its number the product of the numbers of those other terms multiplied together. For example, if the term for an 'animate being' should be imagined as expressed by the number 2, and the term for 'rational' by the number 3, ... the term for 'man' will be expressed by the number 2×3, that is 6.

Moreover, Leibniz had every confidence that this was an easy thing to do and its application to all areas of human reasoning would not long be delayed:

I believe that a number of chosen men can complete the task within five years; within two years they will exhibit the common doctrines of life, that is, metaphysics and morals, in an irrefutable calculus.

The pay-off would be a mind-expanding extension of the power of human thought, making human thought an instrument of the exact sciences, for

Once the characteristic numbers of many ideas have been established the human

* Gödel numbering works by attributing a different prime number to every logical connective, quantifier, and variable. The Gödel Number of the whole statement is the product of the prime numbers associated with each ingredient. Because any number can be expressed as a product of its prime divisors, written in ascending order of size, in only one way, any whole number corresponds to a unique statement and any statement has one and only one prime decomposition.

race will have a new organon,* which will increase the power of the mind much more than the optic glass has aided the eyes, and will be as much superior to microscopes and telescopes as reason is superior to vision.

Leibniz's ideas did not inspire wide interest amongst other mathematicians and philosophers until George Boole's formulations of symbolic logic in 1847. In his book '*An investigation of the laws of thought*' (1856) Boole enters into the spirit of Leibniz's programme by seeking out logical fallacies in well-known philosophical works by casting their arguments into symbolic form. The other interesting pursuit of Leibniz's goal was undertaken by the Italian logician Giuseppe Peano. Besides being a world-class mathematician, Peano was also an expert linguist with a wide and detailed knowledge of classical languages and comparative philology. He devoted a lot of his energy to promoting a new international language, called '*Interlingua*', that he had developed. In the course of this work he had made many attempts to reduce the grammatical structure of languages to mathematical form. Not surprisingly, he became convinced that this was an impossibility because of the vagaries of history which dictate so many quirks of idiom and usage. His '*Interlingua*' was in fact just a simplified version of Latin—a language which, not so very long before, had indeed been a universal language for scholarly discourse. Its vocabulary is that of classical Latin, but there is no grammar at all: nouns are undeclined and verbs are not conjugated; cases and tenses are made plain by the use of prepositions just as in modern conversational usage.

## BOURBACHIQUE MATHÉMATIQUE

*Mathematicians are a species of Frenchmen: if you say something to them they translate it into their own language and presto! it is something entirely different.*

GOETHE

The incompleteness that Gödel demonstrated had profound consequences that have resonated down the corridors of science during the half-century since his demonstrations created a new perspective on mathematics and the Universe in which we live. Until very recently it loomed over the subject of mathematics in an ambiguous fashion, casting a shadow over the whole enterprise, but never emerging to make the slightest difference to any truly practical application of mathematics. In 1981, Jean Dieudonné, a prominent French mathematician who is a modern formalist, happy to continue with the weaving of mathematical symbols in the manner that Hilbert commended, yet shorn of any grandiose plan to

* The 'Organon' was a work of Aristotle laying down systematic rules of logical reasoning. It dominated Western thought and practice on these matters for over two thousand years.

demonstrate completeness, claimed that the demonstration of Gödel had remained irrelevant to the practice of mathematics except in rather refined areas of the theory of sets and logic which had parted company from the rest of mathematics:

> [There is an] optical illusion [that] comes from the more or less unconscious assimilation that many philosophers make of two parts of mathematics: mathematical logic and the theory of sets on the one hand, all the rest on the other. This attitude was justified at the beginning of the twentieth century, for then these two parts were tightly bound to each other, and the greatest mathematicians of the period had a passion for questions of the 'foundations' of mathematics, even if their works concerned neither logic nor the theory of sets. Today, it must be recognized, the situation is radically different; there is almost total divorce ... logic and the theory of sets are become marginal disciplines ... We admire the cleverness and depth of the works that led to the metamathematical theorems of Gödel ... but they exerted *no* influence (positive or negative) on the solution of the immense majority of problems that mathematicians study. Perhaps that is shocking, but I cannot help it; I speak not of opinions but of *facts*.

One of the consequences of this perception was that a neutered version of formalism continued to be practised and can be found even today as a way of viewing mathematics that avoids becoming embroiled in philosophical questions regarding its meaning. It appeals to those who place great emphasis upon technique and who wish to draw a distinction between pure and applied mathematics. The tattered banner of formalism has been carried on most notably by the French consortium of mathematicians known by the pseudonym 'Nicolas Bourbaki', who in the last fifty years have co-authored a series of books about the fundamental structures of mathematics, of which geometry and arithmetic are particular examples. They personify the last hopes of the formalists; axiomatics, rigour, and soulless elegance prevail; diagrams, examples, and the particular are all eschewed in favour of the abstract and the general. The aim of the Bourbaki project has been not so much the discovery of new results but the codification of the known in new and more succinctly abstract ways. They are the ultimate textbooks for the *cognoscenti*.

Jean Dieudonné, the leading propagandist for this approach to mathematics, believes that this formal approach exemplifies what every science should aspire to be, for

> the scientific study of a whole class of objects presupposes that the peculiarities which distinguish these objects from one another are forgotten on purpose and that only their common features are retained. What singles out mathematics in this respect is its uncommon insistence on following that programme to its utmost consequences. Mathematical objects are to be considered as *completely* defined by

the axioms which are used in the theory of these objects; or, in the words of Poincaré, axioms are 'definition in disguise' of the objects with which they deal.

The Bourbaki project began in 1939 and has an amusing background. No one seems to know why the group of French mathematicians who began the project named themselves after a non-existent Frenchman. It seems they may have been inspired by an army officer, General Charles Denis Sauter Bourbaki, who was prominent in the Franco-Prussian war and apparently declined an offer of the throne of Greece in 1862. A decade later his fortunes had reversed and we find him and his soldiers interned in Switzerland where he attempted to shoot himself—but missed. There is a statue of him in the town of Nancy, and many of the Bourbaki collaborators have been connected with the University of Nancy during their early careers. There are many other peculiar stories about the real Mr Bourbaki, most of them created by the Bourbaki group to keep the legend suitably embellished.

Bourbaki's emphasis upon the identification of analogous structures in different branches of mathematics is a new emphasis. Previously, the formalist would probably have highlighted axioms and deductive reasoning as the defining hallmarks of mathematics. But this is viewed as so weak a characterization that it fails to distinguish mathematics from a host of other activities. The Bourbaki spokesmen complain:

> It is a meaningless truism to say that this 'deductive reasoning' is a unifying principle for mathematics. So superficial a remark can certainly not account for the manifest complexity of different mathematical theories, not any more than one could, for example, unite physics and biology into a single science on the ground that both employ the experimental method. The method of reasoning by syllogistic chains is nothing but a transforming mechanism, applicable just as well to one set of premisses as to another; it could not serve therefore to characterize these premisses.

Yet, even within the ranks of mathematicians Bourbaki is strongly criticized by some for its arid 'scholasticism' and artificial 'hyperaxiomatics' which seek to make mathematics into 'a kind of logical theology'. One of its supporters, Laurent Schwartz, tries to justify its approach by drawing a distinction between the processes of discovery and the distillation of knowledge:

> Scientific minds are essentially of two types, neither of which is to be considered superior to the other. There are those who like fine detail, and those who are only interested in grand generalities ... In the development of a mathematical theory, the ground is generally broken by scientists of the 'detailed' school, who treat problems by new methods, formulate the important issues that must be settled, and tenaciously seek solutions, however great the difficulty. Once their task is

accomplished, the ideas of the scientists with the penchant for generality come into play. They sort and sift, retaining only material vital for the future of mathematics. Their work is pedagogic rather than creative but nevertheless as essential and difficult as that of thinkers in the alternative category ... Bourbaki belongs to the 'general' school of thought.

Despite the limits imposed by Gödel's discoveries, the Bourbaki group sought to codify the decidable part of mathematics in a unified fashion, focusing upon the concept of algebraic *structures* created by the different sets of axioms and rules appropriate for the different branches of the subject. They were anxious to organize the disparate pieces of mathematical knowledge into a single corpus so that the similarities between superficially different structures could be appreciated and hence exploited across the whole range of studies. Mathematics for Bourbaki is simply what is produced by the work of mathematicians. It is 'a human creation and not a divine revelation'. It sees mathematics as a living, growing structure that requires an organization to be imposed upon it if it is to avoid a future of chaos and fragmentation:

Mathematics, according to Bourbaki, is similar to a rapidly growing city with many suburbs encroaching the surrounding counties. The expanding city is in an identity-crisis, as it were, since there have been so many random additions inside and outside the city as well as some serious deteriorations at several sections of the city. A few suburbs have grown so fast, in fact, that they may become autonomous cities; meanwhile, there is hardly any traffic among these suburbs ... With little or no communication amid the random proliferation, the city needs a drastic urban renewal which will be 'in accordance with a more clearly conceived plan and a more majestic order, tearing down the old sections with their labyrinths of alleys, and projecting towards the periphery new avenues, more direct, broader and more commodious'. Bourbaki enters here as a grand city-planner and construction-contractor.

Yet, Bourbaki's programme of organizing mathematics into a neat network of logical connections is always haunted by the lack of any proof of self-consistency. This spectre is faced pragmatically by appeal to experience, and a faith in the rarity and innocuousness of the lack of any ultimate proof of consistency. One just has to live a little dangerously:

We believe that mathematics is destined to survive, and that the essential parts of the majestic edifice will never collapse as a result of the sudden appearance of a contradiction; but we do not maintain that this opinion rests on anything but experience. That is not much, it may be said. But for twenty-five centuries mathematicians have been correcting their errors, and seeing their science enriched and not impoverished as a consequence; and this gives them the right to contemplate the future with equanimity.

The hostility that one often finds amongst applied mathematicians to the Bourbaki philosophy springs from their feeling that it divorces the practice of mathematics from physical problems and the real world of things which inspire new ideas. Nonetheless the influence of the Bourbaki project has been very wide-ranging. In the 1960s and 1970s it appears to have been the stimulus for the so-called 'new math' teaching programmes adopted in high-school mathematics teaching in many countries. This approach to the teaching of mathematics departed significantly from the traditional model whose emphasis was upon manipulation and problem-solving. Instead of the old-fashioned emphasis upon arithmetic, calculating interest rates, using logarithms, geometry, and calculus, there was the study of sets, groups, and other abstract mathematical structures. This experiment seems to have been unsuccessful and present mathematical teaching of young children is far less abstract. In this way it exploits the tactile intuition for shapes and structures that they have developed from playing with toys. Much of the outcry against the 'new math' syllabuses by parents was undoubtedly caused by the fact that they found themselves frustrated by finding their children unable to carry out traditional mathematics with much proficiency but then found themselves unable to understand the type of mathematics being taught so they were unable to offer any additional help to their children. In retrospect this educational experiment to introduce abstraction to children at such an early age seems to have been somewhat idealistic in its devotion to the presentation of mathematics as a logical work of art. Emil Artin reveals this deep feeling of the formalists for mathematics as a vast tapestry of interwoven statements pulled taut by the threads of logic into a finished work of art:

> We believe that mathematics is an art. The author of a book, the lecturer in a classroom tries to convey the structural beauty of mathematics to his readers, to his listeners. In this attempt he must always fail. Mathematics is logical, to be sure; each conclusion is drawn from previously derived statements. Yet the whole of it, the real piece of art, is not linear; worse than that, its perception should be instantaneous. We all have experienced on some rare occasions the feeling of elation in realizing that we have enabled our listeners to see at a moment's glance the whole of architecture and all its ramifications.

Ten years ago a survey of working mathematicians revealed 30 per cent of them to be formalists in the Bourbaki mode. One reason for this is the point that Laurent Schwartz makes: most mathematical work is far from its popular image of 'discovery'; rather, it is dominated by the process of refinement so that difficult and complicated proofs are made simpler and shorter until one can claim that their chain of argument is 'obvious' or 'trivial' by which mathematicians mean simply that it appeals to no new type of argument. It is merely cranking through a well-worn set of operations. In order to do this sort of thing it is probably most expedient to act as though one is a formalist even if one might find all the implications of such a cramped perspective rather less attractive if asked to reflect upon it in the armchair over the weekend.

Bourbaki must also respond to the challenge of answering the crucial question, posed here by Einstein:

How can it be that mathematics, being after all a product of human thought independent of existence, is so admirably adapted to the objects of reality?

He sees the true course of the mathematician's work to be the elucidation of the basic structures of logic. If fully explored these will encompass all the interrelationships sanctioned by logic. The world around us is seen as a specialization of some of these structures in order that they can be instantiated or modelled by the particular interrelationships of material things. The fact that formal mathematical structures are meaningless can be turned upon its head: instead of maintaining that they therefore apply to nothing, one can maintain that they apply to all possibilities. The observed Universe is but one of them. Bourbaki writes:

The great problem of the relations between the empirical world and the mathematical world, that there is an intimate connection between experimental phenomena and mathematical structures, seems to be fully confirmed in the most unexpected manner by the recent discoveries of modern physics. We are completely ignorant, however, of the underlying reasons for this fact (provided that one could indeed attribute a meaning to these words) ... But, on the one hand, quantum physics has shown that this macroscopic intuition of reality ('from immediate space intuitions') covered microscopic phenomena of a totally different nature, connected with the areas of mathematics which had certainly not been considered for their applications to experimental science. The axiomatic method, on the other hand, has shown the 'truths' from which it was hoped to develop mathematics were but special aspects of general concepts, significance of which was not limited to these fields. Hence it turned out ... that this intimate connection, the harmonious inner necessity of which we were once expected to admire, was merely a fortuitous contact of two disciplines whose real connections are much more deeply hidden than could have been presumed *a priori*.

## ARITHMETIC IN CHAOS

*Truth is stranger than fiction; fiction has to make sense.*

LEO ROSTEN

Until very recently the chain of discoveries that Gödel initiated seemed to have come to an end. The well of undecidability seemed to have run dry leaving his great discoveries shining like a pinnacle as the tide of mathematics turned to move elsewhere. But then, in the late 1980s, new simpler ways of proving and expressing Gödel's theorems were discovered which recast them into statements about information and randomness.

We can associate a quantity of information with the axioms and rules of reasoning that define a particular axiomatic system. We define its information content to be the size of the shortest computer program that searches through all possible chains of deduction and proves all possible theorems. This approach leads to the conclusion that no number whose complexity is greater than that of the axiomatic system can be proved to be random. If one tries to make good this deficiency by adding additional axioms or rules of inference to increase the information content of the system, then there will always exist even larger numbers whose randomicity cannot be demonstrated. There is a real limit to the power of arithmetic.

Most strikingly, the American mathematician Gregory Chaitin has explored the consequences of this line of thinking within the context of a famous mathematical problem. Suppose we write down an equation linking two (or more) quantities $X$ and $Y$, say $X + Y^2 = 1$. Then, if we do not restrict the values of X and Y to be whole numbers, there are infinitely many pairs $(X,Y)$ which solve it (for example, $X = 3/4$ and $Y = 1/2$). But suppose we are interested in ascertaining the solutions for which $X$ and $Y$ are both positive whole numbers. This is called a 'Diophantine problem' in honour of Diophantus of Alexandria, the greatest algebraist of antiquity,* who was the first to engage in the systematic use of algebraic symbols, using special symbols for unknown quantities, reciprocals, and powers of numbers. These problems have more than one possible solution. In our elementary example of a Diophantine equation, $(X,Y) = (1,0)$ or $(0,1)$ are the only solutions.

We can create another Diophantine equation containing a variable number $Q$, which could have any integral value 1,2,3,..., and so on; for example $X + Y^2 = Q$, which reduces to the example we gave above when

* Diophantus lived and worked around the time AD 250. Metrodorus' *Greek Anthology*, compiled from various sources in about AD 500, records his epitaph. We know little else about him than this apposite biographical puzzle:

This tomb holds Diophantus. Ah, how great a marvel! The tomb tells scientifically the measure of his life. God granted him to be a boy for the sixth part of his life, and adding a twelfth part to this, He clothed his cheeks with down; He lit him the light of wedlock after a seventh part, and five years after his marriage He granted him a son. Alas! late-born wretched child; after attaining the measure of half his father's life, chill Fate took him. After consoling his grief by this science of numbers for four years he ended his life.

From this description, a little calculation should reveal to you that he lived to the age of eighty-four, whilst his son lived for only forty-two years.

$Q = 1$. Now suppose we create a more elaborate Diophantine equation that contains $N$ different variables $X_1, X_2, X_3, \ldots X_N$ instead of just two ($X$ and $Y$) as well as the fixed number $Q$. Chaitin asked whether an equation of this form typically has finitely or infinitely many solutions in whole numbers as we let $Q$ run through all its possible values: $Q = 1,2,3,\ldots$, and so on.

At first this appears to be but a minor variation on the traditional question of whether this equation has a solution in whole numbers for each $Q = 1,2,3 \ldots$ But it is an infinitely harder question to answer. It transpires that there is no way to determine the answer. The answer is random in the sense that it requires more information to resolve than is present in arithmetic. It cannot be computed by reducing it to other mathematical facts and axioms. Chaitin represents this state of affairs by forming a number Ω whose digits are the sequence of binary numbers 0 or 1 chosen as follows. Take each value of $Q$ in turn and write 0 if the Diophantine equation has a finite number of solutions and 1 if the number is infinite. The result is a binary string of ones and zeros, for example,

$$\Omega = 0010010101001011010\ldots,$$

and so on to infinity, which specifies a real number in binary arithmetic. Chaitin showed that Ω is a veritable cloud of unknowing: its value cannot be computed by any computer; no computer program, no matter how complex, can ever do better than determine a finite number of the infinite digits required to specify Ω. These limitations arise because each digit in the specification of Ω arises in a manner that is logically completely independent of any of the others. No machine that merely follows a given rule or program possesses the ingredient of novelty that is required to create the next entry in the list. Finally, Chaitin changed this number by placing a decimal point in front of it, so that the new number becomes a fraction less than one. Hence, with the example we have given above, we would change to

$$\Omega = .0010010101001011010\ldots$$

and this number gives the probability that a randomly chosen computer program with a random input will eventually stop after a finite number of steps.* Its value is always some number between (but not equal to) zero and one. A value of zero would signal that no programs stop whilst the other extreme, of one, would mean that every program would stop.

* Alan Turing first considered the less general question of whether a particular program will stop if given a random input.

This state of affairs has the final striking consequence that if we pick some very large whole number then, for values of $Q$ in a Diophantine equation like that given above, there is no way in which we can ever decide whether the $Q$th binary digit of the number $\Omega$ is a zero or a one. Moreover, this is the situation for an infinite number of possible choices for $Q$. Human reasoning can never get at the answer to this question. It is a question whose answer does not correspond to a theorem in any formal system. Each of the infinite string of digits which define the number $\Omega$ corresponds to an undecidable fact of arithmetic.

Even arithmetic contains randomness. Some of its truths can only be ascertained by experimental investigation. Seen in this light it begins to resemble an experimental science.

## SCIENCE FRICTION

> *A species stumped by an intractable problem does not merely cease to compute. It ceases to exist.*
> SETH LLOYD

After seeing how Gödel's theorem is established one could be forgiven for becoming a little sceptical about its real relevance to science. It appears to emanate from the formulation of a rather artificial linguistic paradox. Far more impressive would be the demonstration that some great unsolved problem of mathematics which has tortured mathematicians for centuries *is* actually undecidable. Or perhaps that some very practical mathematical problem, like 'what is the optimal economic strategy?', is logically irresolvable. In 1982 some 'natural' examples of undecidable mathematical statements were discovered in the course of trying to solve a real problem—they are not artificial concoctions.

Suppose we call a set of numbers 'large' if it contains at least as many members as the smallest number it contains; otherwise, we shall call it 'small'. So, for example, the set of numbers {3,6,9,46,78} is large but the set {21,23,45,100} is small (because it contains fewer than 21 members). Now it can be shown that if you take a *big enough* collection of numbers and give each pair of them a label, either black or blue, say, then you can always find a 'large' set within the collection such that the pairs in this set are either all black or all blue. This is not totally surprising but what is, is the fact that the question 'how big is 'big enough'?' cannot be answered using arithmetic: it is undecidable. Several other examples of undecidable questions of a similar type are now known; they are natural in the sense that they arose in the course of trying to solve other mathematical problems.

Another very interesting aspect of Gödel's theorem is its connection with the idea of randomicity. Superficially this connection is rather surprising, yet it turns out to be rather deep. Not only does it transpire that the question of whether a sequence of numbers is random or not is logically undecidable, but posing this question in the right way leads to an illuminating proof of Gödel's theorem which sheds light on the limitations of axiomatic systems.

Suppose we are presented with two lists of numbers, the first entries of which are

$$\{3,56,6,23,78,....\} \quad \text{and} \quad \{2,4,6,8,10,...\}$$

How can we gauge the extent to which these sequences are random? We would like an answer that does not rely upon our subjective impressions but yet incorporates a little of what our minds are telling us about the presence or absence of order. Ask what is the length of the shortest computer program that can generate each sequence. The length, in computer bits, of this shortest program is called the *complexity* of the sequence. If a sequence is haphazard and contains no special rule for generating one entry from another (as in our first example), then the shortest program can be nothing less than the listing of the sequence itself. But, if the sequence is ordered, then the required program can be much briefer than the infinitely long sequence which is given. In the second of our examples the program will just list the even numbers: {PRINT 2N, N = 1,2,3,...}.

A sequence is *random* if its complexity equals the length of the sequence itself. In this case it requires the entire listing to specify it. So, given any two random sequences of different length, the longer sequence is regarded as the more complex. If you pick a large number of sequences of numbers, say telephone numbers, you will find that most have rather high complexity and it is extremely rare to come across strings of numbers that have low complexity.

Using this notion of complexity, consider giving a computer, whose programs include all the symbols and operations of arithmetic, the following instruction:

Print out a sequence whose complexity can be proved to exceed that of this program.

The computer cannot respond. Any sequence it generates must by definition have a complexity less than that of itself. A computer can only produce a random sequence that is less complex than its own program. We are now able to exploit this quandary to show that there must exist undecidable statements. Simply pick a random sequence, call it *R* say, whose complexity exceeds that of the computer system. A question like

Is $R$ a random sequence?

is undecidable for the computer system. The complexity of the statements '$R$ is random' and '$R$ is not random' in each case is too great for them to be translated by the computer system. Neither can be proved or disproved. Gödel's theorem is proved.

The inevitable undecidability of certain statements that this example demonstrates arises because the logical system of the computer, based upon arithmetic, has too small a complexity to cope with the spectrum of statements that can be composed using its alphabet. There is, as a consequence, no way in which you can decide whether the computer program you are using to perform a particular task is the shortest one that will do it.

This result poses restrictions upon the scope of any approach to the laws of Nature on the basis of simplicity alone. The scientific analogue of the formalist methodology in mathematics is the idea that, given any sequence of observations in Nature, we try and describe them by some all-encompassing mathematical law. There may be all sorts of possible laws that will actually generate the data sequence, but some will be highly contrived and unnatural. Scientists like to take the law with the lowest complexity in the sense described above. That is the most succinct coding of the information into an algorithm. Sometimes such a prejudice is called 'Occam's razor'. We can see that this approach will never allow us to prove that a particular law we have formulated is a complete description of Nature. Just as there must exist undecidable statements that can be framed in its language, so a law can never be proved the most economical coding of the facts. Unfortunately, one can never know whether or not one has discovered the 'secret of the Universe'—whether the remaining compression required to unveil it is a minor one or a vast enterprise.

It has been said that the test of any idea is whether it can be used to build better machines. It is a remarkable feature of Gödel's theorem, brought to light by its modern elucidation in the language of complexity, information, and randomness, that it reveals to us something about the limitations that logical undecidability will place upon the machines of the future. Suppose we take the example of a rather fancy household gas cooker with an electronic control panel. It is equipped with all manner of microprocessors which are heat-sensitive and respond to instructions regarding cooking time and oven temperature. They transmit information about the environment inside the oven to the valves that regulate the flow of gas to the burners. The microprocessors store information temporarily until it is overwritten by new information or instructions. The more efficiently this information can be encoded and stored in the microprocessors, the more efficiently the machine operates, because it minimizes the work done in erasing and overwriting instructions in its memory. But Chaitin's investigations have revealed that Gödel's theorem is equivalent to the statement that we can never tell whether a program is the shortest possible one that will accomplish its task. We can never compute the most succinct program required to store the instructions for the operation of the cooker. So, the microprocessors that we use will always overwrite more information than they need to: they will always possess *some* inefficiency or redundancy. In practice, the amount of redundancy introduced into the microprocessors of a household gas cooker by Gödel's theorem is fantastically minute—billions of times smaller than the improvement that could be achieved just by keeping it clean. Nonetheless, the loss of efficiency is finite and in much more sophisticated machines it could one day prove significant in defining the ultimate powers of artificial minds or supercomputers. This *logical friction* relates the restrictions imposed by Gödel's theorem to the workings of any device that makes use of logic. We could go further and envisage the entire Universe as a computer whose software consists of what we normally call the laws of Nature. The hardware consists of the particles and forms of matter that constitute the observed material world. The shortest representation of the laws of Nature—the Theory of Everything that physicists are currently seeking—would be the most succinct encapsulation of all the laws governing the superficially disparate aspects of the physical world. It is the set of rules which, when discovered, would signal the end of physicists' quest for the ultimate law of Nature. But our new statement of Gödel's theorem reveals that we can never prove that a succinct representation of the laws of Nature is the shortest possible. We may have found the ultimate codification of the laws of Nature, but we would never be able to prove it so to be. All we could do would be to discover a briefer representation. That is, we could falsify a candidate for the most succinct representation of the laws of Nature, but we could never verify one. All we can hope for is that our discovery might be so close to the minimal representation that further abbreviation would tell us nothing of great significance. But we can never be sure.

## MATHEMATICIANS OFF FORM

> *Formalized mathematics, to which most philosophizing has been devoted in recent years, is in fact hardly to be found anywhere on earth or in heaven outside the texts and journals of symbolic logic.*
> REUBEN HERSH

We have taken a long look at the topsy-turvy history of formalism. It was born of a desire for certainty and control over the mathematical creations of the human mind. It began in confidence with no image of how its quest could ever be frustrated save by humans' inability to reach it. Its attainment would enable mathematics to avoid all nebulous questions of 'meaning'. Mathematics was simply the sum total of the strings of inky marks on pieces of paper linked by the prescribed rules of the game starting from consistent initial axioms. Any statement we care to make can be checked against the logical web of formalist mathematics. If it falls within the net of deductions then it is true; if it falls outside, it is false. But 'true' and 'false' are not words that need correspond to any thing outside the logical game: they need not correspond to anything in the physical world. Gödel's discoveries showed these expectations to be unattainable. Nonetheless a residue of the formalist philosophy lives on, treating mathematics as an artificial creation of human minds: a collection of pretty patterns in an abstract mental space. True, it is limited by Gödel's theorems and its practitioners' belief in its ultimate consistency rests upon a faith strengthened by the experience of thousands of years of faultless experience. What can one say about the formalist explanation of what mathematics *is*? Is it an adequate response to the mystery of the harmony between the mind, the physical world, and the symbolic relationships of mathematics? I think that we must find it wanting in many respects. The accumulated circumstantial evidence against it is telling and diverse. The cornerstones of the case one brings against formalism are as follows.

The arid nature of formalism—its divorce from the application of mathematics to the real world and its removal of meaning from mathematics—were high prices to pay. But whilst they might have been palatable in return for the original goals of completeness and consistency, one seems to have ended up with the worst of all worlds: a view of mathematics that sees it as a lifeless game that we must simply hope is consistent. If we are true to its rigid doctrines, then we shall find all mathematics to be shattered into fragments. Pick on some familiar concepts, like that of a point or that of a line; then whenever the formalist sets up a different axiomatic system which includes them they must be regarded as *different* concepts each time, because they are completely defined by the rules of the axiomatic system. Because the formalist does not allow mathematical concepts to have any meaning, say by correspondence with objects in the real world, one cannot permit concepts to be transportable from one axiomatic system to another. If we add an axiom to a system in which 'points' and 'lines' are defined, then we cannot take the enlarged system to contain identical concepts of 'point' and 'line'. We have made all things new.

This dilemma explains why Hilbert laid so much stress upon establishing the consistency of theories. For if we cannot settle the truth or falsity of any claim arising from a certain set of axioms by using those axioms, then nothing can settle the truth of that claim for us. We can always do it by adding extra axioms, but by so doing we create a new and distinct system which is thus incapable of providing answers to questions about the old system.

Formalism doesn't square with mathematicians' experience of doing mathematics. It is more appropriate as a philosophy of double-entry bookkeeping than of mathematics. Indeed, it seems to encourage a sort of intellectual deception. Mathematicians do not think about mathematics as if it had no concrete representations. They draw diagrams, imagine examples, and have insights as to what results might be true. Only afterwards is the formalist philosophy applied to recast the results so obtained into a series of abstract deductions from which all information about the chain of thinking that led to their discovery has been expunged. According to the formalist, mathematical theorems are not really 'proved' or 'discovered' they are just generated by stringing together sequences of symbols in patterns dictated by the rules and axioms. The meaning of any string of symbols can be nothing more than the thread that links it back to its axioms. Some deductions are 'deeper' than others because they are further from the axioms than others or require more effort to be expended to reach them. But Gödel teaches us that we can never prove a chain of deductions, or string of symbols, to be the shortest path to a particular conclusion. Thus the relative 'meaning' of statements classified in terms of their depth is also beset by undecidability.

If formalism were a true and complete picture of what mathematics is, then we are naturally led to ask why we should bother with it. What is it that leads us to develop certain structures rather than others? The answer must be that they are in some sense richer, more interesting, more significant than others. But from whence comes this significance? Why are some of the logical patterns prettier than others? Usually the mathematician will tell you that the most telling ideas are those that appear in quite different branches of the subject and draw out analogies between disparate parts of the subject. But we have seen that the way in which particular concepts and conclusions are tied uniquely to the defining axioms of their sub-branch of the subject makes analogies like this a coincidence that cannot be pursued or milked of any further benefits. They have to be pursued in terms of some overarching 'structure' of which the analogous examples are particular aspects. Again, this is the sort of activity that occurs after the analogies have been identified by other processes of thought.

Formalism seems to be an anthropomorphism that was created because of the appearance of semantic paradoxes in the use of human language. The employment of axiomatic systems began with Euclid but his axioms were extracted from experience of the real world. Subjects like arithmetic existed for thousands of years before retrospective systems of axioms tried to encapsulate them. Arithmetic systems did not spring from formal dictates and the two sit together rather uneasily. We find ourselves confronted with something of a dilemma when we try to restore our inherited sense of correspondence between certain truths of mathematics and those experienced elsewhere. For *any* statement can be shown to be a true deduction from *some* set of axioms. This means that we cannot associate the true statements of mathematics with *the* collection of axioms and rules of reasoning of formal systems. Moreover, we have found that there exist mathematical systems which cannot be reduced to a system of axioms that have a unique interpretation: they generate other non-equivalent mathematical structures as well.

The basis of mathematical knowledge now becomes something of a problem. Hilbert and like-minded formalists held that mathematical knowledge was absolutely certain. If that knowledge rests upon observations of the real world, then it cannot be absolutely certain, just as science cannot be. Instead, Hilbert appeals to a notion of human intuition as providing knowledge about certain mathematical objects. But how does this differ from the first option? The concepts we are able to frame are the result of the particular conceptual and manipulative capabilities of the human brain. The brain is a consequence of the process of human evolution and is partially a summary of information about the environment in which it has developed. Its complexity is a reflection of the complexity of the environment in which we have evolved. Thus these mathematical intuitions of Hilbert have no more claim to infallibility than observations of the material world. Indeed they are merely pieces of information processed by slightly different parts of our brains. As a result the formalist is forced to conclude either that mathematics is inexact or that it rests upon some foundation other than that provided by the human senses or mental processes.

Most mathematicians are not content to accept that what they do is merely follow the chains of symbolic deductions that the formalist condemns them to pursue. In their pursuit of theorems and mathematical insights they often draw diagrams to illustrate the truth of their conjectures. If the formulae were just squiggles on paper devoid of any meaning, then one might wonder why any diagram should provide a representation of any aspect of the system of formulae.

The last objection brings us to the heart of our dilemma. Formalism made the most of the discovery that one was free to define axiomatic systems as one pleases. They did not have to correspond to things in the real world. The axioms did not have to correspond to observed facts about the world as Euclid's did. The sole criterion for their acceptability is their lack of logical inconsistency. One can still extract candidate systems of axioms from reality if one chooses. Indeed, their concrete existence might be a good indication of their non-contradictory character. But we know that large parts of mathematics do work as descriptions of the physical world. The formalist is forced either to ignore this fact completely or regard the Universe as a physical representation of some formal system. We must now ask why the Universe picks out particular sets of axioms to define a particular formal system on which to base its behaviour. Why have our minds evolved to take for granted a two-valued form of logic in which things are either true or false? Presumably they have been found to possess a greater capacity for survival when they have this structure than if they don't because on the scale of everyday experience the world reflects this structure. Although the formalist sees mathematics as mere formulae without meaning we know that something makes these dead formulae take wings and fly.

Formalism has a suspiciously close affinity to the general philosophical notions that were in the air at the time of its conception and early popularity. It is a specific and precise application of the doctrine of logical positivism that was formulated by the Vienna circle of philosophers and which was to become an influential idea in the philosophy of science in the 1940s and 1950s. This philosophy wished to formalize all knowledge in a systematic fashion so that any statement could be tested to discover whether it followed from agreed initial assumptions following the rules of reasoning. In this way, they hoped to banish religion and metaphysics into a limbo of unverifiable statements that could not be deduced logically from the body of certain knowledge. Gödel, incidentally, was an occasional participant in the Vienna circle's early discussions, but opposed their goals of arriving at a unified and complete scheme of knowledge. One detects also something of the Teutonic desire for order and formality in things in the very notion of formalism. The hopes of the logical positivists were dashed and their approach undermined by others, most notably by Karl Popper who stressed the falsifiability (as opposed to verifiability) of statements as a criterion for scientific meaning. The wider philosophy of science and knowledge which complemented formalism has devolved and mutated into other forms as new ideas have arisen to capture the essential character and method of science. By comparison, the philosophy of mathematics has stagnated. Formalism has not adapted to meet the challenges posed by new mathematics and its relationship to the external world and to the activities of mathematicians and computers.

We have already raised the problem that formalism does not square with most mathematicians' experience of doing mathematics. The mathematician relies on all manner of rather vague intuitions and insights, sometimes merely trial and error, to get things right. Sometimes progress is replaced by regress as mistakes are found, misconceptions revealed, and false starts retracted. This experience was enlarged and illuminated most memorably by the thesis of a Hungarian émigré named Imre Lakatos. Lakatos was a senior official in the Hungarian Ministry of Education who fled to Austria during the Hungarian uprising in 1956. Eventually, he came to Cambridge, where he enrolled as a doctoral student working on the history and philosophy of mathematics. His thesis is a famous one, which was published in book form with the title *Proofs and Refutations* only after his sudden death from a brain tumour in 1974, by which time he was a famous figure in the world of philosophy. His study was an attack upon formalism and other dogmatic philosophies of mathematics. By presenting extraordinarily detailed case histories of how particular mathematical problems have come to be formulated and resolved, Lakatos argued that the formalist picture was totally at variance with reality: it was a secondary phenomenon that grew up after the important work of mathematical creation and distillation had been done. Formalism was quite secondary to the informal reasoning that is carried out when mathematicians have a first idea about the sort of theorem that they would like to prove; they attempt to do this, and discover that it can't be done because there are unusual counterexamples to their potential theorem. Their response is to first revise and refine their goal and then try to prove a new theorem. Lakatos showed how some geometrical problems had evolved in this way through a chain of reformulations, counterexamples, and partial proofs until something related to—but quite distinct from—the original proposal was stated and proved. This process, whereby the formal statements of mathematics play a secondary role to the informal changing ones, he saw to be the essence of a mathematics that was a dynamic and evolving discipline rather than an ossified structure churning out meaningless strings of symbols from immutable axioms. Lakatos criticized the formalists for a crucial ambiguity which left them never explaining whether mathematics *ought* to be like a formal system or whether they were merely describing the fact that mathematics did resemble a formal system.

In retrospect, it is interesting to see Lakatos' scheme in its historical and political context. For the Eastern European intellectual fleeing from the terrors of totalitarianism and becoming immersed in the anti-establishment culture of the sixties, an anti-authoritarian approach to the philosophy of mathematics is not entirely unexpected. In Lakatos's mathematics things do not proceed only according to preordained rules of reasoning; they are partially unstructured and free from restrictions.

The most curious feature of formalism is what can be found in that formalism. Suppose that we have a computer that carries out the deductions within our formal system. We know that if the system is rich enough to contain ordinary arithmetic then we can carry out most of the logical deductions that make up the subject of science. It appears that the chain of logical deductions can give rise to very complex structures of the sort that we use in physics, biology, and computer science. Ultimately, it would be possible for something like the phenomenon of human consciousness to be expressed by some very complex formal string of characters. Thus within this formal system there are character strings which can reflect upon their own nature. They can recognize that they reside in a particular systems of laws and structures. At this stage of complexity in the formal structure 'meaning' must inevitably appear unless we deny that anything has a meaning.

Thus we are confronted with the dilemma that there exist very special formal systems: those that correspond to the workings of the physical world and those which can become self-aware in the same way that we are parts of the material world that have become self-aware. The fact that mathematics describes the workings of the most elementary workings of the world in areas totally divorced from those areas of human experience that played a role in our evolutionary development convinces us that mathematics does possess some irreducible meaning. The fact that some mathematics, rather than others, is chosen for deployment in Nature is a fact whose significance we can ill afford to ignore if we are to unlock the deepest secrets at the heart of the Universe.

Formalism is found lacking in two crucial respects. It fails to account adequately for the relationship between mathematical symbols and the minds of mathematicians and it fails to explain the utility of mathematics in describing the workings of the physical world.

# CHAPTER FOUR

# *The mothers of inventionism*

*I have come to believe that the whole world is an enigma, a harmless enigma that is made terrible by our own mad attempt to interpret it as though it had an underlying truth.* UMBERTO ECO

## MIND FROM MATTER

*There is a very good saying that if triangles invented a god, they would make him three-sided.* BARON DE MONTESQUIEU

Despite its forbidding rigidity and sterility, formalism has a human face. It reveals mathematics to be a human activity made possible by the capabilities and complexities of the human mind. It is a man-made universe, even if the 'man' needs to be a competent mathematician. In centuries gone by this would have been interpreted as a welcoming confirmation of man's primary role in Nature as the ruler of the natural world and the pinnacle of its created order. Those were the days when neither the objective character of observed reality nor our unadulterated knowledge of it was questioned. But throughout the nineteenth century we saw the infallibility of ancient beliefs being increasingly questioned and the relative character of our knowledge became the fashionable viewpoint. Darwinian evolution revealed the true place of humankind in the animal world, Laplace's picture of the origin of the solar system from one of many spinning clouds of gas emphasized the fickle nature of the chain of events that gave rise to the treasured environment around us. The sceptical philosophers, like Kant, realized that the very act of understanding adds something to the nature of the reality and so creates a gap between the real world and the perceived world. The possible existence of such a gap, its measure and meaning, have since formed the issues about which so many twentieth-century philosophers have argued. Today, our picture of the evolution of the Universe from an uninhabitable Big Bang to the quiescent galaxy-infested state replete with stars and planets it finds itself in today, and the evidence we have of our own humble origins, provokes us to ponder the origins of the faculty we possess for understanding the world.

Our minds possess the ability to simulate the processes of Nature so that we no longer have to experience events physically in order to learn from them. This ability to invent and project little movies in our mind's eye is a crucial aspect of the complex phenomenon we call 'consciousness'. This ability seems to have evolved rapidly at some critical stage in our past history and enabled us to manipulate our environment and its other living occupants more effectively than any other living thing. Our minds are the current edition of this evolutionary process. Their complexity is a reflection of the complexity of the environment in which we find ourselves. The fact that we have not only survived in a complex and rapidly changing environment but, to a considerable extent, even learnt to control it is a telling fact. It reveals that the mental picture that we have of the world is an extremely faithful one. If our mental picture of the world—of light, of sound, or of motion—were distorted very significantly by the very act of perceiving, listening to, or thinking about the world, then our chances of survival would be negligible. Our eyes tell us something about the real nature of light because they have evolved in response to the real properties of that entity we call 'light'. Our ears tell us things about the real nature of sound because they have evolved by a process of natural selection which renders more advantageous any development of a sound reception system which delivers a faithful record of reality. This recognition of the way in which our faculties have evolved in tune with the need to respond to the environment around us also reveals why we might expect that (almost) all of us have very similar perceptions of the world. If the human mind were adding a large interpretational ingredient to our perception of the world, then we would expect this subjective addition to vary considerably from person to person. Maybe once it did. But a single physical world, with its particular laws of Nature, is shared by all living things, and so we expect there to exist convergence in the perception of the world by all processors of information about the world (like brains) that turn out to have a high survival value.

Despite the compelling nature of this argument, one might worry that there are aspects of the physical world, like the nature of the most elementary constituents of matter or the structure of distant galaxies, black holes, and other astronomical objects, which we have discovered too recently for that process to have had any survival value. So, in these esoteric areas of human understanding, we cannot appeal to the survival value of faithful representations. Yet, ironically it is in precisely these areas that the utility of mathematics displays to us its greatest power. The sceptic could claim that this is an illusion created by the fact that we can only see the tip of the iceberg of complexity that truly defines these esoteric things and that, when we are able to learn as much about them as we can the weather or human behaviour, we will find them every bit as difficult to capture fully with mathematics. One is reminded of the story about an astronomer who began a public lecture about stars with the words 'Stars are pretty simple things...', only to hear a voice calling from the back of the room, 'You'd look pretty simple too from a distance of a hundred light years!'

## SHADOWLANDS

*Between the idea*
*And the reality*
*Between the motion*
*And the act*
*Falls the Shadow*

T.S. ELIOT

A strange thing about people is that they are never content to take things as they come. Human consciousness acts as a buffer which cushions us from raw reality. It creates symbols or pictures of reality for us. This ability to represent the world in a succinct way is necessary for our evolution and survival. In order to learn from experience we need to have ways of gathering and storing information about the environment in which we live. When we recognize that what we perceive is a representation of reality and could differ from it in essential ways, we have invented philosophy. Ever since ancient times philosophers have wrestled with the problem of what the relation is between the world as it really is and the perception of it that our senses offer. We can have some confidence that a substantial part of our picture of the world is a true and faithful one because were it not so we could not be here to talk about the matter.

The difficulty of dealing with mathematics in this light is that we know that many attributes of the things that we observe are ascribed to them by us. If we decide that some of those attributes, the 'mathematical' ones among them, for example, are to be regarded as mind-independent, then we have to know where to draw the line.

Galileo was amongst the most effective users of mathematics to elucidate the structure of the world. From his studies of motion, light, and astronomy he became convinced that Nature was 'a book written in mathematical characters' and if we did not learn this language we would be unable to fathom the subtleties of the Universe. Anything about the world which did not lend itself to a mathematical treatment, he regarded as secondary and subjective, something ascribed to it by the observer's mind rather than something primary and objective. Galileo writes that

> many affectations which are reputed to be qualities residing in the external object, have truly no other existence than in us, and without us are nothing else than names.

As an example, he cites things like smells and tastes and odours which, if their quantitative aspects were taken away, would leave nothing at all

> If the ears, the tongue, and the nostrils were taken away, the figures, the numbers, and the motions would indeed remain, but not the odours nor the tastes nor the sounds, which without the living animal, I do not believe are anything else than names.

Thus Galileo distinguishes subjective from objective properties by distinguishing those which are imagined in some way to reside in the object itself from those which reside in the observer's mind. It is important to recognize that before Galileo most thinkers had regarded man and the natural world to be essential parts of the Universe with man playing the major role. Once mathematics is introduced to describe things, Galileo finds that man is not a very good subject for mathematical treatment, so it is useful to distinguish between the world within man and the 'external' world outside. The latter is thus the real 'objective' part of things, whilst the former constitutes the subjective part. This distinction thus sets up man as a mere reader of the great mathematical book of reality. It was a view taken also by the French natural philosopher René Descartes to establish what is now known as 'dualism'—a sharp distinction between the observer and the observed. Unfortunately, such a simple view of things can no longer be sustained. The discussion we have given above indicates how our minds and bodies are not supernatural; they too are part of the natural world and can be described by the language of mathematics. We cannot view mathematics as either a product of our mental processes or a part of external reality that we discover. We know of no way to sustain such a distinction for our minds bear the imprint of the external world in which they have evolved and in which they must persist.

## TRAP-DOOR FUNCTIONS

> *What goes up a chimney down but won't go down a chimney up?*
>
> ANONYMOUS RIDDLE

The problem of how much the true 'message' of Nature has been distorted or shifted by the process of our perception, assimilation, and thought is one that scientists try to ignore or define away by saying that they are only studying the appearances because that is all we have available for study. The modern study of 'information' has progressed to such an extent that we are able to draw conclusions about the effects of processing or transmitting information of any sort by any means. Although the rules that lead to such conclusions were designed to apply to the transmission of information along wires or the airwaves they can be applied metaphorically to the problem of knowledge—the puzzle of how our understanding of the world degrades the information available to our brains and senses. We view 'understanding' as a process of receiving a message that may be distorted in some known or unknown ways.

First, let us explore this image of 'understanding' as the breaking of a code. We are all familiar with simple codes like exchanging the identities of the letters of the alphabet for numbers. These are simple enough for a child to solve. More secure codes involve the use of keys of which two copies exist, one for the sender and one for the receiver. The cipher can be enormously complex, involving hieroglyphics and all manner of unusual symbols in an effort to defeat an enemy seeking to break the code. But such 'single pad' codes always run a risk of having the key stolen or copied in transit from the original source of the two copies. About fifteen years ago the problem of the secure transmission of coded information was solved in a remarkably simple way. It achieved what at first sounds impossible: the transmission of a message that can be decoded by the recipient without him knowing what the sender's code is. So there is no possibility of the code falling into enemy hands because no deciphering key needs to be sent to the recipient at any time. To get an idea of how this might be possible, let's start with a simple picture to convince you that such a coding is possible.

Suppose I place my secret inside a trunk, padlock the trunk and send it to you. The process of coding is represented by the process of padlocking; decoding would involve unlocking the padlock. If I want you to read the secret message inside the trunk, then I could arrange for a key to be sent to you separately; but that would run the risk of interception by others anxious to discover our secret. Yet there is a simple way to avoid having to do that. I lock my padlock on the trunk and send it off to you. You put your padlock on the trunk (but ignore mine) and return it to me. I remove my padlock with my key and return it to you. You unlock your padlock with your key and open the trunk. Neither of us needed to know anything about each others' keys or padlocks.

If we want to use a trick like this to encode a message, then we exploit the existence of what mathematicians call 'trap-door functions'. These are operations which are very easy to perform in one direction but practically impossible to carry out in the other direction. Threading a piece of cotton through the eye of a needle illustrates the general idea: it is easy to pull the cotton out but very difficult to thread it through the eye of the needle. The ease one has in falling through a trap-door compared to the difficulty of climbing back through gives these operations their name (see Fig. 4.1).

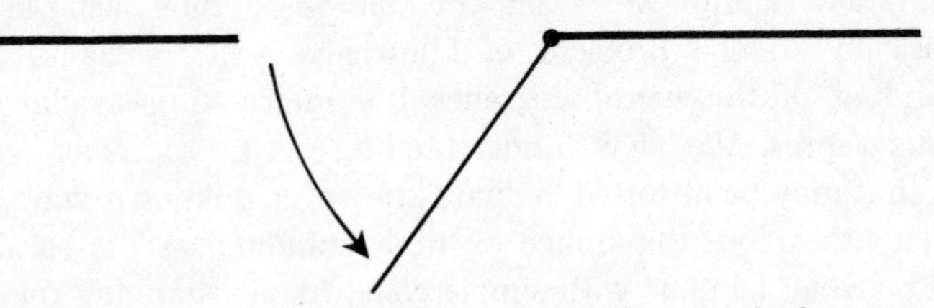

**Figure 4.1** The familiar image of the 'trap-door' through which it is easy to fall through but not so easy to climb back up. This metaphor captures a feature of some mathematical computations that are easy to perform but once done, are practically impossible to reverse.

It is easy to create a simple mathematical operation which has a trap-door-like property. Take two very large prime numbers—that is, numbers like 11 or 23 which cannot be exactly divided by any number other than themselves and 1—so large that they have maybe hundreds of digits in them. Now multiply those two numbers together. The result is a new number with about twice as many digits in it. This multiplication is a simple, although lengthy, operation for any person to perform unaided; a computer could do it very quickly though. But the reverse operation of discovering the two unknown factors of some composite number is something else entirely. It could take the fastest computers on earth thousands of years to perform.

Big prime number × big prime number = very big composite number
This direction is easy ———————>
<——————— This direction is totally impracticable

Now, suppose that we code our message in some way so that it is represented by a large prime number which we shall call $M$. I possess a secret large prime number P known only to me, and I encode the message by multiplying $M$ and $P$ together and send the result, $M \times P$, to the recipient, who has some other secret large prime number $Q$, known only to him. They encode the message they receive using their own code; that is, they multiply the number they receive by $Q$ and return the result, $M \times P \times Q$, to me. I decode the message by dividing it by my number $P$ and then send back the number remaining to the others. They receive $M \times Q$ and decode this by dividing by their own secret number $Q$; they are then left with the message *M*. And neither of us needed to know the large prime number that the other was using. To discover it we would have needed to factor an enormous number into its two prime factors. If the other person suspected that we could have developed some new type of computer which might do this in a matter of months rather than thousands of years, he can simply change the number he uses after each message is decoded.

Many of the world's most secret codes are based upon trap-door operations like this. If a mathematician suddenly discovered a way for a computer to produce the prime divisors of very large numbers with hundreds of digits in a matter of seconds, then there might well be something of an international crisis in military and commercial circles.

We can apply the lesson that we have learnt from these trap-door operations to the problem of unravelling the nature of the world by mental processing. True reality is 'coded' as it were by our perceptive and cognitive apparatus and we come thereby to know an encryption of the true nature of things. We see that it is quite possible for us to know the coding that transforms true reality into perceived reality and yet be unable to invert it and recover the picture of true reality from our perception of it.

We do not know how complicated and trap-door-like is the process which converts reality into perceived reality. In 1948 the American mathematician Claude Shannon discovered an extraordinary mathematical theorem which tells us something about the transmission of any form of information in a situation where there exists the possibility of distortion or dilution of the true signal. This is what engineers call a 'noisy channel'.

Shannon taught us that, somewhat surprisingly, it is always possible to find a way of coding a message so that its decoded form has a distortion smaller than any level you care to specify, regardless of the level of noise and interference. In practice, signals can only possess high levels of resilience to noise if there is a lot of redundancy built into the transmission code, with far more symbols being transmitted than are necessary to specify the message in the absence of any noise. For example, the message could just be repeated many times or it could contain instructions which automatically correct errors.

Shannon's theorem is very general; it refers to information of any sort, whether it be in the form of writing, radio signals, music, impulses in the human visual cortex, illustrations, or human speech, and it applies to any method of communication whatsoever. But there is a price to pay for this generality: although one knows that there exists a way of coding any message so that the distortion to the decoded message is as small as you like, you are not told what that optimal method of coding the message is or how to construct it.

It is interesting to apply this result to the problem of human cognition. Again, let us regard the process of perceiving and understanding the world as the reception of some message, but regard the influences of human bias, misapprehension, human mental processing, incorrect observations as forms of 'noise' which distort the true nature of reality into the one that we contemplate. Shannon teaches us that there exists a coding of the world which minimizes the gap between true and perceived reality. How close is Nature to this optimal situation? We know from experience that the codings with high fidelity have lots of built-in redundancy as insurance against vital pieces of their message being lost or distorted. Moreover, if the same message is sent to different recipients, then they can all decode the same message as accurately as they wish even though they may not all receive the same sequence of symbols (or even the same types of symbol). We could thus regard the entire legacy of human investigation of the Universe as the reception of information in Shannon's sense. Different observers and thinkers are distorted by different types of 'noise'. The fact that many observers of Nature make similar independent discoveries or draw similar conclusions about the world is a measure of the degree of optimal coding of Nature because it is a measure of the amount of repetitive redundancy in the information available to us about the world. Despite the diversity of human investigators into the nature of the Universe, Shannon's theorem tells us that there exists an optimal way for Nature to be packaged so that the distortion introduced by our minds can be made as small as one likes. Perhaps the evidence that the laws of Nature give a consistent picture regardless of so many superficial distorting influences, and in the light of such a diversity of approaches, is indicative of their possessing an optimal coding in some sense. It is interesting to reflect that it might be the language of mathematics that provides this optimal coding by minimizing the amount of irrelevant information that can be carried along with the message and providing a way of identifying errors easily by reference to the intrinsic pattern of logic being employed.

## MATHEMATICAL CREATION

*There was a young man of Milan*
*Whose poems, they never would scan;*
*When asked why it was,*
*He said, 'It's because*
*I always try to get as many words into the last line as I possibly can'.*

ANONYMOUS

Mathematicians *do* mathematics. Some people think that 'doing' involves inventing and creating after the manner of artists and musicians. In some cases this creative energy is directed, like the architect's, by the utilitarian desire to describe some particular goal, but in others it may be motivated purely by aesthetics. Regardless of the intention, there is no denying the feeling of the inventor that invention is taking place. Perhaps mathematics is nothing more than the activity of mathematicians: an invention rather than a discovery. This view we shall dub 'inventionism'. Although it is not often found amongst physicists and other 'hard' scientists, it is occasionally detected amongst mathematicians of a more formalistic persuasion who regard the human mathematician as the free creator of axioms and rules of reasoning which come to define formal systems. But when one enters the realm of the softer social sciences like economics, or the human sciences, one finds this view of mathematics to be endemic. In the social sciences mathematics is used as a tool for achieving certain specific goals. It is natural to treat it as one might any other tool and regard it as a creation motivated by specific human needs. The growth of this view has occurred primarily in the twentieth century. It can be detected all through the physical sciences if one is sensitized to it. If one is not, then it is rather easy to act as an unwitting agent in its propagation. Whereas the library of a nineteenth-century scientist would be full of formidable tomes bearing sombre matter-of-fact titles like *Hydrodynamics* or *The Theory of Sound*, today one would find *Mathematical Models of Fluid Flow* or *Concepts of Sonic Phenomena*. The intrusion of the notion of 'modelling' betrays a departure from the Victorian confidence in the existence of a unique theory of hydrodynamics or sound which we could discover and write down for all to understand. The 'concepts' are ways of representing or visualizing some elements of that part of reality which we have come to call 'sonic phenomena'.

This fashionable perspective looks upon mathematics as a representation of particular intelligible aspects of a very complicated phenomenon which one is not claiming can be understood or described in its entirety. Instead, we isolate aspects of it and describe those by means of whatever bits and pieces of mathematics are the most effective tools for the job. If we find the going too tough using just pencil and paper, then we can bring in the computer to help in the modelling. If we were studying how water flows over a sea barrier, then we could model the water as a collection of millions of little balls which bounce off one another when they collide, just like billiard balls. The computer could follow all the collisions that would occur as the collection of balls was launched towards the barrier and flowed over and around it in a complicated way. Here the real physical quantity is indeed 'modelled' by something else and one exploits certain analogous aspects of the behaviour of the real system and the model. The model bears the same relation to reality as does a model aeroplane to a real one. It possesses some of its properties. It is a representation that is helpful for study of the real. Many social sciences use mathematics in this way. The parts they use are obviously chosen to possess very particular properties unique to the problem they are studying. Moreover, that problem—for example a model of our society—is likely to be very complicated when every detail is included and so it will always be modelled in some simplified edition which ignores many complications and details, just to make the study tractable. Here the mathematical representation of things is very obviously a simple model that the author has constructed. It may not be the best place to seek out the truth about mathematics.

## MARXIST MATHEMATICS

*Culture is an instrument wielded by professors, to manufacture professors, who when their turn comes will manufacture professors.*

SIMONE WEIL

If one believes that mathematics is invented by human beings for a particular purpose, then one is implicitly regarding it as a cultural phenomenon of the sort studied by sociologists. To make such a position credible, it is necessary to answer several immediate challenges that opponents will launch against it. How does one explain the very similar mathematical traditions of widely different cultures? How does one explain the phenomenon of simultaneous independent discovery by different mathematicians of the same concepts? The first question could be countered by pointing out the differences in the development of the major mathematical civilizations: the Chinese did not develop geometry like the Babylonians and the Greeks. The inventionist would argue that mathematics was subject to laws of development and diffusion between different cultures just like the arts and language. The simultaneous discovery of different concepts by different mathematicians working in different cultures would have to be attributed to diffusive influences or to the appearance of similar social and practical problems, which arise as a consequence of having a civilization, whose solution must necessarily be found the same by different mathematicians in those cultures. Moreover, one often finds that within a culture there can be an evolution of thinking to a certain critical point after which several people independently make essentially the same discovery. The American sociologist of science Robert Merton maintains that multiple discovery is actually the rule rather than the exception in science:

When a cultural system grows to the point where a new concept or method is likely to be invented, then one can predict that not only will it be invented but that more than one of the scientists concerned will independently carry out the invention.

Predictably, there exists a Marxist picture of mathematics that lays prime stress upon its outgrowth from human needs and activities. It sees mathematics as just another fallible human activity and mathematicians as foremost social beings whose activities contribute to the overall growth of society and are likewise shaped by its evolution. Interestingly, the vast majority of the sociological studies of mathematics that have been conducted have been by Marxists or by thinkers sympathetic to a materialist picture of mathematics and its origins. For orientation we might take an extract from Karl Marx's* *Das Kapital*:

> The existence of things *qua* commodities, and the value-relation between products of labour which stamps them as commodities, have absolutely no connexion with their physical properties and with the material relations arising therefrom. There it is a definite social relation between men, that assumes, in their eyes, the fantastic form of a relation between things. In order, therefore, to find an analogy, we must have recourse to the mist-enveloped regions of the religious world. In that world the productions of the human brain appear as independent beings endowed with life, and entering into relations both with one another and the human race. So it is in the world of commodities with the products of men's hands. This I call Fetishism which attaches itself to the products of labour, so soon as they are produced as commodities, and which is therefore inseparable from the production of commodities.

What lies behind these sentiments is the belief that in a civilization like early Greece that is seen to have nurtured many aspects of modern science and mathematics, the existence of distinct slave and leisured classes brought about a distinction between thinking and doing. But that thinking was necessarily coupled to the commercial activities of day-to-day affairs and so when it gave rise to abstract mathematical notions they were subtle outgrowths of that activity. The mathematical notions of the early Greeks are seen therefore as having arisen from financial dealings, land surveying, and the like. However, the argument can be taken further into more interesting territory where it amplifies some of the ideas about the early development of counting that we made in Chapter 2. It is claimed that the bartering and exchange of goods for other goods or money is a source of abstract thinking and exact computation whereby the things themselves are ascribed some value which is independent of their precise

* Marx wrote quite extensively about mathematics at a technical level. In his work *Mathematical Manuscripts* there is an extensive discussion of the merits of various definitions of the derivative of a function in an attempt to avoid the use of logical steps which could not be explicitly implemented in reality.

physical identity. Subsequently, the idealization of this abstraction was developed by the Greeks in other contexts and used to introduce mathematical ideas in the abstract devoid of specific reference to tangible things. The subsequent de-emphasis of this origin is associated with the general development of the capitalist perspective. Ideas, like wealth and property, come to be viewed as entities existing outside the milieu of human activity and independent of human affairs. This distortion is then seen as the origin of the mistaken view that abstraction is a mental process by which we discover truths which exist independently of our human activities.

What can one say about this thesis that monetary exchange is the specific source and the ultimate origin of those abstractions that dominate our mathematical thinking? It is an argument that parallels Marx's more well known characterization of religions as the outgrowth of particular social processes. However, one might equally well turn the tables and argue that religion was the origin of abstract thinking. One only has to look in the right place to find what one really wants to find. The early Hebrews had a religion which was unusual, perhaps unique, in the ancient world in its taboo upon a representation of God in any way, shape, or form. Any form of graven image was forbidden and indeed early Jewish artefacts offer no representations of any living things. Thus one finds here that the notion of God was deliberately abstract. Aside from this approach one might take issue with the idea that mathematical abstraction arose from commercial activity because the latter is already quite a sophisticated affair that requires its practitioners to have at their fingertips a number of 'mathematical' notions. They need to have ideas about relative quantity and value; they also need some idea of ordering. Moreover, a more detailed investigation of how the Greeks regarded their coinage makes the general Marxist thesis difficult to sustain for them. Some have claimed that it is just as common for coins to be referred to as worth so many cattle or sheep as for animals to be spoken of as worth so many coins. The bulk of mundane daily business was carried out by slaves rather than philosophers and they do not appear to have been greatly interested in the concept of monetary exchange.

Arguments of this sort regarding the origin and cultural development of mathematics run into difficulty when they confront the utility of mathematics. It is easy to understand why mathematics turns out to be useful for solving problems or understanding observations of the world which were known at, or before, the invention of the mathematical concepts concerned. What we cannot understand is why mathematical concepts invented for purely aesthetic reasons, or as extensions of other mathematics invented for practical purposes, turn out to be, if anything, still more accurate in their description of the workings of the world. Cultural traditions are contexts for the development of mathematics but they cannot be its universal causes.

## COMPLEXITY AND SIMPLICITY

*My theology, briefly, is that the universe was dictated but not signed.*

CHRISTOPHER MORLEY

The different attitudes of the sociologist and the physicist to the nature of mathematics betray a difference in the nature of their subjects which results in a different perspective upon the world. We want to explore the reasons for this because, at root, this different perspective upon the world has a strong influence upon their views of and the importance they attach to mathematics.

Stop a particle physicist in the street and you will soon find yourself hearing how simple, symmetrical, and altogether elegant is this thing we call the Universe. All the diverse materials around us reduce to the permutations of a handful of microscopic building-blocks obeying elementary rules governing who can do what to whom and with whom. Yet when we get back to the workaday world it is surely nothing of the sort. Our daily lives, the workings of our businesses, national economies, local ecologies, or weather systems are anything but simple. Rather, they are a higgledy-piggledy of complexity governed by a concatenation of interlinked processes that possess neither symmetry nor elegance. As complexity becomes more organized so the range of phenomena that issue from it blossom and grow with unpredictable subtlety. And indeed, if it had been a biologist you had stopped along the way, you would have been told nothing about simplicity and symmetry. He would have waxed eloquent about the interlinked complexity of the outcomes of natural selection that we see around us. Neither planned nor guided, there is no reason for them to be simple: after all, there are so many more ways to be complicated. Their primary characteristic is persistence, or stability, rather than simplicity. So what are we to make of this apparent dichotomy: Is the world simple or is it complicated? The answer is important because the physicist, impressed by the simplicity of the world, will find it neatly encapsulated by abstract mathematical descriptions of pattern and structure and will be greatly persuaded of the intrinsic mathematical nature of things. The social scientist will find all his problems too hard for mathematics to solve. He will need to model, to approximate, and to guess in order to quantify what he sees. He will ascribe no elevated status to mathematics; instead he will be persuaded that the world cannot be captured by its simple structures.

Ever since the early Greeks began the serious contemplation of natural things there have existed two different emphases in thinking about what is important for our understanding of the Universe. The 'Platonic' thread is a tradition that emphasizes the timeless and unchanging aspects of the world as being the most fundamental. For Plato himself these were the 'forms', the invariant blueprints, of which all observed things were merely shadowy examples. The observed happenings were therefore less fundamental than the unchanging blueprints that governed them. For scientists of the post-Newtonian era these unchanging aspects were the conserved quantities of physics—energy, linear momentum, angular momentum, electric charge, and the like. This Platonic emphasis seeks to expunge the process of change and the notion of time from the description of things.

This desire can be realized to a surprising extent. The traditional laws of Nature dictating change can always be recast as equivalent statements that some 'conserved' quantities remain unchanged in all physical processes. Thus the conservation of linear momentum is equivalent to the fact that the laws of Nature must be the same in all places, the conservation of energy to the requirement that they be the same at all times, the conservation of angular momentum to their sameness in all directions of space. This Platonic approach has reached its zenith during the last fifteen years in the study of elementary particle physics. For whilst one can replace laws of Nature governing changes in space and time by statements that certain quantities remain unchanging, these statements of solidarity can in turn be replaced by the dicta that certain patterns or 'symmetries' be preserved in Nature. Particle physicists have developed this connection between laws and invariance still further in the creation of 'gauge theories' in which the symmetry that is preserved is of a more abstract geometrical character so that laws of Nature can remain the same under arbitrary changes in space of the particles involved. The requirement that such powerful invariances be preserved turns out to demand the existence of the forces of Nature and to dictate the way in which particles interact with each other. They allow us to offer answers to questions like 'why do certain forces of Nature exist?' rather than merely provide descriptions of *how* they act in the Universe.

All the known forces of Nature—gravity, the strong nuclear force, the weak force, and electromagnetism—are described by varieties of gauge theory founded upon the immutability of some pattern when any change occurs. They are founded therefore upon the Platonic assumption that symmetry is fundamental and the ultimate expression of that faith is the search for a 'Theory of Everything' within which all these separate theories of the four different forces of Nature can be subsumed and unified into a single description of the ultimate symmetry, or law of Nature, from which all else follows. Whether that ultimate theory is a conventional gauge theory or a superstring theory, in which the most fundamental entities are lines or loops of energy rather than points, makes no difference to the appeal to symmetry. The primary appeal of superstring theories is that the further requirement that all calculable quantities be finite is sufficient to pin down just one or two possible all-encompassing symmetries within which those respected by the individual gauge theories of the forces of Nature can be embedded as pieces in the complete kaleidoscopic pattern of things. This success in beating a path towards one all-embracing symmetry—the Theory of Everything—through the complex jungle of experience lies behind claims of physicists that the Universe is simple and deeply symmetrical.

But there is a second tradition in the study of Nature that, until recently, has been less popular than the Platonic search for the invariants of Nature. The Aristotelian perspective laid emphasis upon the observable happenings in the world rather than the unobservable invariants behind it. As a result the process of temporal change was regarded as fundamental. It is no accident that the original advocates of such an emphasis drew their intuition more from the study of living things than the purposeless pendula of the physicist. For the advocates of this approach the world looks complicated and messy and one need not expect to explain away all aspects of that complexity by appeal to simple 'laws' acting behind the scenes.

To understand the real difference between the simple Platonic view of the world and the complicated Aristotelian perspective we need to appreciate one important fact about the world: symmetrical laws of Nature need not have outcomes which possess the same symmetries as those laws. If we place a pencil in a vertical state and allow it to fall then it will fall in *some* direction. The laws governing the fall of the pencil do not have any special preference for one direction over any other, but the pencil must fall in some particular direction and, in so doing, the underlying symmetry of the governing law of Nature is broken in the observed outcome of the law. Were this not so, then every outcome of the law of Nature would have to carry the full invariance of the laws. We could not be sitting in the spot we happen to occupy at the moment unless the laws of Nature showed a special favouritism for that spot. Thus we see that outcomes are much more complicated things than laws of Nature.

Moreover, we do not observe the laws of Nature: we observe only the outcomes of those laws and from the heap of broken symmetries before us we must work backwards to reconstruct the pristine laws behind the appearances. Sometimes this is very easy to do, but often it is impracticable because of the sensitivity of the direction of the symmetry-breaking to the whims of the environment. But we have learnt one important lesson. This process of symmetry-breaking explains how we can reconcile the existence of observed complexity with underlying laws of Nature that are simple.

Particle physicists earn their living by studying the laws of Nature and their claims for the simplicity and symmetry of the world point to the economical forms that can be found for the laws of Nature. The life scientist, or the economist, by contrast, troubles himself not at all with any 'laws' of Nature. The focus there is entirely upon the complicated outcomes of the underlying laws. This state of affairs is perhaps responsible for the surprising lack of success that accomplished mathematical physicists so often have when they turn their attention to the problems of the life sciences. Accustomed to pristine symmetry and mathematical beauty they discover that the higgledy-piggledy results of natural selection possess neither of those desirable features. Instead, they are faced with understanding outcomes that are separated from the underlying 'simple' laws of physics by a long sequence of hidden symmetry-breakings.

We have seen therefore that the world can be both simple and complicated in important ways and the aspect that impresses you most will depend upon whether you are more concerned with the laws of Nature or their outcomes. An interesting historical example where this division was clear but unrecognized can be found in the eighteenth and nineteenth centuries. There one finds examples of 'design arguments' for the existence of God from those examples of order and apparent contrivance in Nature from which humanity seems to benefit. There were always two varieties of such arguments. The older, which was turned on its head by the insights of Wallace and Darwin concerning natural selection, pointed to the existence of specific situations in the natural world—the design of the eye, or of the hand, or the way in which animal habitats appeared tailor-made for their inhabitants, for example—where the outcomes were advantageous, as evidence of divine providence. The other style of design argument, popularized first by Newton's followers, pointed to the invariant laws of Nature as the primary evidence for a deity. In these two examples one sees design arguments based, in the first case, upon the *outcomes* of the laws of Nature, stressing particular examples of complex symmetry-breakings, and in the second, upon the simplicity, invariance, and symmetry of the underlying *laws*.

During the twentieth century the Platonic approach has dominated fundamental physics. Since the mid-1970s, when gauge invariance and symmetry was found to be a master-key with which to unlock the secrets of the elementary particle world, the laws of Nature have been regarded as more interesting than their outcomes. This is not altogether surprising: for laws are simpler to study and one might imagine that once in possession of laws one could understand and predict their outcomes. But in the last few years physicists, mathematicians, and computer scientists have realigned their focus of attention upon the outcomes, having come to appreciate that there exist sequences of events which cannot be replaced by timeless invariants in the Platonic manner.

The Aristotelian perspective has re-emerged in the study of complexity in the abstract, that is, as a general phenomenon not necessarily tied to a particular complicated physical situation. Suppose we consider some sequence of outcomes (numbers issuing from a computer, for example): what do we mean by saying that the sequence is ordered in some way? It means that our minds have picked upon some pattern which enables us to abbreviate the sequence in our minds. If it is possible to store the information in a sequence in an abbreviated form shorter than the sequence itself then the sequence is regarded as non-random and we call it *compressible*. If no such abbreviation exists, then the sequence admits no representation other than the complete explicit printout of itself. In this case we call it *incompressible*. These are ideas that we took a first look at in the last chapter in our discussion of randomness.

The existence of incompressible sequences means that the Platonic approach is of no use in their analysis. The lack of an abbreviated representation means that there exists no symmetry or invariance whose simple preservation is equivalent to the data content of the sequence. For that would be a compression of the sequence. The outcomes contain a level of complexity that requires nothing less than their explicit listing to capture their full information content.

Besides elevating the study of outcomes to something that is not necessarily included within the study of natural laws this notion of compressibility gives simple ways of characterizing many of our intellectual activities. We recognize a possible new definition of 'science' as being simply the search for compressions: the laws of Nature are the compressions of our sense data. The discovery of a Theory of Everything would be the ultimate compression. Moreover, the apparent success of this process hinges upon two superficial features of things: the physical world that we observe seems to be surprisingly amenable to compression, and the brain is remarkably good at effecting compressions when presented with events. In a predominantly incompressible world we would not have scientists but archivists who simply recorded every observed event. The compressibility of many aspects of the world saves us from this 'Bureaucracy of Everything'. We can use a simple law of motion to describe the motion of heavenly bodies instead of having to keep a record of their positions and velocities at all times. Yet, clearly, this compressibility and the brain's remarkable ability to make sense of complicated things is an important necessary condition for our own existence. We could not survive as 'intelligent' observers and readers of a book like this in a world where no compressions were possible or with brains that produced imaginary or erroneous compressions. A certain level of predictability and innate predictive power is required for the successful evolution and survival of living things. However, we must beware of the fact that our brains are altogether too good at finding compressions. It has clearly proved efficacious to overdevelop our pattern-recognition capability (presumably because if you see tigers in the bushes when there are none your friends will merely call you paranoiac, whereas if you fail to see tigers in the bushes when there are, then your continued survival must be rather doubtful). As a result we see canals on Mars and all manner of exotic things lurking in inkblots. Yet, the brain cannot gather all the information potentially on offer to it; that would be as impractical as gathering none—would we really want to receive information about every last electron orbital when we looked at a painting? It overcomes this problem by storing only a part of all the information available to the senses. Our physiological make-up helps to effect this truncation by placing limits on the intensities of light and sound that we can respond to. However, this serves to warn us that the brain would effect a compression of the observed information even if one did not truly exist. This is a modern way of restating something first stressed by Kant in the eighteenth century. Furthermore, we know that many aspects of the scientific enterprise set out to truncate the information available to us in order to effect a compression, for example by random sampling to obtain a representative opinion poll.

This teaches something more about different sciences. In the so-called 'hard' sciences the most important characteristic of their subject matter, that encourages compression, is the existence of simple idealizations of complicated situations which can underpin very accurate approximations to the true state of affairs. If we wish to develop a detailed mathematical description of a star like the Sun, then it serves as a very good approximation to treat the Sun as being spherical with the same temperature all over its surface. Of course, no real star possesses these properties precisely. But many stars are such that some collection of idealizations like this can be made and a very accurate description still results. Subsequently, the idealizations can be relaxed slightly and one can proceed step by step towards a more realistic description that allows for the presence of small asphericities, then to further realism, and so forth. By contrast, many of the 'soft' sciences which seek to apply mathematics to such things as social behaviour, prison riots, or psychological responses fail to produce a significant body of sure knowledge because their subject matter is far less compressible and does not readily provide obvious and useful idealizations from which one can proceed towards better and better approximations to reality.

What we have learnt from this digression into the relation between laws of Nature and their outcomes is that different types of scientist focus upon different aspects of the physical world and as a result will not be equally impressed by the role played by mathematics.

## MATHS AS PSYCHOLOGY

> *I know that when I was in my late teens and early twenties the world was just a Roman candle—rockets all the time ... You lose that sort of thing as time goes on ... physics is an otherworld thing, it requires a taste for things unseen, even unheard of—a high degree of abstraction ... These faculties die off somehow when you grow up... profound curiosity happens when children are young. I think physicists are the Peter Pans of the human race ... Once you are sophisticated, you know too much—far too much. Pauli once said to me, 'I know a great deal. I know too much. I am a quantum ancient.'* ISIDOR RABI

The inventionist pitch really amounts to the claim that mathematics is a branch of human (and maybe also to a lesser extent animal) psychology. It is an invention of the human mind. But we have already seen that this terminology is slightly misleading. The common idea of an invention is of something that is suddenly created into a vacuum. If someone 'invents' the motor car, we have in mind the situation that one day we have the idea, or even the material entity, of a motor car when the previous day we did not. However, we have seen that the evolutionary process has tied us to the environment around us so that our physical capabilities and our sense-perceptions of the world are influenced by the fact that truer, more accurate, representations of the world will have a greater survival value and so over huge periods of time will tend to be preserved. So, even if we lean towards the idea that mathematics is a human invention we should not imagine that it has necessarily been invented out of nothing. It may be a creation that has arisen because of certain inputs into the human mind, or because inputs are processed in a certain way by the brain.

This line of thinking leads us to consider how the human mind comes to develop the notion of number during our early childhood. We have already looked at the history of the development of the notions of number and counting in ancient cultures. Whilst this may be closely related to the development of human cognition in early childhood, we shall not try explicitly to connect the two here. However, some parallels with the pattern of historical development mapped out in Chapter 2 will often be apparent.

The Swiss psychologist Jean Piaget developed an influential picture of how mathematical ideas evolve and come to be entrenched in a child's mind during its early formative years. Whilst the extension of this picture to cover the evolution of all other forms of cognition, as Piaget intended, is somewhat precarious and one would be nervous of extending it too far from the European cultures in which Piaget's observations were made, its picture of development is persuasive.

Piaget's underlying view was that everything we know about the world and the routes by which we come by that knowledge, derive in their earliest stages from our physical actions on things—handling, touching, and manipulating objects. In the nursery, children less than two years old handle objects and learn to recognize them again after being separated from them. They develop personal attachments to things, but only after about eighteen months do they develop a sense that a thing is the same when it moves somewhere else or when they see it again at a different time. This is a key part of their psychological development: it shows that they are beginning to appreciate that objects have some sort of existence that is independent of their own actions upon them. This allows them to think about them as objects in their own right and to compare them with other objects. This leads to the ability to group similar things together. All the cars or all the furry animals can be gathered together into a collection. This grouping ability shows that one has gained the notion of a set or a class of similar things. Having appreciated the concept of a collection one can move on to the idea of some collections being larger or smaller than others. At first this will be a rather impressionistic appreciation. A child shown two collections of chocolates can be induced to choose the one that contains fewer members if it is arranged to cover a larger area or appear 'bigger' in some other more impressive fashion. Only a general notion of quantity and an appreciation of small numbers is apparent. There is no notion of a uniform sequence of quantities due to the successive addition of a single quantity. As this ability grows it is at first primarily a linguistic ability to learn numbers off by heart. Not until about four or five years of age does this learning of numbers by rote start to be linked to the earlier recognition of collections and sets of objects. When this happens the child begins to learn that the sequence of numbers can be mentally transported to correspond to an arrangement of objects so that the last number counted in the sequence gives the total number of objects.* Moreover these operations do not depend upon other properties of the things being counted. By age six or seven, more sophisticated notions can come into play. The child can count two different collections and, unlike younger children, compare them and identify the one containing more objects unambiguously without being misled by other measures of size. This process is novel because it means that two images have been created in the mind which can be compared even if the real collections are no longer seen side by side. Following this step more complicated manipulations can be carried out, transferred to other situations, or employed upon collections of real objects. At this stage the foundation for mathematical reasoning is being laid. It has grown out of the manipulation of ordinary objects, but the process has gradually become internalized in the mind so that one can remember it or recreate it rather than merely react to it.

After this stage of mastering and internalizing concrete operations on things, there is a growing realization of certain necessary truths about the nature of reality rather than merely experience of the properties of collections of things. One learns that if one member is removed from each of two equal collections then they remain equal. Two collections are either equal in number or they are not. The order in which things are counted doesn't affect the total one obtains. As one reaches the age of nine or ten, it appears that this recognition is transferable to less concrete notions about things. We see here an explicit source of mathematical intuition in the material objects in the world and their interrelationships. Gradually, in the early teenage years it becomes possible to carry out sets of mental operations upon representations of things. They are replaced by symbols and these collections of symbols can be manipulated by the mind. The previous range of necessary truths about operations like subtracting and adding becomes applicable to the symbols that represent quantities. A subject like algebra becomes possible where a symbol like the letter $x$ can represent any number which can be added to either side of an equation just as equal numbers of coins can be added to equal collections. This step is the essence of all future mathematics. Later on, it will be possible for the mind to invent new rules for manipulating symbols which are not tied to any empirical set of operations that can be carried out with

* Whilst this can apply to the situation in English and other Indo-European languages, it does not hold elsewhere. In Japanese, for example, the numbers used to count are not the same as those those employed to describe the total number in a collection being counted. It is as if one could count up to twelve but the word one should use to describe a set of twelve must always be 'dozen'.

real objects. At this stage the internal mental manipulation of symbolic representations of concrete things has taken off and flown like a dragonfly leaving the mundane chrysalis of past experience behind. It is no longer constrained in any way by our experience of concrete manipulations, only by the scope of our imagination to come up with sets of rules for the manipulation of symbols. The only requirement we place upon those inventions is that they be 'consistent' after the manner of the formalists that we met in the last chapter.

This, in a nutshell, is Piaget's picture of our mental evolution of mathematical intelligence. It originates in a child's activities with things in the world; shuffling, sorting, and comparing them; the notion of quantity is discovered and is then internalized in the mind whereupon it becomes available as a way of representing things in symbolic form; these symbols are then manipulated after the manner of the things themselves; the rules for their manipulation thereupon become the essential features of the activity, taking over from the things in themselves.

Animals lack the mental complexity to pass through this sequential development. Experiments reveal that some animals possess a rudimentary notion of quantity although not the notion of counting. They can distinguish small and large collections of things so long as the numbers involved are small and they are not, as very young children also were, confused by counter-impressions of size like a smaller number of things spread out over a larger area. More particularly, we know that even small domestic animals like dogs and cats are able to detect a small change in quantity. Experiments with birds reveal that in general they can choose the larger of two piles of seeds if the variation in quantity is 3 and 1, 4 and 2, or 4 and 3, but generally lose track when the choice is between piles of 4 and 5. The historian Tobias Dantzig tells the following amusing story which illustrates the limit of a crow's number sense and reveals that it is not dissimilar to our own immediate intuitive grasp of small quantities which we discussed on p. 40:

> A squire was determined to shoot a crow which made its nest in the watch-tower of his estate. Repeatedly he had tried to surprise the crow, but in vain; at the approach of man the crow would leave its nest. From a distant tree it would watchfully wait until the man had left the tower and then return to its nest. One day the squire hit upon a ruse: two men entered the tower, one remained within, the other came out and went away, but the bird was not deceived; it kept away until the man within came out. The experiment was repeated in the succeeding days with two, three, then four men, yet without success. Finally, five men were sent: as before, all entered the tower, and one remained while the other four came out, and went away. Here the crow lost count. Unable to distinguish between four and five it promptly returned to its nest.

There is no evidence that any animals have graduated from merely having a number sense to actually being able to count. This is undoubtedly a fairly elaborate mental operation and distinguishes our mental abilities from those of even the most intelligent animals in a dramatic way. If our consciousness has indeed evolved from a more primitive state then we might expect to discover some evidence of a more limited number sense in the earlier phases of human evolution. One memorable discussion of this question was that by the Princeton psychiatrist Julian Jaynes in his book entitled *The Origin of Consciousness in the Breakdown of the Bicameral Mind*. He claims that one can determine the period around three thousand years ago when consciousness became fully developed amongst early humans. He argues that at far earlier times the emerging structure of consciousness resulted in people 'hearing voices' in their heads and supports this by drawing attention to the way in which literature and art become more fantastic in earlier times when visions and dreams seem to have had a common place amongst stories and accounts of events. This world of vivid imaginings is, for Jaynes, the outflowing of the preconscious human world. This is a wonderful idea but I just can't bring myself to believe it on the basis of the evidence provided by Jaynes; most of it is consistent with too many other, more prosaic, alternatives.

One can see from this discussion that there is an uneasy tension between mathematicians and psychologists. They are not really interested in the same things. The psychologist wants to determine the origin of the particular skills that underpin mathematical dexterity and constructs his studies and experiments with that clear end in view. The mathematician, in contrast, is more attracted by the desire to discover the origin of certain primitive intuitions which seem to be presupposed by number. In that quest he tends to be guided more by rational deduction than by experimental studies. Little more than a hundred years ago one would have found a great gulf dividing these two approaches. The mathematicians would have assumed mathematical practice to be the discovery of a Platonic world of absolute truth. The growth of the study of human thinking and how it is linked to other aspects of our physiology and environment throughout the twentieth century has led to a growing belief that many aspects of mathematical reasoning are founded upon constructs that are peculiar to the make-up of human minds and so can be studied in detail by psychologists. As we shall see later on, there are particular interpretations of mathematics which are quite contrary to this view. Some regard the ordinary natural numbers as a completely irreducible basic concept. This is unacceptable to someone like Piaget because it implies that the concept of number must suddenly spring up fully grown in the human mind, which is contrary to his observations of the gradual growth and elaboration of the number concept over many years in young children.

Piaget's ideas about the gradual acquisition of tactile intuitions as a prerequisite for mathematical conceptualization have a number of implications for the educational process. In the distant past, when toys were limited in variety, most children would have had rather similar experiences of the manipulative skills offered by objects of different shapes and sizes. Today this is no longer the case. The diversity of children's toys is vast and one can see how carefully designed toys can help important cognitive skills to develop. Construction kits and train-sets exhibit geometrical features that encourage an intuitive feel for how the world is, how things fit together, properties of shapes and angles, curves and right angles; all this informal development of how things work and fit together is an important preparation for more formal instruction about such things—although, of course, if it is good instruction it will continue to get pupils to do practical things and to learn by experience, not merely by rote. If conceptual understanding requires these early experiences of real-world properties, then one might speculate that some individuals who just cannot seem to understand certain mathematical concepts might simply have missed out on some essential aspects of their early informal experience of the world. A particular worry in this respect, that many educational psychologists must have debated, is the extent to which the traditional difference between girls' toys and boys' toys might play a role in their subsequent aptitude for mathematical thinking. Until comparatively recently, boys' toys were dominated by train-sets and construction kits, Meccano sets and cars, all of which are replete with latent information about the geometrical aspects of the world and the fact that there are right or wrong ways of putting things together. Girls' toys, by contrast, tended to be dominated by cuddly toys, dolls, and things that create a personal emotional attachment. If Piaget's picture of the development of mathematical intuition is part of the story of human psychological development, then it is likely that these sexual differences in the objects of play have some impact upon the later aptitude of children for mathematical concepts and their appreciation of the nature of the external world. Of course, during the last decade we have seen a significant change in the objects of attention of children of school age. The small computer and electronic games machine now have an all-pervasive influence in Western cultures. It remains to be seen what the long-term implications of this change might be.* It is easy to focus on the new

* However, the sceptic might argue that a child's constant computer-game playing is about as likely to create knowledge of real computing as hanging around pool and snooker halls used to give teenagers an improved intuition for Newtonian mechanics.

aptitudes acquired by exposure to computers but it is less easy to evaluate the aptitudes lost by the lack of exposure to the traditional activities whose place the computer game has usurped.

## PRE-ESTABLISHED MENTAL HARMONY?

*Poets do not go mad; but chess players do. Mathematicians go mad, and cashiers; but creative artists very seldom. I am not, as will be seen, in any sense attacking logic: I only say that this danger does lie in logic, not in imagination.*
G.K. CHESTERTON

The ease with which humans pass through their early development to emerge in possession of an aptitude for numerical things that far outstrips that of any other living creature points towards some rather special aspect of the mind's architecture that facilitates it. Many psychologists look to some particular type of hard-wired neural circuitry of great complexity that enables abstract concepts like mathematics to be manipulated. For example, the American psychologist Howard Gardner summarizes some interesting neurophysiological findings which he believes form a rationale for the neural organization of mathematical ability:

> The ability to carry out logical-mathematical operations commences in the most general actions of infancy, develops gradually over the first decade or two of life, and involves a number of neural centers that work in concert. Despite focal damage, it is usually the case that these operations inhere not in a given center but in a generalized and highly redundant form of neural organization. Logical-mathematical abilities become fragile not principally from focal brain disease but, rather, as a result of more general deteriorating diseases, such as the dementias, where large portions of the nervous system decompose more or less rapidly. I think that the operations studied by Piaget do not exhibit the same degree of neural localization as [others] that we have examined ... and that they therefore prove relatively more fragile in the case of general breakdowns of the nervous system. In fact, two recent electrophysiological studies document considerable involvement of both hemispheres during the solution of mathematical problems. As one author puts it, 'Each task produces a complex, rapidly changing pattern of electrical activity in many areas in front and back of both sides of the brain.' In contrast, abilities like language and music remain relatively robust in the case of general breakdowns, provided that certain focal areas have not themselves been especially singled out for destruction.'

Where Piaget and psychologists like him claim that linguistic ability and mathematical intuition are just parts of the general learning and adaptation process of the brain, there are others, like Chomsky, who argue that the ability to acquire mathematical or other forms of language derives from an innate and unique attribute of the human mind. There is certainly a strong case to be made for such an idea. Very young children seem to possess a natural propensity for language acquisition—have you ever heard of a person who was unable to speak any language?—after only a few years of exposure. This skill seems to fade as we age though. Piaget's theory makes that difficult to understand. We would expect to become better at it. But perhaps our minds become too crowded with other things or simply lose a certain flexibility?

What is most interesting about our problem is how the study of mathematics and counting might tell us something about the structure of the mind. Recently, Roger Penrose has attempted to proceed in the same direction, arguing that the human mind arrives at judgements which supersede the capabilities of mechanical devices that follow the dictates of a single algorithm. This leads to the further (and to this author's mind rather unlikely) suggestion that the microscopic workings of the brain are directly influenced by quantum uncertainties which render its action beyond emulation by man-made machines. If true, this would be a most dramatic example of how the process of doing mathematics by human minds leads to new speculations and insights into the detailed workings of the human mind. However, the heart of Penrose's argument is the claim that our ability to judge one of Gödel's theorems to be true is proof of a non-algorithmic element in human thinking. It is curious that such a basic deduction about the nature of thinking should rest upon the thinker's apprehension of Gödel's demonstration. It implies that animals and young children who cannot yet carry out such mental gymnastics really think in a qualitatively inferior way and their more predictable thought processes would be simulatable by a computer algorithm. We shall be taking a closer look at this claim in the last chapter.

A traditional response to the question 'what can mathematics tell us about the mind?' is that there exists some special faculty of the mind which channels its operations into the mathematical mode. This aspect of the 'wiring' of the mind is then held to be responsible for creating certain laws of thought which seem to be automatic. Although mathematicians have invented logics in which statements need not be either true or false, this type of two-valued logic seems to be intuitively built into the way that we think. We regard it as self-evidently correct and if any schoolchild were to query its validity we would be at an impasse because no full justification could really be given.

One of the difficulties that must be faced if one is to push further into the psychology and even the neurophysiology that underlies mathematical thinking is that the basic intuitions regarding counting and ordering things that form the basis of studies like those of Piaget are focused upon collections of fairly randomly chosen children. They are not mathematicians and many of them will never carry out the sort of very abstract and peculiar thinking that characterizes the work of professional mathematicians. Moreover what the ordinary person in the street regards as mathematics is usually nothing more than the operations of counting with perhaps a little geometry thrown in for good measure. This is why banking or accountancy or architecture is regarded as a suitable profession for someone who is 'good at figures'. Indeed, this popular view of what mathematics is, and what is required to be good at it, is extremely prevalent; yet it would be laughed at by most professional mathematicians, some of whom rather like to boast of their ineptitude when it comes to totalling a column of numbers. Unfortunately, this image of mathematics is what is being studied in most psychological studies because they often focus—for very good educational reasons—upon the process by which children become numerate. Yet, when we come to face the problem of what mathematics really is and the mystery of its applicability to the real physical world, it is not the mathematics of the accountant that is of most interest. Rather, it is the status of much more abstract structures and logical relations between entities which are often far removed from everyday intuition and experience. They may have been invented merely to go one step further in creating abstract patterns of great intricacy but later find application in the description of the inner working of Nature at the heart of matter; they may even be contrary to those kindergarten intuitions that we gain from touching and seeing things, yet describe the workings of things in the world that we cannot touch or see. Thus, just as we have seen how it is possible to have a 'number sense' without being able to count, or to be able to speak having no knowledge of grammar, so it is possible to be able to count without having any knowledge or sense of mathematical structures.

As a result of this divide between numbers and mathematics one would do well to be suspicious of explanations of mathematics or of mathematical reasoning that tried to lump it all under one heading: trying to show that *all* mathematical knowledge has been abstracted from the physical world, or that there is a single type of mathematical intelligence whose cultivation and form is responsible for all aspects of human mathematical manipulation and creation.

Any attempt to trace intellectual attributes from the evolution of other human traits is a difficult one because of the vast number of possible scenarios and the paucity of evidence to discriminate between them. The evolution of mathematical intelligence is at root an aspect of the evolution of language and in the study of that problem unusual possibilities have been raised which, if correct, might apply to the development of mathematical reasoning also. We mentioned in Chapter 2 that Noam Chomsky and Stephen Jay Gould have argued that language acquisition may have evolved merely as a by-product of an evolutionary process, like increase in brain size, that selected for other entirely different traits and abilities. It is not our intention to try to defend such a view in opposition to the standard picture of the evolution of language by a process of gradual specialization honed by natural selection. We mention it just to indicate that what we regard as a fundamental feature of the human mind might well turn out to have rather humble beginnings of no special significance at all.

We see that there is every reason to believe that the complex structure of our brains permits the type of abstract and symbolic reasoning that mathematics requires. When we look outside the Western cultural heritage, we have to look harder for evidence of such abstract reasoning ability, but that is usually because we do not know where to look. If we merely test primitive tribesmen to see how they score on tests and trials designed by our own standards, we will not learn very much. But in almost all primitive non-Western cultures one finds a common core of numeracy that underpins basic operations in their societies. Buying, selling, and bartering reveal individuals who are adept at calculation and who are at home with symbolic representations. Legal systems are often well-developed bodies of logical deduction based upon certain agreed premisses. In this sense they are not unlike formal mathematical systems in which particular conclusions are reached from agreed axioms. The aim of the legal system is to try every claim and show it to be true or false. An interesting case history has been studied by the anthropologist Edwin Hutchins, who carried out a study of land disputes in the Trobriand Islands. He discovered that two litigants disputing the ownership of a piece of land would each have to produce logical histories which terminate in the state in which they possess the sole right to the land in question. In the process he will usually try to show that there can exist no such logical history which culminates in his rival having the right to the property. Hutchins comments that

> the problem-solving task of the litigant is akin to theorem-proving in mathematics or logic. The cultural code provides the axioms or implicit premises of the system. The historical background of the case, and especially the state in the past at which the litigants agree on the disposition of the garden, provides the explicit premises of the problem. The theorem to be proved is a proposition which represents the litigant's own rights in the land.

Another situation where mathematical intelligence is evident in most cultures is in the practice of games similar to nim or chess. These games of strategy, like games of chance, gradually encourage the development of logical and strategic thinking for material gain or enhanced status in the group. Likewise, the development of religious or social ritual can generate intricate hierarchies and an appreciation of geometry and symmetry. Some have even argued that the historical origin of counting and geometric sense is to be found in ancient ritual. The fact that we find quite separate ancient societies developing similar systems of counting is an important fact. It makes us suspect that there exists some common mental propensity in human thinking processes and/or some aspect of our environment or anatomy (ten fingers?) which produces a common response. This, we shall investigate more fully in the next chapter.

Another version of the inventionist approach is to point to human culture, rather than human psychology, as the root of mathematical ideas. One recent study of mathematics in history which takes this line begins with a forceful declaration which reveals the strong motivation for taking such a line:

> The assumption made in the present work is that the only reality mathematical concepts have is as cultural elements or artifacts. The advantage of this point of view is that it permits one to study the manner in which mathematical concepts, as cultural elements, have evolved and to offer some explanation of why and how concepts are created from the syntheses produced by cultural forces in the minds of individual mathematicians. Moreover, the mysticism that creeps surreptitiously into most forms of idealistic attitudes toward mathematical existence disappears...

By attributing mathematical ideas to human experience it is imagined that one can safely ignore the awkward question of what the mathematical ideas really are and whether they exist in some disembodied form. The mathematician Raymond Wilder was a fervent advocate of mathematics as a cultural creation. One motivation for such a view, as we have already seen, is the way in which alternative geometries and logics were discovered and it might be claimed that some were merely outgrowths of the cultures involved in their discovery. Against this, one could raise the objection that there are examples of multiple discovery wherein totally separate cultures 'discover' the same mathematical ideas.

Kant's theory of our knowledge of the world is a more sophisticated version of these notions. He argues that our minds do indeed possess a structure which ensures that our cognizance of the world is filtered through particular 'categories'. For a Kantian, like Joong Fang, the remarkable adaptation of mathematics to the workings of the natural world

> is possible because our knowledge of the 'reality' is not of the reality in itself, but the reality already adapted to our cognitive faculty. The feeling of admiration is of

our own making, and so is the 'adaption'. Precisely because of such man-made adaptions the jig-saw puzzles of our studies on the objects of reality fit together at all and, when fitted together, tell us only how they appear to us and not how they really are in themselves.

The standard form of fiction disclaimers may be employed here, too, namely: Any similarity between reality and appearance in the mathematico-physical models of the physical world are purely coincidental. The similarity, then, depends solely on the ingenuity and competence of mathematicians and physicists. Mathematics, as such, is the study of ideal models, preferably adaptable to real structures.

### SELF-DISCOVERY

> *Imagination is dependent for its activity on the quantity and quality of its available material. There is nothing in imagination that has not been in sensation.*
>
> MONTGOMERY BELGIAN

We have explored the case for regarding mathematics as a human invention, shaped primarily by the structure of the human mind and its particular ways of processing and organizing information, and responsive to the ways of human society and culture. This approach to mathematics has similarities with formalism, which also regarded mathematics as a human creation, but does not share its aims of consistency and completeness. Indeed, how could it? The products of human thinking must necessarily be fallible at some level. On this picture we do not discover mathematics 'out there'; it need not exist in the absence of mathematicians and the form it takes is strongly associated with our own genetic make-up. If mathematics is studied on Alpha-Centauri, we would expect it to differ from our own variety in many respects. We could not assume that we shared a common knowledge of some absolute truth that we could both discover. The idea that extraterrestrials might possess an alien way of thinking and reasoning had occurred to Frege in the nineteenth century. His reaction indicates the extent to which mathematicians of that time regarded this possibility as bizarre:

> But what if beings were even found whose laws of thought flatly contradicted ours and therefore frequently led to contrary results even in practice? The psychological logician could only acknowledge the fact and say simply: those laws hold for them, these laws hold for us. I should say: we have a hitherto unknown type of madness.

Like Frege we also feel that inventionism falls short of explaining the whole mystery of mathematics. The mind that invents has evolved out of the physical Universe and inherits its propensity for the mathematical from the adaptive advantage to true reality that such an adaptation bestows. To pursue the inventionist philosophy is to make mathematical truth dependent upon time and history. We are forced to an anti-Copernican stance which sees mathematical truth changing with the evolution of the human mind. Inventionism is a wonderful philosophy for the arts and humanities where we seen the fruits of imaginative subjectivity; their nature and practice contrast so drastically with that of mathematics that the objective element seems to have failed to be adequately incorporated into this view of mathematics. Inventionism fails to provide insight into the fact that Nature is best described by our mental inventions in those areas furthest divorced from everyday life and from those events that directly influence our evolutionary history. In the end, one cannot help but feel that humanity is not really clever enough to have 'invented' mathematics. In the next chapter we shall see what happened when we had a go at doing just that.

# CHAPTER FIVE
# *Intuitionism: the immaculate construction*

*Ah! what is man?*
*Wherefore does he why?*
*Whence did he whence?*
*Whither is he withering?*

DAN LENO

## MATHEMATICS FROM OUTER SPACE

*In Ireland the inevitable never happens and the unexpected constantly occurs.*

JOHN PENTLAND MAHAFFY

There is great excitement at NASA today. Years of patient listening have finally borne fruit. Contact has been made. The first evidence of extraterrestrial intelligence has been found. Soon the initial euphoria turns to ecstasy as computer scientists discover that they are eavesdropping not upon random chit-chat but a systematic broadcast of some advanced civilization's mathematical information bank. The first files to be decoded list all the contents of the detailed archives to come. Terrestrial mathematicians are staggered: at first they see listings of results that they know, then hundreds of new ones including all the great unsolved problems of human mathematics. They recognize those great insolubilia that we call Fermat's 'last theorem', the Riemann hypothesis, Goldbach's conjecture; on and on they go. This news alone stimulates renewed interest in these problems around the universities of the Earth. Buoyed by the knowledge that these problems are decidable, maybe someone can come up with a proof before the chance is gone for ever and the answer is revealed when the radio telescopes finally receive the extraterrestrial archives. Predictably, these last-ditch efforts fail. Soon, the computer files of the extraterrestrials' mathematical textbooks begin to arrive on earth for decoding and are translated and compiled into English to await study by the most distinguished representatives of the International Mathematical Congress. Mathematicians and journalists all over the world wait expectantly for the first reactions to this treasure chest of ideas. But odd things happened: the mathematicians' first press conference was postponed, then it was cancelled without explanation. Disappointed participants were seen leaving, expressionless, making no comment; the whole atmosphere of euphoria seemed to have evaporated.

After some days still no official statement had been made but rumours had begun to circulate around the mathematical world. The extraterrestrials' mathematics was not like ours at all. In fact, it was horrible. They saw mathematics as another branch of science in which all the facts were established by observation or experiment. They had used their fastest computers to check that every even number was equal to the sum of two prime numbers case by case through the first trillion examples. They found it so in every case and therefore regarded this as a general truth (which we call Goldbach's conjecture) established by experiment to a particular level of statistical confidence. The notion that there might be a counterexample waiting upstream amongst the uninvestigated numbers, they regarded in the same way that Newton would have reacted to an objector to his law of gravity who argued that he could not establish a universal law because he had not observed every falling apple—there might, after all, be one that levitates! Their advanced capability in electronics gave the extraterrestrials amazing powers of enumeration so they could check the truth of their mathematical conjectures through millions of possibilities in a split second. They had established definite statistical criteria for confidence in statements and they were regarded as true when no exceptions had been found amongst trillions of cases.

Needless to say, terrestrial mathematicians were shell-shocked ('gob-smacked' according to one unlearned journal's front-page headline). Expecting the deepest logical insights, they found nothing but results established by empirical methods or generalization from special cases. Hoping to gauge the intelligence of the extraterrestrials by their powers of deduction, they found their achievements in some sense incommensurable with our own. Where they sought the insight of a Nobel laureate, they found merely the winner of 'Master Mind'.

Now the dust has settled, one can see to the heart of the problem. What was all so unexpected was the fact that the extraterrestrials lacked any notion of *proof* in our sense. They did not follow Euclid's example by laying down initial postulates and agreed rules of reasoning before proceeding to *deduce* all the conclusions that follow from those assumptions. True enough, they used some initial assumptions which they regarded as self-evident, but they would then merely set about *confirming* their conjectures by experimental discoveries about what was true of these systems. As they became technologically expert so they could investigate possible truths by the enumeration of special cases very quickly and thoroughly. They seemed to have a subdiscipline that we would have called the philosophy of mathematics, although its principal concern seemed to be with the methodology of mathematics. There seemed to be some groups who thought that one should program computers to search thoroughly through all possible relationships between quantities to find trends and patterns which could then be explored systematically. A large part of their knowledge of statistics and probability seemed to derive exclusively from a series of very sophisticated experiments. They constructed curious devices which produced sequences of events of various types. The trends detected in sequences of events were compared with the outputs from their collection of standard devices and named after them. If a new trend was discovered, they simply produced a miniature version of it and added it to their library of standards.

What, in retrospect, is so fascinating about the development of mathematics in this culture, and which seems to have depressed our own mathematicians so much, is the speed and confidence with which it progressed. It had no worries about logical paradoxes or concerns about Gödel's theorems but had access to a vast area of truth which Gödel taught us must be out of reach of deduction. Most surprising of all was the evident fact that science in this extraterrestrial culture had not been impeded at all by their approach to mathematics. In fact, their philosophy books had footnotes about a suggestion that had been made that they develop some way of ensuring that truths of mathematics held for all cases and not just the very large number tested by enumeration. But this approach rapidly became a backwater as mathematicians were unwilling to take a step backwards and redefine their subject in such a way that many of the results they regarded as true were no longer to be regarded as such. To give up their method of confirmation would be like fighting with one hand tied behind their back. So, it appears that they certainly knew about our method of proof, but it was just not competitive with their process of confirmation. They were aware of its intrinsic limitations in deciding the truth of all statements. They knew that the length of deductive proofs of very deep truths might be longer than even their best computers could reach in the lifetime of their civilization. Reliance upon deductive proof would have crippled their pace of scientific development, and so this notion that everything should be proved just faded away as a philosophical curiosity. The most interesting objection to it was that any such move would create a new subject that differed from science in a useless way. The study of the Universe was based upon a finite number of observations from which generalizations could be made. But those generalizations could never be totally certain. Thus it was entirely right that the mathematical language in which such experimental facts were represented should have just the same foundation. Because their mathematics was essentially experimental, it was not regarded as separate from science at all. Indeed, as far as our search of their archives can ascertain, there is no evidence that anyone had ever posed such a question as 'why does mathematics work?' Mathematics was just part of the natural workings of the world guided along particular pathways. Computers were just particular configurations of natural materials whose subsequent history defined those processes which we are in the habit of calling mathematics. Their development of computer technology with techniques to manipulate numbers according to all manner of rules was highly advanced and was adequate to do all the things that their science required.

While the impact of these discoveries was still sinking into the minds of terrestrial mathematicians, somebody remembered how, early in the twentieth century, there had been an Indian mathematical genius who had grown up without any knowledge of the modern idea of proof but who had, by a strange mixture of pure intuition and experimental enumeration, come up with scores of deep and unproven formulae about numbers which surprised and in some cases defeated the ablest mathematicians of the day.

## RAMANUJAN

> *I remember once going to see him when he was lying ill at Putney. I had ridden in taxi-cab No. 1729, and remarked that the number seemed to me rather a dull one, and that I hoped it was not an unfavourable omen. 'No,' he replied, 'it is a very interesting number; it is the smallest number expressible as a sum of two cubes in two different ways.'* G.H. HARDY

Srinvasa Ramanujan died in India of tuberculosis on 26 April, 1920. Thirty-three years earlier he had been born into a poor middle-class family in the Tanjore district of Madras. As a child he was recognized as quite extraordinary, with exceptional powers of memory and calculation. He was introduced to elementary mathematics by a book entitled *A Synopsis of Elementary Results in Pure and Applied Mathematics,* written by the English mathematical tutor George Carr. Interestingly, this is a book designed to assist candidates intending to sit the entrance and scholarship examinations for the University of Cambridge mathematics course. It contains a long list of important and interesting formulae (with 'interest' judged by the likelihood of appearing as an examination question) without proofs. Thus it introduced Ramanujan to what was interesting in mathematics without biasing his intuition with the traditional methods of proof. In retrospect, this might be seen as a piece of good fortune. He absorbed Carr's book of formulae quickly and used what he found there as a springboard for the development of his own ideas, pursuing mathematics to the exclusion of his other school studies. This imbalance prevented him passing the entrance examinations to college and he was left at the age of twenty-two seeking employment to support himself and his new wife. As chance would have it, this search for employment brought him into contact with a keen mathematician living near Madras. Ramachandra Rao describes how he was visited by an impoverished but impressive young man clutching two fat notebooks of mathematical discoveries and a burning desire to find some financial support so that he could continue his mathematics. Rao recalls how the ragged Ramanujan took his frayed notebooks from under his arm,

> opened his book and began to explain some of his discoveries. I saw quite at once that there was something out of the way; but my knowledge did not permit me to judge whether he talked sense or nonsense. Suspending judgement, I asked him to come over again, and he did. And then he had gauged my ignorance and showed me some of his simpler results. These transcended existing books and I had no doubt that he was a remarkable man.

Rao helped Ramanujan in various ways and as a result he was encouraged to write to Trinity College, Cambridge, where Hardy and Littlewood, two of the world's foremost mathematicians, were Fellows. The letter explained his circumstances and history—'I have not trodden through the conventional regular course...I am striking out a new course for myself ... I would request you to go through the enclosed papers'. The work enclosed consisted of about 120 formulae establishing mathematical results. Some of them are reproduced in Fig. 5.1. Hardy managed to prove some of them, although it took him considerable effort even in areas where he considered himself an expert. Others were related in ways that made Hardy suspect that his mysterious correspondent was in possession of far more general results of which even the 'difficult and deep' ones he was seeing were but particular simple instances. But of the most impressive formulae, Hardy 'had never seen anything in the least like them before ... they could only be written down by a mathematician of the highest class. They must be true because, if they were not true, no one would have had the imagination to invent them.' The rest, as they say, is history. Ramanujan came to Cambridge in 1914, and was soon elected to a Fellowship of Trinity and then of the Royal Society at a remarkably early age. But despite his mathematical success Ramanujan did not really fit into Cambridge life. There were many social, family, and dietary problems

(1.1) $1-\frac{3!}{(1!\,2!)^3}x^2+\frac{6!}{(2!\,4!)^3}x^4-\cdots$

$$=\left(1+\frac{x}{(1!)^3}+\frac{x^2}{(2!)^3}+\cdots\right)\left(1-\frac{x}{(1!)^3}+\frac{x^2}{(2!)^3}-\cdots\right)$$

(1.2) $1-5\left(\frac{1}{2}\right)^3+9\left(\frac{1\cdot3}{2\cdot4}\right)^3-13\left(\frac{1\cdot3\cdot5}{2\cdot4\cdot6}\right)^3+\cdots=\frac{2}{\pi}$

(1.3) $1+9\left(\frac{1}{4}\right)^4+17\left(\frac{1\cdot5}{4\cdot8}\right)^4+25\left(\frac{1\cdot5\cdot9}{4\cdot8\cdot12}\right)^4+\cdots=\frac{2^{3/2}}{\pi^{1/2}[\Gamma(\frac{3}{4})]^2}$

(1.4) $1-5\left(\frac{1}{2}\right)^5+9\left(\frac{1\cdot3}{2\cdot4}\right)^5-13\left(\frac{1\cdot3\cdot5}{2\cdot4\cdot6}\right)^5+\cdots=\frac{2}{[\Gamma(\frac{3}{4})]^4}$

(1.5) $\int_0^\infty \frac{1+\left(\frac{x}{b+1}\right)^2}{\left(1+\frac{x}{a}\right)^2}\cdot\frac{1+\left(\frac{x}{b+2}\right)^2}{1+\left(\frac{x}{a+1}\right)^2}\cdots dx=\frac{1}{2}\pi^{1/2}\frac{\Gamma(a+\frac{1}{2})\Gamma(b+1)\Gamma(b-a+\frac{1}{2})}{\Gamma(a)\Gamma(b+\frac{1}{2})\Gamma(b-a+1)}$

(1.6) $\int_0^\infty \frac{dx}{(1+x^2)(1+r^2x^2)(1+r^4x^2)\cdots}=\frac{\pi}{2(1+r+r^3+r^6+r^{10}+\cdots)}$

(1.7) If $\alpha\beta=\pi^2$ then

$$\alpha^{-1/4}\left(1+4\alpha\int_0^\infty\frac{xe^{-\alpha x^2}}{e^{2\pi x}-1}dx\right)=\beta^{-1/4}\left(1+4\beta\int_0^\infty\frac{xe^{-\beta x^2}}{e^{2\pi x}-1}dx\right)$$

(1.8) $\int_a^\infty e^{-x^2}dx=\frac{1}{2}\pi^{1/2}-\frac{e^{-a^2}}{2a+}\,\frac{1}{a+}\,\frac{2}{2a+}\,\frac{3}{a+}\,\frac{4}{2a+}\cdots$

(1.9) $4\int_0^\infty\frac{xe^{-x\sqrt{5}}}{\cosh x}dx=\frac{1}{1+}\,\frac{1^2}{1+}\,\frac{1^2}{1+}\,\frac{2^2}{1+}\,\frac{2^2}{1+}\,\frac{3^2}{1+}\,\frac{3^2}{1+}\cdots$

(1.10) If $u=\frac{x}{1+}\,\frac{x^5}{1+}\,\frac{x^{10}}{1+}\,\frac{x^{15}}{1+}\cdots$, $v=\frac{x^{1/5}}{1+}\,\frac{x}{1+}\,\frac{x^2}{1+}\,\frac{x^3}{1+}\cdots$

then

$$v^5=u\,\frac{1-2u+4u^2-3u^3+u^4}{1+3u+4u^2+2u^3+u^4}$$

(1.11) $\frac{1}{1+}\,\frac{e^{-2\pi}}{1+}\,\frac{e^{-4\pi}}{1+}\cdots=\left\{\sqrt{\frac{5+\sqrt{5}}{2}}-\frac{\sqrt{5}+1}{2}\right\}e^{2\pi/5}$

(1.12) $\frac{1}{1+}\,\frac{e^{-2\pi\sqrt{5}}}{1+}\,\frac{e^{-4\pi\sqrt{5}}}{1+}\cdots=\left[\frac{\sqrt{5}}{1+\sqrt[5]{5^{3/4}\left(\frac{\sqrt{5}-1}{2}\right)^{5/2}-1}}-\frac{\sqrt{5}+1}{2}\right]e^{2\pi/\sqrt{5}}$

(1.13) If $F(k)=1+\left(\frac{1}{2}\right)^2k+\left(\frac{1\cdot3}{2\cdot4}\right)^2k^2+\cdots$ and $F(1-k)=\sqrt{210}\,F(k)$

then

$$k=(\sqrt{2}-1)^4(2-\sqrt{3})^2(\sqrt{7}-\sqrt{6})^4(8-3\sqrt{7})^2(\sqrt{10}-3)^4$$
$$\times(4-\sqrt{15})^4(\sqrt{15}-\sqrt{14})^2(6-\sqrt{35})^2$$

**Figure 5.1** A page from Ramanujan's letter to Hardy displaying a selection of his beautiful mathematical discoveries. Some, like the formulae (1.7), (1.8), (1.9), Hardy recognized as familiar to him and his colleagues; others, like (1.5) and (1.6), he managed to confirm after rather a lot of work even though he was the world's foremost authority on the solution of such problems. But (1.10), (1.11), and (1.12) he found deeply mysterious and difficult, announcing that 'they must be true because, if they were not true, no one would have had the imagination to invent them'.

that Hardy should have foreseen, but didn't. Tuberculosis was contracted. He died little more than seven years after sending those first formulae to Hardy.

The remarkable contributions he made to number theory both alone and in collaboration with Hardy were evidences of Ramanujan's remarkable intuition about the properties of numbers. Yet these intuitions lacked almost any formal mathematical training of the type that would have been given even to good high-school students of the time. After rhapsodizing on his unique familiarities with the interrelationships of numbers, and his partial rediscovery of whole areas of mathematics for himself despite huge gaps in his armoury of concepts and techniques, Hardy confesses:

> His ideas as to what constituted a mathematical proof were of the most shadowy description. All his results, new or old, right or wrong, had been arrived at by a process of mingled argument, intuition, and induction, of which he was entirely unable to give any coherent account ... It was his insight into algebraical formulae, transformations of infinite series, and so forth that was most amazing. On this side most certainly I have never met his equal ... He worked, far more than most mathematicians, by induction from numerical examples in a way that [was] often really startling [and] without a rival in his day.

Hardy's colleague, Robert Carmichael, gives the following perspective upon the problem of inducting him into the methods and ethos of modern mathematics following his achievements in isolation:

> In some directions his knowledge was profound. In others his limitations were quite startling ... But notwithstanding the fact that he had never seen a French or German book and that his command of the English language was meager, he had conceived for himself and had treated in an astonishing way problems to which for a hundred years some of the finest intellects in Europe had given their attention without having reached a complete solution. That such an untrained mind made mistakes in dealing with such questions is not remarkable. *What is astonishing is that it ever occurred to him to treat these problems at all.*

The emphasis we have placed on the final lines of these remarks highlights the curious fact that Ramanujan was led to develop mathematical ideas in the same directions as others who had different backgrounds, motivations, and approaches to mathematics. Ramanujan displayed no interest in anything but the purest relationships between numbers. He had no interest in the application of mathematics to scientific problems, nor in extracting mathematical ideas from the physical world. He had intuition about the structure of formulae. This intuition seemed to draw more naturally from the idea of confirmation rather than proof. He seems to be a formalist who doesn't use logic; an experimentalist who doesn't look at the physical world: an explorer of numerical realms.

## INTUITIONISM AND THREE-VALUED LOGIC

*To be or not to be: that is the question.* WILLIAM SHAKESPEARE

What our two stories, the first invented, the other true, teach us is the possibility that our most basic notions regarding 'proof' may be questioned. In the last chapter we explored the idea that mathematics may in some sense be merely an abstraction from the physical world that is invented by our minds and possesses the form it does largely because of the way in which our minds are constituted and pre-programmed by the process of evolution. In the early years of the twentieth century this idea was developed in an extreme and novel form by the Dutch mathematician Luitzen Brouwer and his school of 'intuitionists'. It took its cue from a belief that one cannot inquire into the foundations and nature of mathematics without delving into the question of the operations by which the mathematical activity of the mind is conducted. If one failed to take that into account, then one would, in Brouwer's opinion, be left studying only the language in which mathematics was represented rather than the essence of mathematics. Whereas Russell and his school had sought to show that mathematics was just a manifestation of logic, Brouwer sought to do the opposite: to show that logic is built upon mathematics, but in such a way that one cannot take the usual principles of logic to be of universal application.

These ideas had many precursors and many consequences. But before we explore them, we should try to state the nature of Brouwer's philosophy of mathematics in as simple a fashion as possible. Because of his concern about the uncertain and subjective influence of the mind upon our mental constructions, he sought to found mathematics in as conservative a manner as possible, upon the smallest and surest island of those intuitions which he believed we all unarguably share. For Brouwer, this island consisted of the 'natural' numbers 1,2,3,... and simple counting processes. From this basis he defined mathematics to be the edifice that can be constructed from them by step-by-step deductions using a finite number of steps.

This sounds an innocuous conception, but a mathematics defined in this 'intuitive' manner turns out to be a much smaller thing than the edifice that his contemporaries understood by 'mathematics'. It outlawed familiar concepts like 'infinity'; it removed cornerstones of logical reasoning; for only statements which could be established explicitly by a finite sequence of constructive steps are now included in mathematics. This outlaws the ancient logical device of the *reductio ad absurdum*, or proof by contradiction, in which one proves something to be true by showing that a logical contradiction arises if it isn't. This traditional form of logical deduction does not construct the true statement step by logical step; rather, it shows that a logical contradiction would arise were it to be false. This sounds acceptable but we need to recognize that it is founded upon the stipulation that a statement is either true or false. Brouwer denied this 'principle of the excluded middle' as it was known, allowing a third limbo status of 'undecided' to exist for statements whose truth or falsity had not been constructed by following a finite number of deductive steps. This distinction is somewhat reminiscent of the difference between English and Scottish law. In English law the defendant must be found either guilty or not guilty whereas in Scotland there exist not only these two verdicts but a third option of 'not proven'. This last verdict differs from that of 'not guilty' in that it permits the defendant to be retried on the same charge in the future. English law does not permit such a retrial.*

The consequences of following Brouwer's demand that mathematics consist only of those statements that can be constructed in a finite number of steps from the properties of the natural numbers are very great. It produces a mathematics that is far smaller in extent, far more limited in power, and far more predictable than the conventional mathematics which employed a two-valued logic in which every statement was either true or false. The mathematics according to the intuitionists was just a part of the ocean of mathematical truths that were accepted by other mathematicians. Any truth of intuitionism would be a truth of traditional mathematics but not necessarily vice versa.

Three-valued logic produces a host of changes to traditional assumptions. Tautologies of two-valued logic need not be tautologies in three-valued logic. An obvious example is the principle of the excluded middle itself (that is, 'a statement is either true or false'). One interesting development by Kolmogorov was the realization that if we reinterpret the conventional logic that applies to entities like 'statements' or 'sets' as applying instead to problems then the principle of the excluded middle is not necessarily true. Thus, if $A$ and $B$ are problems, 'the negation of $A$' is

* The reader may recall that the essence of the plot in Agatha Christie's first novel *The Mysterious Affair at Styles* was that a villain, having committed a murder by some means, deliberately laid false clues to make it appear that he had committed the murder in a different way. He had a cast-iron alibi against having committed the murder in the manner suggested by this trail of false clues, and so if he could be charged and tried for the murder on the basis of the false evidence then the revelation of his alibi at the last moment in court would result in his acquittal with no possibility of him being retried even if the true means of the murder were to be discovered in the future.

interpreted as meaning that *A* is insoluble; '*A* and *B*' as meaning that both of the problems *A* and *B* are soluble; '*A* or *B*' to mean that either problem *A* or problem *B* is soluble; '*A* implies *B*' to mean that if problem *A* is solved then so is problem *B*. Hence, there are some deductions that hold just as in two-valued logic. For example, *A* 'implies' *B* 'and' *B* 'implies' *C* taken together 'imply' that *A* 'implies' *C*. However, it is no longer necessarily true that problem *A* is either soluble or insoluble.

Whilst restricting mathematical truth to what can be constructed explicitly, without the use of the *reductio ad absurdum*, has dire consequences for mathematics, it will also have ramifications for science. The application of mathematics to the natural world assumes that two-valued logic holds with the exception of some attempts to interpret quantum uncertainty by the use of the assumption that three-valued logic prevails in the world of elementary particles. There are many classic scientific results which make use of the *reductio ad absurdum* in either of its two forms. The most famous are the 'singularity theorems' which lay down conditions which are sufficient to guarantee that the Universe had a 'beginning'.* These theorems do not explicitly construct the origin of the Universe. Rather, they demonstrate that a logical contradiction would arise with other physically reasonable assumptions about the Universe were there to be no beginning. Now, the intuitionist allows you to deduce that something is false by showing that its truth would produce a contradiction, but will not allow you to deduce that something is true by showing that its falsity would produce a contradiction. The singularity theorems of cosmology follow the first of these strategies, but some other proofs, like the demonstration that black holes cannot split up, follow the second.

The application of mathematics to the physical world has to *assume* a particular logic applies to its workings. At present there is no reason to believe that it has to be the standard two-valued logic. A full description of the Universe—a Theory of Everything—should reveal either that the logical structure is inevitable or that it is part of initial defining characteristics of our Universe that could be different. Perhaps one could show that 'observers', like ourselves, could only exist in realities governed by two-valued logic. Any full explanation of why the Universe is well described by mathematics needs to offer some explanation for the underlying logical system that is adopted. Presumably, the development of mathematics over thousands of years, before it became formalized, enshrined two-valued logic because this was extracted from the nature of the physical world around us. But the part of the world that impinges upon human experience is really

* Most of these conditions can be tested by observation. At present it is not believed that they are obeyed by the material content of the Universe.

very small. How do we know that there are not aspects of the microscopic and astronomical dimensions of reality that march to a different logical tune?

## A VERY PECULIAR PRACTICE

*The point of philosophy is to start with something so simple as to seem not worth stating, and to end with something so paradoxical that no one will believe it.*
BERTRAND RUSSELL

'The integers were made by God; all else is the work of man.' Although this after-dinner remark by Leopold Kronecker has been taken by some critics to reveal more about his early career in banking than his philosophy of mathematics, it is a succinct summary of the intuitionists' position. All mathematical statements of greater complexity than the natural numbers must be constructed from them explicitly by a finite number of deductive steps. There is no ready-made mathematical world waiting to be discovered. The meaning of mathematics is nothing more nor less than the sequence of steps that are required to construct its statements from the natural numbers.

One can readily identify the affinity that this idea has with the so-called 'operationalist' philosophy so popular in the early years of the twentieth century. This was an attempt to found all knowledge upon a firm foundation of experience so that the meaning of something was reduced to the finite sequence of operations that would have to be conducted in order to construct or measure it. Einstein was greatly influenced by this style of thinking in the formulation and presentation of his special theory of relativity and, in particular, its feature that the intuitively reasonable notions of simultaneity, or length, or mass, or time, turn out to have no absolute meanings independent of the state of motion of the observer.* The Nobel-prize-winning physicist Percy Bridgman was an enthusiastic propagandist for this philosophy of the physical world and was elated to discover that the mathematicians had the option of an analogous constructivist philosophy to interpret their own subject. Although Einstein was impressed by this approach in his early work on special relativity, he

* It is very common to see the emergence of this type of operationalist philosophy in science whenever unusual new concepts emerge which run strongly counter to common sense or natural scientific intuition. It represents a way of sticking very close to things one knows and proceeding with a minimum of logical leaps into the unknown. It is very interesting that the physicist Russell Stannard, who is actively involved in trying to explain quite advanced scientific ideas (like relativity) to young children, maintains that the teaching of science to children is most effective if it is done in this constructive mode. Thus, rather than dwelling upon the nature of concepts and the logical flow of ideas as one does when talking to adults, the young child's mind seems to appreciate most an explanation of what is possible by a step-by-step prescription for how it would be implemented.

seems to have turned away from this philosophy after about 1915 when he became increasingly impressed by the power of mathematical formalisms. He was eventually completely captivated by their potential to produce a unified mathematical description of the physical world and one sees no trace of his penchant for simple thought-experiments, which so characterized his brilliant early work, after he achieved so much success with the abstract mathematical formulation of the general theory of relativity.

Henri Poincaré believed that the intuitive sense of the natural numbers is the most rudimentary mental notion that we possess and we use it to generate whole sequences of things and ideas by the inductive process. This sense arises from our conscious experience of time in a linear sequence. We feel distinct thoughts and observe individual aspects of the world. The resulting sequence reveals that the natural numbers, the notion of a sequence, is implicit in our consciousness. As we become more practised at reflecting upon and interpreting our conscious experience, we may be able to order it in more elaborate ways. Many of these ideas of Kronecker, Poincaré, and Brouwer were in stark opposition to the philosophy underlying Hilbert's formalist programme. Hilbert spelt out his opposition to their ideas in one fell swoop:

> Mathematics is a science without presuppositions. For its establishment, I need neither our dear Lord as Kronecker, nor the assumption of a specific faculty of our understanding attuned to the principle of mathematical induction as Poincaré, nor Brouwer's primitive intuition, nor finally, as Russell and Whitehead, the axioms of infinity, reducibility, or completeness.

Hilbert did have a place for mathematical intuition though. He certainly did not maintain that one could found it upon logic alone. Rather, he claims,

> something is already given to us in the imagination as a preliminary condition for the application of logical inferences and for the performance of logical operations: certain extra-logical concrete objects which are present in intuition before any thinking has come about. If logical inference is to be certain, then these objects must be in all parts completely open to examination, and their presentation, their distinction, their succession or juxtaposition must be immediately given to intuition at the same time as the objects themselves, as something that cannot be reduced to something else or that requires such a reduction. This is the basic philosophical attitude which I deem requisite for mathematics as well as, in general, for any scientific thinking, understanding, and communication. In mathematics, specifically, the object of our consideration is formed by the concrete symbols themselves whose shape, by virtue of our attitude, is immediately clear and recognisable. This is the scantiest presupposition with which no scientific thinker can dispense and which therefore everyone must hold, be it consciously or unconsciously.

Whereas Brouwer was arguing that there was some particular formal system which is naturally evolved in harmony with our patterns of thought, Hilbert maintained that there was no 'special relationship of our knowledge to the subject-matter concerned but rather it is one and the same for every kind of axiomatics'. So, Hilbert sees that some intuitive component of our understanding must complement evidence of a logical nature, but the focus of that intuitive component he believes to be quite different to that claimed by Brouwer. Whereas Brouwer tries to safeguard every single deductive step that a mathematician makes from some illegal intuitive deviation, Hilbert, as we saw in Chapter 3, separates the formal manipulation of mathematical symbols from the intuitive business of talking *about* those processes and their properties, choosing to call the latter subject 'metamathematics'.

The earliest trace of this constructivist philosophy of things is actually to be found in Johannes Kepler's astronomical work, in the early years of the seventeenth century, where he makes the curious remark that we have no right to assert that a regular heptagon exists until someone has prescribed the geometrical method by which it can be constructed with a ruler and a pair of compasses. In more recent times the earliest supporter of the constructivist view seems to have been Kronecker, who was then a distinguished German mathematician working at the University of Berlin in the nineteenth century. He was strongly influenced by one of his famous colleagues, Weierstrass, who had attempted to show that all mathematical quantities could be generated by appropriate sequences of operations with natural numbers, generalizing the obvious cases like the construction of fractions which consists of one natural number divided by another, to numbers which are not of this type. Kronecker took a more extreme view, maintaining that there were actually no mathematical entities other than the natural numbers. He sought to expunge all other quantities from mathematics, believing that fractions, irrationals, and complex numbers were illusory concepts that had arisen merely through some misguided application of mathematical logic to the artefacts of the physical world. Eventually, he believed, a way would be found to recast those subjects into their most natural and elementary form wherein only the natural numbers would appear. Of the situation with regard to algebra and other branches of pure mathematics that appeared to be founded upon logical operations more extensive than counting processes, Kronecker wrote in 1887 that

I also believe that we shall succeed some day in arithmetising the total content of all these mathematical disciplines, that is in grounding them on the number concept taken in its narrowest sense, and thus eliminate the modifications and extensions of this concept which were for the most part occasioned by applications in geometry and mechanics.

Kronecker set about prosecuting this programme in very definite ways. All simple equations of the form 1 - 2 = 5 - 6 were recast into equivalent forms which avoided the appearance of negative numbers. Because the equation cited has the form minus one equals minus one, Kronecker regarded it as meaningless because on both sides of the equation there was a quantity which could not have a real existence. There is indeed a way to achieve this goal but it is most complicated and impractical.* One wonders how often Kronecker employed it to represent the overdrafts of his customers at the bank. No doubt in this context he was able to impress upon his clients the reality of negative quantities without too much difficulty. Needless to say Kronecker's ill-conceived programme did not capture the imagination of mathematicians.

Just a few years later Poincaré came out in favour of a view that we must regard the natural numbers as undefinable given things from which the whole of mathematics must be obtained by a step-by-step process which we recognize as intuitively given by our conscious experience.

Despite these interesting precursors the intuitionist school of constructive mathematics is most closely associated with the name of Luitzen Brouwer who became its leading advocate, committed to showing that it could be worked out in rigorous detail to define a new and rock-solid foundation for a mathematics that made no use of non-intuitive notions like 'infinity' or non-constructive arguments like proof by contradiction.

One must remember that Brouwer was strongly influenced by the crisis of confidence that had affected mathematics at the turn of the century. The discovery of logical paradoxes had threatened the entire basis of the subject and had been totally unsuspected by many leading mathematicians. The entire formalist programme arose as one response to this state of affairs; Brouwer's intuitionism was another. A not dissimilar state of affairs had existed in philosophy a hundred years earlier. David Hume and Immanuel Kant had made sceptical attacks upon the naive but dogmatic assumptions that were held about the process by which we come to know things.

Brouwer's campaign against the traditional picture of what mathematics was, and should be, began with the completion of his doctoral thesis at

* For the aficionado it is enough to say that he advocates transforming an equation like 1 - 2 = 5 - 6, which results in negative quantities, into a so-called 'congruence relation' $1 + 2x = 5 + 6x$ to be solved in arithmetic with base $x + 1$. However, it was then pointed out that this only works if the variable quantity is finally assigned a value by taking $x + 1$ equal to zero, but this entails $x$ taking the value minus one—exactly the situation the entire device was designed to avoid!

the University of Amsterdam in 1907. He pursued it vigorously until, as we shall shortly see, some extraordinary events effectively damped down his appetite for mathematics and mathematicians.

## A CLOSER LOOK AT BROUWER

*The Groups of Wrath* COSGROVE

Brouwer was a strange and brilliant man. Self-centred, catatonic, and misanthropic, he seems at odds with everyone he had contact with, choosing to work for long periods in total isolation at a small 'hut' in the country. Despite his lack of interest in others, he was acutely sensitive and throughout his life was engaged in endless disputes, real and imagined, which grew out of his deep pessimism and suspicion of the motives of others. These tendencies were present in tell-tale forms even during his school-days. At the age of seventeen he went through the traditional form of public confirmation into the Dutch Reformed Church at Haarlem. This requires a series of responses to particular doctrinal questions, but also involves a number of free statements about the foundations of the candidate's faith and thinking. Here we find Brouwer saying that 'in the world surrounding me and part of me I am struck by its loathsomeness and that I want to remove this, also from the human world. I can hardly call this love-of-my-neighbour, for I don't care twopence for most people. Hardly anywhere in human society do I recognize my own thoughts and inner life. The human spectres around me are the ugliest part of my world of images ... I consider religious forms to be good for the stupid masses, to be kept in reverent ignorance ...'.

After going through a period of deep depression as a student, during which he considered abandoning his mathematical studies altogether, he fell under the influence of a charismatic Dutch mathematician, Gerrit Mannoury, whose stimulating lectures and wide interests rekindled his enthusiasm for mathematics. Moreover, Mannoury made him aware of the crisis that existed in the foundations of arithmetic: the positions of Frege, Russell, and Whitehead, and the formalist programme of David Hilbert. Mannoury delivered a weekly lecture on this subject at the University and we know that Brouwer attended every one of them.

Ambition restored, and engaged to be married to a divorcée twelve years older than himself, Brouwer graduated from Amsterdam University in 1904 with highest honours in mathematics and chose to pursue a doctorate in mathematics under the supervision of Diederik Korteweg, a Dutch mathematician renowned for his work on the application of mathematics to fluid flows and other aspects of the real world. Brouwer took the rather unusual step of pursuing this programme of work full-time. Normally it would have been the part-time occupation of an experienced schoolteacher seeking a higher qualification in order to obtain a junior teaching post at a university. Korteweg soon discovered that Brouwer was an awkward and unmanageable research student but, to his great credit, he allowed Brouwer to pursue research into the foundations of mathematics despite his strong fears that Brouwer would end up doing philosophy rather than mathematics. Korteweg's own philosophy of mathematics was a straightforward one. He believed that we had discovered mathematics from the physical world and so its applicability there was just following the stream back to its source. Brouwer's idea of seeking the origin of mathematics in the human mind would have been something entirely foreign to his natural way of thinking, despite his familiarity with, but lack of interest in, Kant. Korteweg's fears were compounded by his own style of mathematics which was traditional and focused upon solving practical problems where there would be definite practical applications; but one must also remember that philosophy was a more or less non-existent academic subject in Holland at that time and, to cap it all, his student was involved in some pretty intemperate public speaking on philosophical and social issues. He appears to have been something of an 'angry young man'. Many of these pieces of oratory were condemnations of those around him, especially women, whom he seems to have regarded as the lowest form of life and a danger to the position of men because 'the usurpation of any work by women will automatically and inexorably debase that work and make it ignoble'. In 1905 he published a compendium of these rather ill-considered pieces of invective under the title *Life, Art and Mysticism*. One might have expected him to repudiate all this youthful iconoclastic nonsense in later life, but on the contrary, he just seems to have developed it into more extreme and sophisticated forms and invariably referred back to this first publication as the source of his thinking.

Brouwer set himself up in isolation in a hut he built for himself near the little village of Laren, in an environment that attracted artists and radical thinkers from the arts. He began to take on board the work of Russell, Hilbert, and Poincaré, as well as the later critical philosophy of Kant, still finding time to produce the occasional research paper on other mathematical problems. By the autumn of 1906 Brouwer had completed the development of his thoughts about the nature of mathematics and laid out the plan of his thesis in six chapters:

I The construction of mathematics
II The genesis of mathematics related to experience

III The philosophical significance of mathematics
IV The foundations of mathematics on axioms
V The value of mathematics for society
VI The value of mathematics for the individual.

The first chapter planned to lay out his constructivist programme, with its basis in our intuition of the natural numbers, but was amplified by the solution of some unsolved mathematical problems by the use of his constructive methods. Much of the ensuing material was non-mathematical in character, outlining his pessimistic personal philosophy, and introducing Kant's ideas about the basic human intuitions of time and space which dictate our perceptions of the world.

Not surprisingly, Korteweg was less than happy with the structure of this thesis when it was presented to him for comment in draft form. It didn't look like any mathematics thesis he had ever seen before! After much strained argument and revision Korteweg succeeded in persuading Brouwer to remove a good deal of the early philosophical discussion, increase the mathematical content and moderate his dogmatic tone in much of what remained. Eventually a compromise was arrived at, sympathetic examiners were appointed and Brouwer successfully defended his thesis. He was even awarded his doctorate 'with distinction'.

The next stage in Brouwer's career is most remarkable. He attended international conferences on mathematics where he saw and heard many of Europe's greatest mathematicians like Hilbert and Poincaré, but his letters show that he had set his heart upon gaining a professorship with the financial security it offered. So, as a result of the frosty reception that his thesis work (written in Dutch) on the foundations and meaning of mathematics had received further afield, he did not speak publicly on this subject until after 1912 when he was appointed to a professorship in Amsterdam. Instead, he worked with enormous energy on topology and group theory, producing an astonishing total of fifty-nine research papers in the period from 1907 to 1912.* Much of this work was deep and difficult—he virtually created the modern subject of 'topology' during this period—and it established him as a mathematician of international reputation.

Safely appointed to his professorship in Amsterdam on the strength of this penetrating mathematical work, he returned to his heart's first desire: the intuitionist philosophy of mathematics. Beginning with his inaugural

* A further one hundred and six papers would follow in the period from 1912 to 1928 despite a slackening of output due to the intervention of the First World War. But only thirty-six papers were published from 1928 until his death in a road accident in 1966 at the age of eighty-five.

address, he spoke vehemently (and frequently) of his opposition to the whole foundation of mathematics as it was done by others, advocating its re-establishment upon the firm foundations of step-by-step construction from the natural numbers, banishing all infinities or meaningless deployment of the assumption that a statement is either true or false before it is explicitly shown to be one or the other. Eventually, he converted another great mathematician to his viewpoint. Hermann Weyl, Hilbert's most brilliant student in Göttingen, shared Brouwer's views on mathematical foundations and for a while considered accepting an offer of a professorship in Brouwer's department in Amsterdam, but Hilbert managed to persuade him that his future lay in Göttingen

During this period Brouwer also became involved in strange movements for social and political reform, where he manifested all his talents for irascibility and unreasonableness. This, coupled with analogous disputes within the academic world and with the media, drained more and more of his time and energy and exacerbated his inability to work harmoniously with others. The biggest consequence of this was the rift with Hilbert which opened in 1919. First, Brouwer turned down an invitation by Hilbert to leave Amsterdam and join his famous department in Gottingen; then, his advocacy of intuitionism began to be seen as a revolutionary threat to mathematics by Hilbert. At first, he probably thought it little more than a youthful flirtation with philosophy which the young Brouwer would grow out of during the productive phase of his research after the defence of his thesis. But the defection of Weyl to Brouwer's camp with talk of 'a revolution' occurring in mathematics, together with the increasingly influential lectures that Brouwer was giving around Europe struck a raw nerve in Hilbert. Hilbert launched a counter-attack against the intuitionist doctrine. He announced:

> What Weyl and Brouwer are doing is precisely what Kronecker did. They seek to save mathematics by throwing overboard all that is troublesome ... If we were to follow the kind of reform they suggest we would risk losing a great part of our most valued treasures.

Later, we shall see what extraordinary events ensued because of this. For now, it is enough to note the analysis of Brouwer's biographer Walter van Stigt, who concludes that, at this stage,

> neither Brouwer nor Hilbert were temperamentally capable of keeping mathematical controversy at the level of a detached professional debate. Brouwer in particular needed the stimulus of a personal challenge to stir him into action; he was a fighter, who needed a personal enemy on whom to concentrate his attack. Even if Hilbert's and Brouwer's views on mathematics could hardly be described as

antipodal in every respect, the foundational debate now became polarised into a battle between Intuitionism and Formalism with international leadership as the prize.

## WHAT IS 'INTUITION'?

*Intuition is reason in a hurry.* HOLBROOK JACKSON

The first question one should ask about 'intuitionism' is: Why is it called intuitionism? At first this seems obvious—surely it just focuses upon the role of intuition in the process of mathematical discovery. But a closer look does not seem to reveal such an idea at all. To begin with, it is much more specific. It focuses upon the primitive feeling and recognition that humans have for quantity in the form of the natural numbers 1,2,3,..., or whatever they might choose to call them. The intuitionists believed this concept to be an irreducible one that is universally shared. Some of Brouwer's followers support this contention by claiming that this intuition of the natural numbers arises from the experience of observation and perception of things:

> In the perception of an object we conceive the notion of an entity by a process of abstracting from the particular qualities of the object. We also recognize the possibility of an indefinite repetition of the conception of entities ... The concepts of an abstract entity and of a sequence of such entities are clear to every normal human being, even to young children.

So we are imagined to acquire the notion of an individual object by our in-built ability to perceive something without taking in every single aspect of its structure; in effect, we idealize it by perceiving a sort of rough sketch. This is abstraction. Having arrived at the notion of 'one' in this way it is then claimed that the feeling for 'many' is obvious and is also, as Kant claimed, built in to our minds as part of our perception of time. Our entire temporal outlook creates a notion of one being split into two by the experience of having our perceptions divided into 'before' and 'after'. It is this rather questionable idea that forms the basis of Brouwer's idea of intuition and its source. Brouwer's definition of mathematics in terms of this basic and primitive form of human intuition is the starting point of his philosophy:

> Mathematics is an essentially languageless activity of the mind having its origin in the perception of a move of time. This perception of a move of time may be described as the falling apart of a life moment into two distinct things, one of which gives way to the other, but is retained by memory. If the twoity thus born is divested of all quality, it passes into the empty form of the common substratum of all twoities. And it is this common substratum, this empty form, which is the basic intuition of mathematics.

From this description we can see already that Brouwer does not see mathematics as some sort of 'theory' or collection of squiggles on pieces of paper that have to follow certain rules without contradiction in order to meet the requirements of existence. For him mathematics is a human activity that emerges from the way in which our minds make sense of all experience, by organizing it into a sequence of single pieces.* Mathematics is now just a rather specialized part of psychology (or even neurophysiology if one is a reductionist). Already, we can detect a dangerous loop waiting for us in the future. We find that mathematics—whatever it is—is useful and accurate in describing those particular aspects of the world that we identify with the subject matter of chemistry, biology, and physics. But if we probe to the limits of those subjects to see how the mind works we must, on Brouwer's view, find a complex collective phenomenon which can be described by mathematics yet which also produces mathematics because of the way that evolutionary processes have fashioned the neurophysiological responses to events around us.

Of course, just about every mathematician would be happy to subscribe to the view that mathematics is a human activity, a 'natural' activity even. But where most would disagree with Brouwer would be in their view of where the exact and rigorous part of mathematics then resides. The formalists would point to collections of formal statements written and spoken by mathematicians. These will not be just some type of raw intuition; they will have been fashioned by trial and error into the neatest and most expedient of forms. They have some intuitive element perhaps, but sometimes one might proceed further by abandoning that element explicitly, taking the very opposite notion on board as a starting-point or axiom. Brouwer, by contrast, thinks that basic human thought patterns—not all the fancy refinements that mathematicians then conduct—give rise to the true mathematics. Mathematics is nothing more, and nothing less, than the exact part of our thinking.

Because it is hard to convey the passionate compulsion that Brouwer clearly felt to think in this way about mathematics, it is useful to trace a few of the connections that exist between his thinking and the earlier philosophy of Kant which was so influential amongst continental scientists

* There is an interesting aside here. Niels Bohr, the creator of quantum theory along with its unusual interpretation of reality in a probabilistic fashion, used to point to a work of Danish fiction by Poul Møller which he read in his youth. This is a story of a young student whose consciousness can split into separate tracks which may then interfere or reconvene in the future. This young man does not have the sort of intuition that Brouwer assumes us all to have. Møller's story has many parallels with the picture of the wave function that Bohr adopted to describe quantum reality. Ironically, those who sought to ascribe a different type of logic to quantum reality actually picked Brouwer's intuitionistic three-valued logic in which the excluded middle does not hold.

and mathematicians of the nineteenth and early twentieth centuries. Brouwer traces his own thinking back to Kant's idea that the human mind had certain forms of perception built into it. Numbers for him were not 'things' which exist external to the mind; rather, they exist in the minds of people who take the trouble to construct them by the processes of their thought. Kant identified what he thought were the two categories of human perception which were built into our thinking processes. One was a type of spatial perception, the other a temporal one. He associated geometry with that spatial intuition, and arithmetic and sequential counting with our knowledge of temporal perception. Unfortunately, Kant tied his ideas about geometry to Euclidean geometry, believing it to be an inviolate truth of the Universe created necessarily by the very nature of our perception. For Kant the axioms and structure of geometry and arithmetic were synthetic *a priori* (i.e. necessary) truths independent of experience. It was entirely unthinkable that they could be disproved in any way. Unfortunately, this view was made to look not a little silly with the discovery of the non-Euclidean geometries which we discussed at length in Chapter 1. Because of that, Brouwer rejected Kant's ideas about geometry and in-built spatial intuition, but he homed in upon the Kantian idea of counting as an in-built aspect of human understanding that lies at the root of our mental processes.

## THE TRAGEDY OF CANTOR AND KRONECKER

> *My theory stands as firm as a rock; every arrow directed against it will return quickly to its archer. How do I know this? Because I have studied it from all sides for many years; because I have examined all objections which have ever been made against the infinite numbers; and above all because I have followed its roots, so to speak, to the first infallible cause of all created things.*
>
> GEORG CANTOR

The popular picture of the mathematician and his subject is one of austere and impersonal arguments, impenetrable to outsiders, but settled entirely objectively by insiders. There is no room for propaganda or bluster, prejudice or innuendo; mathematics goes on its logical way whether we like it or not. But clearly some of the issues we have been exploring do not entirely live up to that unambiguous objective standard. There seems a surprising amount of room for human preference and intuitive feeling regarding the foundations of the subject. Such a situation, transparently common in many other walks of life, academic or otherwise, is very likely to give rise to disagreements between the participants. In mathematics these quarrels are comparatively rare but when they do occur they seem to exhibit an extraordinary ferocity and longevity with all manner of unforeseen results. One of the curiosities about the advent of the intuitionist interpretation of mathematics, first in simple form by Kronecker, and then in all-encompassing form by Brouwer, is that both stages were characterized by long-running disputes between advocates of the more conventional 'formalist' picture of mathematics and the perceived revolutionaries who sought to impose a new definition of mathematics upon everybody else. The first of these battles was between Kronecker and Georg Cantor, the other between Hilbert and Brouwer, although others were peripherally involved in both cases. These clashes reveal interesting things about the participants and their beliefs and, at the time, played an important role in accelerating the development of the constructivist philosophy of mathematics. In retrospect, the unstable personalities of some of the participants played a major role in preconditioning mathematicians against the views they developed and espoused. The first of these episodes is the story of Kronecker and Cantor. Its telling will also enable us to understand what was going on in mathematics in the 1880s that created such vehement opinions and placed the whole question of the meaning of mathematics in a central position, one from which it has probably never successfully recovered.

Leopold Kronecker was born with a proverbial silver spoon in his mouth in 1823 in Prussia. The son of prosperous Jewish parents, he grew up in an atmosphere of finance and business. After graduating in mathematics at the age of 22, following the death of his parents and his uncle, he found himself charged with running the family's estates and banking interests. This he did for many years with enormous success. Yet at the same time he was an active and highly respected mathematician who taught voluntarily at the University of Berlin, finally taking employment as professor only in 1883, after the retirement of his former mentor Ernst Kummer. During his youth Kronecker had come under the philosophical influence of his father and Kummer, both of whom adopted rather sceptical attitudes to the scope and power of human understanding which were somewhat reminiscent of David Hume's. This influence, coupled with Kronecker's natural inclination towards arithmetical deductions and calculations, and an aversion to geometry, gave rise to his extremely pragmatic and conservative attitude towards mathematics. Despite living in a period when famous mathematicians like Weierstrass were establishing whole realms of the subject upon principles that involved the notions of infinite series of quantities and infinitesimally small quantities, Kronecker became a 'doubting Thomas'. He would only accept as true mathematics those pieces of work which explicitly constructed the quantities of which they spoke in a finite number of steps. We have already seen that he took this philosophy to an extreme by outlawing mention and manipulation even of negative numbers, let alone the more esoteric beasts that had recently entered the mathematicians' menagerie. He believed that anything but the natural numbers 1, 2, 3,... were a baseless and futile attempt to improve upon the works of the Creator. At first Kronecker's main opponent in this enterprise was his contemporary Weierstrass. They argued vigorously and frequently, but remained friends and colleagues in Berlin. Weierstrass' attitude to Kronecker's attempts to truncate mathematics and exclude all those parts of it which he himself had spent so much of his career creating was certainly not mild, as this extract from a private letter to the mathematician Sonya Kowalewski in 1885 reveals:

> But the worst of it is that Kronecker uses his authority to proclaim that *all* those who up to now have laboured to establish the theory of functions are sinners before the Lord. When a whimsical eccentric like Christoffel says that in twenty or thirty years the present theory of functions will be buried ... we reply with a shrug. But when Kronecker delivers himself of the following verdict which I repeat *word for word*: 'If time and strength are granted me, I myself will show ... a more rigorous way. If I cannot do it myself those who come after me will ... and they will recognise the incorrectness of *all* those conclusions with which *so-called* analysis works at present'—such a verdict from a man whose eminent talent and distinguished performance in mathematical research I admire as sincerely and with as much pleasure as all his colleagues, is not only humiliating for those whom he adjures to acknowledge as in error and to forswear the substance of what has constituted the object of their thought and unremitting labour, but it is a direct appeal to the younger generation to desert their present leaders and rally around him as the disciple of a new system which *must* be founded. Truly it is sad, and it fills me with a bitter grief, to see a man, whose glory is without a flaw, let himself be driven by the well-justified feeling of his own worth to utterances whose injurious effect upon others he seems not to perceive.

Kronecker's affinity for the natural numbers is a strident development of a tendency one can detect in early Greek thought as an outgrowth of their philosophical concerns. Aristotle had drawn attention to the contrast between the continuity of many aspects of the physical world and the discrete nature of those elements of human reason, like counting, that we seek to employ in its evaluation. Even in Newton's day his teacher, Isaac Barrow, voiced concerns about the non-constructive nature of some of Euclid's geometrical proofs.

The *primitive intuitionism* which Kronecker advocated was founded upon these four precepts:

1. Natural numbers and their addition are a secure foundation for mathematics because they are anchored in our intuition.

2. Any definition or proof should be 'constructive'. That is, it must start from the natural numbers and construct the resulting mathematical entity or relationship in a finite number of steps. Mathematical existence means construction. The only acceptable proofs give the explicit recipe for the construction involved. Indirect proofs which prove non-existence, or which employ devices like proof by contradiction (or *reductio ad absurdum*) are unacceptable as proofs.
3. Logic differs from mathematics. A logical argument may use the *reductio ad absurdum* and be in accord with a set of rules of reasoning laid down, yet not count as a piece of valid mathematics.
4. One cannot consider actual, or completed, infinities. We may contemplate the construction of a set that can be added to without limit as a *potential infinity,* but we may not view this operation at any stage as having produced a member of some *completed infinite* collection. The production of this completion would have required the illegal notion of an infinite number of operations in its production.

Kronecker appears to have been a fairly forceful character who, although he did not often write about his philosophy of mathematics or the failings he saw in the mathematics used by others, was very vociferous in opposing non-constructive proofs in his lectures and public pronouncements. He also held powerful positions within the academic world and was in a position to control the sort of mathematics that was published in the journal which he edited and to influence the content of others. All this, one would regard as a fairly unexceptional little episode in the ongoing saga of human frailty. The private attitude of Karl Weierstrass which we recounted a little earlier, who on occasions laughed it all off as a joke, seems an entirely reasonable one. However, Kronecker's tendencies were soon to come into conflict with those of a far more intense and paranoid personality, whose name is remembered where Kronecker's is now largely forgotten.

Georg Cantor was born in St Petersburg in 1845. His father was a successful businessman but his family circle contained a diversity of successful musicians and artists. The family was close-knit and dominated by a devout Lutheranism. When Georg was eleven, his family moved to Germany for the sake of his father's health; here he completed his schooling, exhibiting a rare talent for mathematics. Georg was keen to pursue a career in mathematics despite his father's wish for him to train as an engineer. A deep crisis of loyalty was eventually avoided when his father, recognizing his son's true desire, relented and gave him permission to pursue his mathematical career. Cantor was overwhelmed with gratitude to his father and for the rest of his life displayed a touching filial responsibility to live up to his father's expectations.

Cantor studied in many of Europe's best mathematical schools and spent some time as a student of Leopold Kronecker. Their relationship appears to have been amicable, but Cantor never managed to obtain a position at one of the good German universities. Instead of his heart's desire, a position at the University of Berlin, he found himself first as a schoolteacher and then teaching in a poor department at the little-known University of Halle, where he was stuck adrift from the centres of intellectual life for more than forty years.

Cantor's early work was of a sort that violated all of Kronecker's key precepts. His first paper on sets appeared in 1874. It was non-constructive; it used actual infinities; it made liberal use of the *reductio ad absurdum*; it was also fundamentally new and innovative. And Kronecker presented it as a siren call to the younger generation of mathematicians, luring them away from the straight and narrow road defined by the natural numbers into a madhouse of meaningless infinite entities. Kronecker missed no opportunity to oppose Cantor's ideas. He described it as 'humbug' and a breed of 'mathematical insanity'. Cantor, already resentful of his lowly position and menial salary, seeing any chance of his promotion to a post in Berlin blocked by Kronecker's presence there, became increasingly embittered and paranoid. He wrote to his sympathetic mathematical friends that he saw everywhere conspiracy inspired by Kronecker and his colleagues in Berlin and had decided to write directly to the Ministry of Education in order to antagonize his opponents:

> For years Schwarz and Kronecker have intrigued terribly against me in fear that one day I would come to Berlin. I regarded it as my duty to take the initiative and turn to the Minister myself. I know precisely the immediate effect this would have: that in fact Kronecker would flare up as if stung by a scorpion, and with his reserve troops would strike up such a howl that Berlin would think it had been transported to the sandy deserts of Africa, with its lions, tigers and hyenas. It seems that I have actually achieved this goal!

This move didn't succeed in getting him a job in Berlin, but it certainly succeeded in annoying Kronecker. Whilst Kronecker was no doubt a fanatic, he was a calm, measured, and calculating one. His response seems to have been calculated to goad Cantor rather then confront his ideas in some public way. Cantor had hoped that a public debate would enable mathematicians to judge once and for all whose ideas were the more profitable. For just as Kronecker saw Cantor as a dangerous revolutionary, so Cantor saw Kronecker as a dangerous reactionary and an encumbrance to the growth and freedom of mathematics, who was attempting to impose a strait-jacket upon the creative development of the entire subject. Kronecker's doctrines would lead to interesting mathematics becoming extinct.

To understand Kronecker's response, one must be aware that there were just two journals where Cantor could publish his work. One was closed to him because Kronecker was the editor; the other, *Acta Mathematica*, was edited by Gösta Mittag-Leffler, someone sympathetic to Cantor's work on infinities, and so his work was published there. Kronecker wrote to Mittag-Leffler saying that he was toying with the idea of writing an article for his journal showing 'that the results of modern function theory and set theory [the subject of Cantor's papers] are of no real significance' because they make use of non-constructive methods and actual infinities. Cantor got wind of this and, persuading himself that this would end up being a polemical attack upon himself, thought Mittag-Leffler had betrayed him by entertaining such a possibility. Imagining another conspiracy, he threatened to have nothing further to do with the journal, thereby cutting himself off from all means of publishing his work in mathematics journals. In the end Kronecker did not write the article and one suspects that maybe he never intended to.

These events, coupled with Cantor's anxiety and failure to solve one of his hardest problems, led not entirely unexpectedly to a nervous breakdown the following spring, in 1884. He seemed to recover after a month of treatment and, recognizing the conflict with Kronecker as a source of the problem, attempted a reconciliation with him. Kronecker replied in a friendly manner to Cantor's letter, but things soon started to go awry again. Mittag-Leffler rejected an article which Cantor submitted to his journal. Cantor was deeply offended and decided never to publish there again. He retreated from mathematics, recognizing that his intense concentration upon particular mathematical questions was beginning to affect his mental equilibrium. He began to explore the philosophical and theological ramifications of his work on infinities and published in philosophical journals. There his work was welcomed, especially by Catholic theologians who believed it to be of fundamental importance to their conception of God and the Universe and were anxious to fulfil Pope Leo XIII's call to incorporate the insights of science into the Catholic world-view. Cantor was a passionate theologian with a detailed knowledge of his subject and, predictably, deep worries that the Church would fall into serious error if it ignored his discoveries. In this more relaxed period away from his mathematical colleagues, Cantor also extended his literary interests and delved in detail into the long-standing dispute over the true authorship of Shakespeare's plays and made significant contributions to the study of the early history of mathematics in India.* Meanwhile, Kronecker continued to voice his opposition to Cantor in his lectures, but the two only seemed likely to collide at the convention of German mathematicians planned for 1891. Cantor seemed to be hoping that there would be some clash of views here so that 'many who were previously blinded would have their eyes opened' to the real Kronecker. But, what turned out to be a fatal accident to his wife prevented Kronecker attending and then Kronecker himself died not long afterwards that same year.

Although Cantor's principal opponent had departed and many younger mathematicians became strongly influenced by his revolutionary treatment of infinite quantities, things did not end happily. Cantor suffered a series of mental breakdowns over the ensuing years. They culminated in the tragic events of 1899 which saw first the death of his younger brother and then that of his twelve-year-old son in the week before Christmas. He was relieved of his teaching duties at the University and four of the next ten years were spent in hospitals and sanatoria until finally he died of heart failure in the clinic at Halle on 6 January 1918.

The conflict between Cantor and Kronecker was later seen by mathematicians of the 1920s and 1930s as a sad spectacle and it came to be viewed as a legacy of the constructivist philosophy. This was strongly reinforced by the account given by Schoenflies in 1927 which presented an analysis that suggested that Cantor's mental breakdown was caused by Kronecker. This view was perpetuated by the very widely read (but historically very unreliable) popular history of mathematics of Eric T. Bell, *Men of Mathematics*, first published in 1937. Bell was clearly strongly influenced by the fashionable Freudian trends in psychoanalysis of the day and tried to explain Cantor's problems with Kronecker in terms of suppressed Freudian conflicts with his father in early life and a special type of rivalry that only exists between academics who are Jews, claiming 'There is no more vicious academic hatred than that of one Jew for another when they disagree on purely scientific matters.' Absurd as this generalization is, Cantor was not even Jewish!

After this sad story the reader is probably wondering just what it was that Cantor had been up to that created such strong feelings.

* Cantor was involved in some of the earliest scholarly investigations of the relationship between Greek and Indian geometrical knowledge. He was one of the first to study the way in which mathematical knowledge spread from one culture to another, rather than assume independent invention in separate places. He also carried out studies of the geometrical significance of Vedic altar designs which we described in Chapter 2. Cantor's early (1877) opinion was that Indian geometry derived from Alexandria and Egypt, but twenty years later he renounced the idea that Alexandrian geometry was merely given theological form by Indian culture. He was eventually convinced of the great antiquity of Indian peg and cord constructions (pre-2000 BC) and of the independence of geometrical concepts arising in the *'Sulba-sûtras*.

## CANTOR AND INFINITY

*Bachelors and Masters of Arts who do not follow Aristotle's philosophy are subject to a fine of 5 shillings for each point of divergence.*

FOURTEENTH-CENTURY STATUTE OF OXFORD UNIVERSITY

In 1638 the famous Italian scientist Galileo noticed a puzzling thing about numbers. If you make a list of all the natural numbers 1,2,3,... and so on, then one can place each of these numbers in direct one-on-one correspondence with its square, that is, the number obtained by multiplying each natural number by itself. This list of squares contains 1×1=1, 2×2=4, 3×3=9, 4×4=16, and so on. The correspondence is shown below:

$$\begin{array}{ccccccc} 1 & 2 & 3 & 4 & 5 & \ldots & N \\ \downarrow & \downarrow & \downarrow & \downarrow & \downarrow & \ldots & \downarrow \\ 1 & 4 & 9 & 16 & 25 & \ldots & N^2 \end{array}$$

Now, what Galileo noticed was that there must be the same number of numbers in both of the lists because there is a direct one-on-one correspondence between the entries in the two lists. Every number in the bottom row is linked by an arrow to one and only one of the natural numbers in the top row. But the paradox seems to be that every number that occurs in the bottom row must also be somewhere in the top row since these are natural numbers as well. So this means that the top row contains the bottom row together with many other numbers, like 2,3 and 5 for instance, which are not squares and so do not appear in the bottom row. Surely, this means the top row contains more numbers than the bottom row, not an equal number as we just argued!

Galileo presents this observation in the course of a dialogue between Simplicius and Salviati. Salviati raises the problem and concludes that

> I see no other decision that it may admit, but to say, that all Numbers are infinite; Squares are infinite; and that neither is the multitude of Squares less than all Numbers, nor this greater than that; and in conclusion, that the Attributes of Equality, Majority and Minority have no place in Infinities, but only in terminate [ie finite] quantities.

What Galileo is more or less telling his readers is that there is something a bit fishy about infinite quantities: they don't behave like collections of finite quantities and I'm not going to worry about them any more. And he didn't.

Two hundred and fifty years later, Cantor not only accepted this 'paradox' at face value, he exploited it to *define* an infinite set as any collection of objects which can be put into a one-to-one correspondence with a part of itself. With a little thought you can easily convince yourself that this is never possible for a set which has a finite number of members since the number of its members must always be greater than the number of members in any part of it. This is straightforward.

What Cantor then set out to do was create exact notions of what it means for an infinity to be equal to, greater than, or less than another infinity. The resulting arithmetic of the infinite, or *transfinite arithmetic*, was a dramatic and controversial departure from the past attitudes to actual infinities by mathematicians who had regarded them as a concept for the theologians. They were happy to use the term 'infinite' as a figure of speech to characterize some unlimited quantity (like the list of all the natural numbers) that could grow as large as one chooses, what they would have called a 'potential' infinity; but actual infinities were anathema. Carl Friedrich Gauss gives the typical early nineteenth-century attitude to that sort of thing with his statement:

> ... the use of infinite magnitude as if it were something finished; such use is not admissible in mathematics.

Although these words were written in 1831, the attitude they espouse was still universally shared by mathematicians fifty years later when Cantor started to announce his new ideas. Whereas earlier mathematicians had convinced themselves, as Galileo did, that infinities just produced paradoxes when you applied the rules of ordinary arithmetic to them, and so they should be outlawed from the subject, Cantor realized that this state of affairs signalled only that they might require a new type of arithmetic:

> All so-called proofs of the impossibility of infinite numbers begin by attributing to the numbers in question all the properties of finite numbers, whereas the infinite numbers, if they are to be thinkable in any form, must constitute quite a new type of number.

His success in doing this arose naturally out of his attempts to produce an exact definition of a set and notions associated with sets. His definition of a set was 'any collection into a single whole of definite well-distinguished objects of our intuition or of our thought'. He required there to be a definite and unambiguous rule that determined which things were members of a set and which were not. This was uncontroversial, but what was not was its application to both finite *and* infinite sets.

Cantor's transfinite arithmetic is very simple: two infinite sets are equal if they can be put in one-to-one correspondence with each other. Sets which can be matched to each other in this sense are then said to have the same *cardinality*. From our example above we see that the natural numbers and their squares have the same cardinality. The set of all the even numbers {2,4,6,...} also clearly has the same cardinality since its members can be put in a one-to-one correspondence with the natural numbers via the association of 1 with 2, 2 with 4, 3 with 6, and so on forever. Such sets are said to be *countably infinite* and their cardinality is denoted by the Hebrew letter aleph with a subscript nought, $\aleph_0$, and referred to as 'aleph-nought'.

We can now try to carry out some transfinite arithmetic in order to create some other infinite sets having the same cardinality as the natural numbers. If we were to add one new member to a countably infinite set, what would happen? For example, we might start with the list of even numbers 2,4,6,... and add zero to it. We can still put the new set into a one-to-one correspondence with the natural numbers; we just shift the correspondences along one notch as shown in Fig. 5.2.

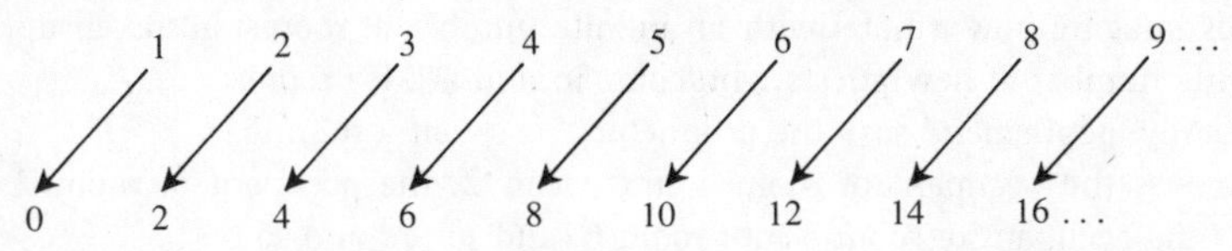


**Figure 5.2** The one-to-one correspondence between the natural numbers and the even numbers after zero has been added to them.

Since the top row contains all the natural numbers and there are $\aleph_0$ even numbers matched one to one with them, the bottom row must contain $\aleph_0 + 1$ members. Therefore, we have the curious result that, in transfinite arithmetic,

$$1 + \aleph_0 = \aleph_0.$$

Likewise, if we had added *any* finite number of extra members to the collection of even numbers, then we could still match the total with the natural numbers. For example, suppose we added four extra members –6, –4, –2, and 0. Then we would match them one-on-one to the natural numbers as shown in Fig. 5.3. This same trick works for any finite number of extra members, and so we have the rule that

$$\aleph_0 + f = \aleph_0$$

for any finite quantity *f*. What is most unusual about this type of addition that it can clearly *never* be true when applied to finite quantities.

There is an amusing illustration of the consequences of Cantor's exploitation of the fact that parts of infinity may be equal to the whole of infinity. It was a little story devised by Hilbert that has become known as 'Hilbert's Hotel'.

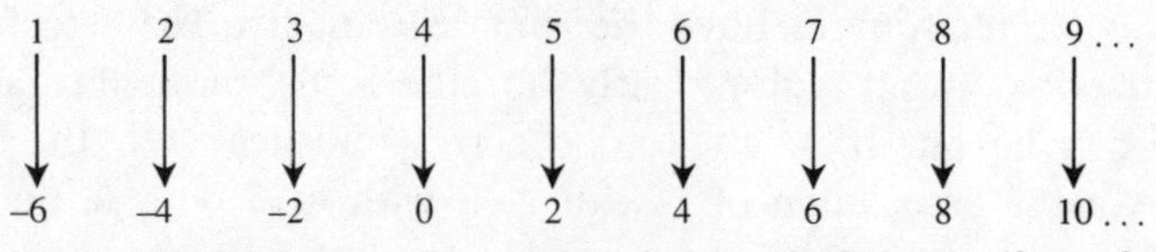


**Figure 5.3** Another one-to-one matching with the natural numbers.

Let us imagine a hotel with a finite number of rooms, and assume that all the rooms are occupied. A new guest arrives and asks for a room. 'Sorry,' says the proprietor, 'but all the rooms are occupied.' Now let us imagine a hotel with an *infinite* number of rooms, and all the rooms are occupied. To this hotel, too, comes a new guest and asks for a room.

'But of course!' exclaims the proprietor, and he moves the person previously occupying room 1 into room 2, the person from room 2 into room 3, the person from room 3 into room 4, and so on ... And the new customer receives room 1, which became free as the result of these transpositions.

Let us imagine now a hotel with an infinite number of rooms, all taken up, and an infinite number of new guests who come in and ask for rooms.

'Certainly, gentlemen,' says the proprietor, 'just wait a minute.'

He moves the occupant of room 1 into room 2, the occupant of room 2 into room 4, the occupant of room 3 into room 6, and so on, and so on ...

Now all the odd-numbered rooms become free and the infinity of new guests can easily be accommodated in them.

This may explain why, in my experience, the strangeness of hotels increases with their size!

There are other curiosities of transfinite arithmetic to come. What happens if we add two countably infinite sets together? For example, add all the even numbers to all the odd numbers. The result is just the complete set of natural numbers which is countably infinite, by definition, so we find that

$$\aleph_0 + \aleph_0 = \aleph_0.$$

Next, Cantor considered the set of all fractions, numbers like 1/2 or 5/7 which are constructed by dividing one natural number by the other. The whole collection of these could be listed in rows by taking first, all the fractions which have the number 1 on the top, then those with 2 on the top and so on, as shown in Fig. 5.4. Each of these rows contains as many entries as there are natural numbers, that is $\aleph_0$, and since there will be a new row for every natural number, the total number of fractions will just be $\aleph_0$ times $\aleph_0$, or what we write as $\aleph_0^2$. A fraction is just specified by a pair of numbers but there are as many infinite pairs of numbers as there are numbers and so we expect to find that $\aleph_0^2 = \aleph_0$. We can easily prove this by showing that it is possible to put the array of fractions into

$$\frac{1}{1}, \frac{1}{2}, \frac{1}{3}, \frac{1}{4}, \frac{1}{5}, \frac{1}{6}, \frac{1}{7}, \ldots$$

$$\frac{2}{1}, \frac{2}{2}, \frac{2}{3}, \frac{2}{4}, \frac{2}{5}, \frac{2}{6}, \frac{2}{7}, \ldots$$

$$\frac{3}{1}, \frac{3}{2}, \frac{3}{3}, \frac{3}{4}, \frac{3}{5}, \frac{3}{6}, \frac{3}{7}, \ldots$$

$$\frac{4}{1}, \frac{4}{2}, \frac{4}{3}, \ldots$$

$$\vdots$$

**Figure 5.4** Cantor's systematic listing of all the fractions, row by row, where all those with 1 on top lie in the first row, all those with 2 on top lie in the second, and so on.

an one-to-one correspondence with the natural numbers so long as you set about counting them in the right order.

The arrowed path shown in Fig. 5.5 enables all the rational fractions laid out in the infinite array to be counted one by one without any being omitted. This shows there exists a direct one-to-one correspondence

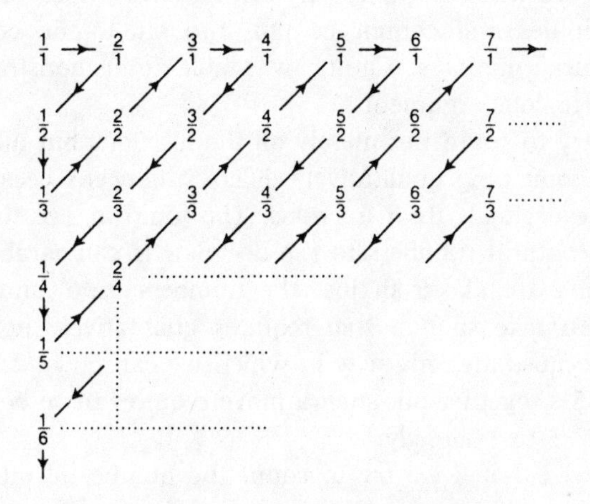


**Figure 5.5** The arrowed path traces out the systematic way to count all the rational fractions. This ensures that each one of the infinite collection of fractions is counted once and only once and establishes the one-to-one correspondence with the natural numbers which label each step in the path that is taken.

between $\aleph_0$ and all the rational fractions through the sequence 1/1, 2/1, 1/2, 1/3, 2/2, 3/1, 4/1, 3/2, 2/3, 1/4, 1/5, 2/4, 3/3, 4/2, 5/1, 6/1, 5/2, 4/3, ... and so on *ad infinitum*. Hence, in this precise sense, the rational numbers are an infinite set of the same size as the natural numbers. At first sight this is a surprising result since natural numbers are rather sparsely distributed whereas there seem to be rational fractions packed everywhere in between them, so a counting process ought to find many more fractions than integers. But this intuition focuses too much upon the *order* in which the numbers appear, whereas the one-to-one correspondence that we have set up does not need to follow the order in size in which the fractions occur in between the natural numbers. This confirms, *en passant*, another of Cantor's rules of transfinite arithmetic:

$$\aleph_0=\aleph_0^2.$$

If we were to multiply both sides of this equation by $\aleph_0$ we would be able to conclude that

$$\aleph_0^3=\aleph_0^2=\aleph_0.$$

and if we kept on multiplying by $\aleph_0$ like this *r* times, we would see that

$$\aleph_0^r=\aleph_0.$$

After this experience with the mysteries of countable infinities, we might be tempted to ask whether there can exist infinities which are bigger than countable infinities and cannot be put into one-to-one correspondence with the natural numbers. Cantor was able to demonstrate this by a remarkably ingenious argument.

If now we try to count not merely all the fractions but all the decimals as well, then something qualitatively different happens because there are so many more decimals than fractions. The jump in size that marks the step from the natural numbers to the decimals is comparable to the step one would have to take from just the numbers 'zero' and 'one' to the larger ones. To take such a step requires qualitatively new conceptual information because the only way in which we can make 2 from 0 and 1 is to add two 1's together but such a move requires us to be in possession of the concept of 'two' already.

Cantor showed that if we try to count the number of infinite decimals —the so-called '*real numbers*'—then we fail. They are of a higher cardinality than the natural numbers, and so cannot be placed in a one-to-one correspondence with them. This he showed by a powerful new argument. It involves the notion of a diagonal number. For illustration, suppose we have four numbers of four digits in length

**1**234
5**6**78
90**1**2
345**6**

Then the diagonal number 1616 is not one of the four numbers listed. What Cantor showed was that if we make this array of numbers infinitely large then there is always a way of concocting a diagonal number that is not one of the infinite list of numbers lined up to make the array. Suppose we just look at the real numbers between 0 and 1 (it does not make any difference to the basic argument if we add all the others as well) and suppose that we count all the infinite decimals. This, Cantor showed, leads to a contradiction. Suppose we could write down all possible infinite decimals and align them one-to-one with the natural numbers and that the list begins as follows,

1 → 0.**2**34566789...
2 → 0.5**7**5603737...
3 → 0.46**3**214516...
4 → 0.846**2**16388...
5 → 0.5621**9**4632...
6 → 0.46673**2**271...

and so on forever. Now take the 'diagonal number' with decimal places composed of the bold digits. It is

0.273292...

Next, alter each digit by adding one to it to get the new decimal

0.384303...

Then this new number cannot appear anywhere on the original list because it differs by one digit from every single horizontal entry. It must at least differ from the first entry in the first digit and from the second entry in the second digit and so on forever. So, contrary to our original supposition, the list could not have contained all the possible decimals. Hence, our original assumption that all the infinite decimals can be systematically counted was false. The reals possess a higher cardinality than the natural numbers. They are *uncountably infinite* and are denoted by the symbol *C* for *continuum*.

Just as there were unusual trans-arithmetic properties of aleph-nought so there are of the continuum. The continuum is made up of all the rational numbers, which are countably infinite, plus the irrationals—those quantities like the square root of 2—which cannot be expressed as the ratio of two natural numbers.

Therefore the irrationals must be uncountably infinite, and so because the irrational numbers plus the rational numbers equals the continuum, we have the relation

$$C + \aleph_0 = C$$

One of things this reveals is that, as numbers go, the irrationals are actually the norm. Contrary to what the Pythagoreans first thought thousands of years ago, it is the rationals that are peculiar—indeed, they are infinitely so.

Another curiosity is that any finite stretch of the real numbers, say from 0 to 1, contains the same number of points as the entire expanse of real numbers from minus infinity to plus infinity. This is most easily demonstrated by a simple picture (see Fig. 5.6). Suppose we take our piece of straight line from 0 to 1 and curve it around to make the bottom half of a circle. If we draw lines from its centre down to the straight line below it, then every single point on the semicircle will be matched with every single point on the line by a line which cuts them both. This tells us that the cardinality of all the real numbers is $C$, just the same as that of any part of them.

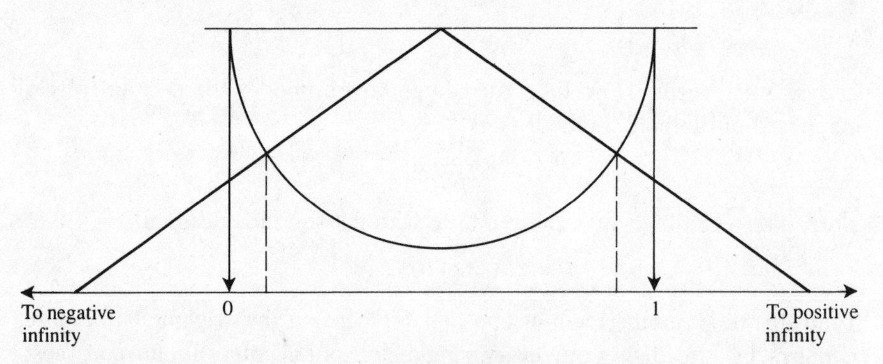


**Figure 5.6** The one-to-one correspondence between the line of length one from 0 to 1 and the entire line from negative infinity to positive infinity. Take any point on the line between negative and positive infinity. Join it by a straight line to the centre of the semicircle as shown. Where the line cuts the semicircle, follow the dotted line vertically downwards to a point on the interval between 0 and 1. Every point on the original line ends up at only one point between 0 and 1.

If we take an infinity of intervals of the real numbers, each from 0 to 1 again, and stack them up on each other until the stack extends up from 0 to 1, then we will have created a square area as shown in Fig. 5.7. The square therefore contains $C \times C = C^2$ points. Each point in the square has a grid reference that is specified by two decimals, for example (0.3333333333..., 0.4567893438...). Now we can match each of these

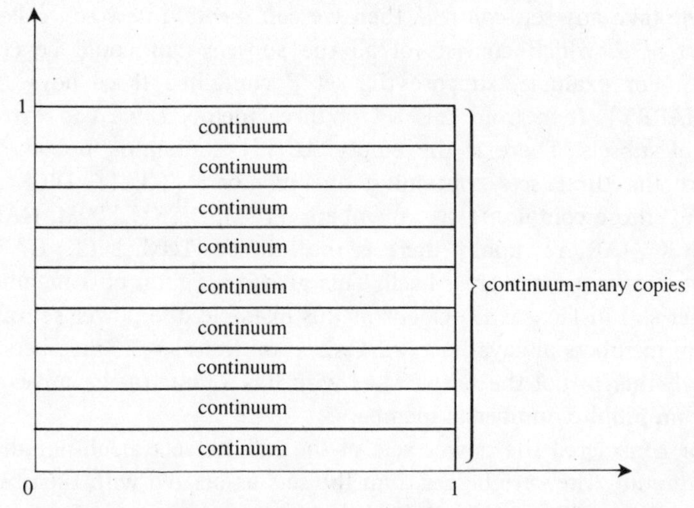


**Figure 5.7** Continuum-many copies of the continuum stacked up to form a square. Each point in the square can be located by a map reference which gives its horizontal and its vertical distance from 0 as two decimal numbers.

specifications with a single decimal number simply by constructing such a single decimal by alternating the entries in the two which specify the grid reference of each point. So, in our example, the point's two coordinates are replaced by the single label 0.34353637383933343338... But the cardinality of all the decimals, of which this is now a member, is just $C$; so we have proved that

$$C = C \times C = C^2,$$

and multiplying through by $C$ we would have $C = C \times C \times C = C^3$ and in general $C = C^p$ for any number $p$. This particular result astonished even Cantor, and when he reported it to one of his correspondents he exclaimed, 'I see it but I do not believe it.'

This was all so totally new and counter-intuitive that many mathematicians found it 'repugnant to common sense' and it was this manipulation of actual infinities that drove Kronecker into his state of studied opposition. Others found it profound, hauntingly beautiful, the crowning glory of mathematics. Hilbert called it

> the most admirable fruit of the mathematical mind and indeed one of the highest achievements of man's intellectual processes ... No one shall expel us from the paradise which Cantor has created for us.

Cantor did not stop at the order of infinity determined by the continuum of real numbers. He realized that there were even larger infinities than this. If we take any set, call it $S$, then we can create a new set, called the *power set of S*, which consists of all the subsets that could be created within $S$. For example, suppose the set $S$ contained three boys {TOM, DICK, HARRY}, then from this set of three members we can extract a variety of subsets. There is the empty set { } containing nobody, then there are the three sets containing one member: {TOM}, {DICK}, and {HARRY}; those containing two members: {TOM, DICK}, {TOM, HARRY} and {DICK, HARRY}; finally there is the full set {TOM, DICK, HARRY}, which is always a member of itself. This gives us a total of 8 members of the power set. In fact, as is evident in this example, the power set of a set having m members always has 2×2×2×2 ... (*m* times) = $2^m$ members. This is true whether or not the set we start with has a finite (as in our example here) or an infinite number of members.

Cantor considered the power sets of the infinite sets aleph-nought and the continuum. They are bigger than the sets associated with them; so the power set of $\aleph_0$ is bigger than $\aleph_0$ and that of $C$ is bigger than $C$, and so we have an ascending hierarchy. We have found a way of generating bigger and bigger infinite sets from ones that we already have. There is no limit to this escalation. If we label the power set of $C$ by $D$, then we could create an infinitely larger set from it by forming the power set $2^D$ of $D$, then its power set, and so on *ad infinitum*. By this means we can create an ever-ascending staircase of infinities, each of higher cardinality than the previous one, whose power set it is. There is no end to this inconceivable infinity of infinities. It can have no greatest member.* As Ian Stewart once remarked, 'it's enough to make Kronecker turn in his grave—infinitely often'.

These are some of the successes of Cantor's work; but there was one problem that he sought constantly to solve. He was never successful and many of his periods of mental disturbance and depression seem to have been exacerbated by his failures. Later, we shall have more to say about this problem. It is far harder, and its implications far deeper and wider, than Cantor ever suspected in his troubled lifetime. It is called the 'Continuum Hypothesis'.

Cantor raised the intriguing question of whether there exist infinite sets which are intermediate in size between the natural numbers and the real numbers; that is, infinite sets with cardinality in-between $\aleph_0$ and $C$. Cantor

* Cantor's transfinite world does have its own paradoxes: 'Cantor's Paradox' is the realization that the set of all sets is its own power set which means that its cardinality is bigger than its cardinality! Cantor didn't regard this as a paradox but he never published it. He mentioned it in a letter to Dedekind. He thought there were two sorts of sets—'consistent' and 'inconsistent'—the first were the true sets whilst the latter were not true unified entities, but they could be useful in carrying out proofs in certain circumstances.

thought that there could not be, but was unable to prove it. The problem remains unsolved to this day. Nonetheless, Kurt Gödel and a young American mathematician Paul Cohen demonstrated some deep and unusual things about it. Gödel showed that if we merely treat the continuum hypothesis as an additional axiom and add it to the conventional axioms of set theory, then no logical contradiction can result. But then in 1963 Cohen showed that the same thing would happen if we added the negative of the continuum hypothesis to the axioms of set theory. Therefore the continuum hypothesis is independent of the other axioms of set theory (just as Euclid's parallel postulate was eventually shown to be independent of the other axioms of plane geometry) and therefore can be neither proved nor refuted from those axioms.

Paul Cohen's demonstration of the insolubility of the continuum hypothesis has a remarkable history. Cohen was a brilliant and confident young student who upon entering graduate school at Stanford gathered a group of his fellow students together to ask them if they thought he would become more famous by solving one of Hilbert's remaining unsolved problems or by showing that the continuum hypothesis is independent of the axioms of set theory. His colleagues voted for the latter. Whereupon Cohen went off, learnt about the relevant areas of mathematical logic and invented a new proof procedure that enabled him to solve the most difficult unresolved question in mathematics in less than a year. Local mathematicians knew there was only one way to know whether the proof was correct and so soon Cohen found himself knocking on the door of Gödel's residence in Princeton. He was clearly not the first such caller. Gödel opened the door just wide enough for Cohen's proof to pass, but not so wide as to allow Cohen to follow. But two days later Cohen received an invitation to tea with the Gödels. The proof was correct. The master had given his *imprimatur*.

Before closing our description of Cantor's heroic efforts, we should mention that they did create the paradox that we have already met in the last chapter. Cantor's seemingly innocent definition of a set allows us to form the creature that is the set of all sets, and having released that genie from its bottle we must face up to the problem of the set of all sets that are not members of themselves, a set of which one is a member only if one is not a member. Perhaps this was the set that Groucho Marx was hoping to get in with when he wrote that 'I don't want to belong to any club that will accept me as a member'!

As a result of Cantor's developments, one could divide the mathematical community into three sorts. There were the finitists, typified by the attitudes of Aristotle or Gauss, who would only speak of potential infinities, not of actual infinities. Then there were the intuitionists like Kronecker and Brouwer who denied that there was any meaningful content to the notion of quantities that are anything but finite. Infinities are just potentialities that can never be actually realized. To manipulate them and include them within the realm of mathematics would be like letting wolves into the sheepfold. Then there were the transfinitists like Cantor himself, who ascribe the same degree of reality to actual completed infinities as they did to finite quantities. In-between, there existed a breed of manipulative transfinitists, typified by Hilbert, who felt no compunction or need to ascribe any ontological status to infinities but admitted them as useful ingredients of mathematical formalism whose presence was useful in simplifying and unifying other mathematical theories. 'No one', he predicted, 'though he speak with the tongues of angels, will keep people from using the principle of the excluded middle.'

## THE COMEDY OF HILBERT AND BROUWER

*Absurdity of absurdity of absurdity is equivalent to absurdity.*

LUITZEN BROUWER

The fraught personal relations between Kronecker and Cantor which characterized the first variety of intuitionism to be proposed seemed to create something of a tradition for the intuitionists. For the next era was characterized by a long period of tragi-comical relations between Hilbert and Brouwer. It is now rather irreverently known as the War of the Frogs and the Mice after Einstein's characterization of it as the 'frog–mice battle' ('Frosch-Mäusekrieg').*

Hilbert admired Cantor's creative work and listed his continuum hypothesis amongst his great list of 'Problems' presented to the International Congress of Mathematicians in 1900. As a corollary, he saw Kronecker's philosophy as a danger to the entire future of mathematics. It threatened to decimate the subject of so much that had been added to it by non-constructive methods of proof; it sought to build the subject upon a subjective 'intuitive' foundation that tied mathematical meaning to step-by-step construction and the real material world. Hilbert required only logical self-consistency to underwrite mathematical meaning and existence. As a result he was outspoken in his opposition to the intuitionists, whether of the old style, like Kronecker, or the new breed represented by Poincaré, Weyl, and Brouwer:

* The term arises from an early Greek poem (*Batrachomyomachia*, *c.*500 BC) whose author is unknown and a later medieval German version by Rollenhagen. The general sense of the title will become clear after reading our account of the modern version starring Brouwer.

What Weyl and Brouwer are doing is mainly following in the steps of Kronecker. They are trying to establish mathematics by jettisoning everything which does not suit them and setting up an embargo. The effect is to dismember our science and to run the risk of losing part of our most valuable possessions. Weyl and Brouwer condemn the general notions of irrational numbers, of functions ... Cantor's transfinite numbers etc, the theorem that an infinite set of integers has a least member, and even the 'law of the excluded middle' ... These are examples of [to them] forbidden theorems and modes of reasoning. I believe that impotent as Kronecker was to remove irrational numbers (Weyl and Brouwer do permit us to retain a torso), no less impotent will their efforts prove today. No! Brouwer's programme is not a revolution, but merely the repetition of a futile *coup de main* with old methods, but which was then undertaken with greater verve, yet failed utterly. Today the State [i.e. Mathematics] is thoroughly armed and strengthened through the labours of Frege, Dedekind and Cantor. The efforts of Brouwer and Weyl are foredoomed to futility.

So much for Hilbert on Brouwer; what of Brouwer on Hilbert's formalist programme founded upon consistency alone?

Nothing of mathematical value will be attained in this manner; a false theory which is not stopped by a contradiction is none the less false, just as a criminal policy unchecked by a reprimanding court is none the less criminal.

Brouwer made his first extensive statement of the intuitionist doctrine in his doctoral thesis in 1907. At first he termed this development of Kronecker's doctrine as 'neo-intuitionism' but he eventually gave up the attempt to preserve a distinction. We shall do the same. Before seeing what Brouwer and his followers advocated, we shall take a look at what ultimately happened as a result of Hilbert's worries about Brouwer's proposals which seemed to have gained some distinguished supporters, albeit supporting a more moderate edition of the original, and their uneasy personal relations.

In the 1920s the foremost mathematical journal of the day was the German *Mathematische Annalen*, or the *Annalen* as it was known. Founded in 1868, its publication was taken over by Springer in 1920, and they inherited a journal with an unparalleled reputation for excellence that rested upon the standing of its board of editors who assessed all papers that were submitted to it for publication. Only those found to be without error and presenting new and important insights were accepted for publication. Being asked to be an editor was a mark of great distinction for a mathematician. In the 1920s the roster of editors included Einstein, Brouwer, and Hilbert, and ten others. Hilbert was effectively the chief because of his pre-eminent reputation and expertise across the whole spectrum of mathematics, whilst Brouwer had been an editor since 1915 and took his role very seriously, examining papers in meticulous detail, although sometimes taking a very long time to do so.

In October 1928 Brouwer received a short letter from Hilbert. It read as follows,

Dear Colleague,

Because it is not possible for me to cooperate with you, given the incompatibility of our views on fundamental matters, I have asked the members of the board of managing editors of the *Mathematische Annalen* for the authorisation, which was given to me by Blumenthal and Carathéodory [two other editors], to inform you that henceforth we will forgo your cooperation in the editing of the *Annalen* and thus delete your name from the title page. And at the same time I thank you in the name of the editors of the *Annalen* for your past activities in the interest of our journal.

Respectfully yours,
D. Hilbert.

Personal relations between Hilbert and Brouwer had begun well enough. Brouwer greatly admired his more senior colleague and made many trips from his native Amsterdam to visit Hilbert's famous institute at Göttingen. Although he had been very critical of Hilbert's formalist programme in his thesis of 1907, this criticism was not widely known because it was written in Dutch, and Hilbert probably knew almost nothing about those aspects of its contents. After recommending Brouwer for a professorship in Amsterdam in 1912, he even offered Brouwer a similar position in Gottingen seven years later. But Brouwer refused and this seems to have initiated the souring of their relationship.

That process was greatly accelerated by Brouwer's active campaigning on behalf of the intuitionist definition of mathematics throughout the 1920s. Hilbert was clearly worried. Brouwer lectured all over Europe to mathematicians and philosophers and was well received. Wittgenstein claimed that he took up philosophy again as a result of hearing Brouwer's lectures on these matters in March 1928 in Vienna. This activity and the challenge that it presented to Hilbert's views on the foundations of mathematics constituted the second and principal source of friction between Hilbert and Brouwer.

The third source of friction was a curious political matter concerning the participation of German mathematicians in the Bologna International Conference of Mathematicians in 1928. There was an international campaign against their participation which Brouwer supported, but not through any post-war animosity towards the Germans (such as some undoubtedly felt); rather because he thought, like some of the Germans from Berlin, that they were being tolerated as less than equal participants in the meeting. Moreover, Brouwer was active in other boycotts against French mathematicians because their country supported harsh anti-German measures. Hilbert was furious at these interventions because they were opposed to his own views and interfered with his assumed authority over all matters concerning German mathematics. Hilbert took this meddling as a personal insult.

Both Hilbert and Brouwer had strong personalities, but Brouwer was very highly strung, unpredictable, and vitriolic. Under stress he often fell victim to fits and fevers of a nervous origin. Hilbert even refers to him as a 'psychopath' in correspondence to Einstein who regarded him as a classic case of the fine line between genius and insanity. All these matters seemed to come to a head in 1928 at a time when Hilbert imagined that his health was failing and so began to fear for the future of the *Annalen* after his death, lest it fall under the influence of the ambitious Brouwer and his intuitionist views. It was in this bizarre climate that Hilbert began his campaign to remove Brouwer from the board of the *Annalen.*

The arrival of Hilbert's letter sank Brouwer into a period of nervous illness and led him to write a disastrous letter to another of the editorial board's members, Constantin Carathéodory, making it appear that he questioned Hilbert's state of mental health at that time:

Dear Colleague,

After close consideration and extensive consultation I have to take the position that the request from you to me, to behave with respect to Hilbert as to one of unsound mind, qualifies for compliance only if it should reach me in writing from Mrs Hilbert and Hilbert's physician.

Yours,

L.E.J. Brouwer

There ensued a lengthy battle of claim and counter-claim as each tried to draw on support from other members of the editorial board, most of whom were anxious to keep as far away from the matter as they possibly could. Eventually, Brouwer appealed directly to the publishers claiming that 'Hilbert had developed a continuously increasing anger against me' because of the three matters we have described already. When Brouwer met with the publisher, Ferdinand Springer, he threatened legal and other actions against both the journal and Hilbert to discredit all concerned. Threatened in this way and facing the resignation of several other editors, Springer realized that the best escape in the face of such an 'embittered and malicious adversary' might be to found a new journal and invite Brouwer to be its editor, thereby saving everyone's faces.

Much activity followed as other members of the editorial board tried to smooth things over and prevent the matter creating a split among German mathematicians. It was at this stage that the physicist Max Born, an ally of Hilbert's, attempted to recruit Einstein to his side, whilst Brouwer and Blumenthal tried with equal lack of success to draw him into their camp.

But Einstein was determined to remain totally neutral and clearly viewed the entire business as a ridiculous tragi-comedy, replying to Born:

> If Hilbert's illness did not lend a tragic feature, this ink war would for me be one of the most funny and successful farces performed by that people who take themselves deadly seriously.
>
> Objectively I might briefly point out that in my opinion there would have been more painless remedies against an overly large influence on the managing of the *Annalen* by the somewhat mad Brouwer, than eviction from the editorial board.
>
> This, however, I only say to you in private and I do not intend to plunge as a champion into this frog–mice battle with another paper lance.

And then to Brouwer, a little more acidly,

> I am sorry I got into this mathematical wolf-pack like an innocent lamb ... Please ... allow me to stick to my role of astounded contemporary.
>
> With best wishes for an ample continuation of this equally noble and important battle, I remain
>
> Yours truly,
>
> A. Einstein.

By this stage Hilbert himself had withdrawn from the battle, on the grounds of failing health, leaving Harald Bohr, the brother of Niels Bohr, and Richard Courant to represent him in any legal matters. After further claim and counter-claim the publishers decided to dissolve the editorial board and reconstitute it in December 1928 with a reduced membership of three, with Hilbert in charge.

Brouwer had lost. After a period of bitter protest and recrimination he almost completely withdrew from playing any active role in mathematics again. He wrote in despair to a friend:

> All my life's work has been wrested from me and I am left in fear, shame, and mistrust, and suffering the torture of my baiting torturers.

Intuitionism continued without the direct contributions of Brouwer and in 1948 the University of Amsterdam appointed Arend Heyting, one of Brouwer's own students, to the professorship of mathematics. Heyting and, later, others like Errett Bishop in the United States continued the development of the constructivist philosophy of mathematics without the mystical elements which Brouwer sought to attach in his desire to trace the origins of mathematical intuition in the human mind.* It became the study of a formal system governed by three-valued logic in which only

* Heyting's attitude towards the free creation and formalization of mathematical systems which are not constructed from the natural numbers is nicely expressed by his retort that 'while you think in terms of axioms and deductions, we think in terms of evidence: that makes all the difference. I do not accept any axioms which I might reject if I chose to do so.'

constructive methods of proof were permitted. In the final years of his life Brouwer appeared to mellow slightly and appeared on the international lecture circuit again, to tell a new generation of young mathematicians about the intuitionist philosophy of mathematics and its programme of reconstructing mathematics according to his own criteria of truth. But the last ten years of his life before his sudden death in 1966 were beset by increasing paranoia and personal problems. His younger colleagues and former students had tried hard to please him, but he refused the special volume they prepared to present on the occasion of his retirement and he stopped Heyting from preparing a publication of his complete works. He died on 2 December 1966, aged eighty-five.

Following Brouwer's withdrawal from mathematics, interest in the intuitionist doctrine soon faded. Brouwer worked only sporadically on the subject with few associates and gradually the entire programme became marginalized. He became active in matters of local politics and commercial investment. Other supporters, like Weyl, were less vociferous in its support and were committed to many different research projects and interests which involved physics. As a result their concern with such fundamental questions tended to be compartmentalized and no real attempt was made to integrate the intuitionist philosophy into the entire spectrum of mathematical concepts and their applications.

There were some little-known exceptions. Besides the Dutch school of constructive mathematics which Heyting led, throughout the 1950s there grew up a Russian school of constructivism led by A.A. Markov in Moscow. What, in retrospect, is interesting about their contribution is the emphasis that they placed upon the concept of an algorithm in the days before computers were used in mathematical research. In 1954 Markov stated that

> The entire significance for the mathematician of rendering more precise the concept of algorithm emerges, however, in connection with the problem of a constructive foundation for mathematics. On the basis of the more precise concept of algorithm we may give the definition of constructive validity of an arithmetical expression. On its basis one may set up also a constructive mathematical logic ... Finally, the main field of application of the more precise concept of algorithm will undoubtedly be constructive analysis—the constructive theory of real numbers and functions of a real variable.

In fact Markov defined a particular type of algorithm, called a 'Markov algorithm' which constructed allowed mathematical statements from the primitive ingredients rather as one might compose words from letters following particular rules of composition. For Markov, the mathematical 'existence' of a quantity meant something quite unambiguous: it can be produced by a Markov algorithm.

In the West it was the publication in 1967 of a work entitled *Foundations of Constructive Analysis* by Errett Bishop that rekindled serious interest in constructive mathematics. The reasons are interesting. First, Bishop was a respected mathematician who had been active in mathematics using traditional methods of proof. He presented his ideas in a clear and readable fashion that any professional mathematician could understand. Second, his book gave the lie to claims that constructivism would empty mathematics of many of its most treasured possessions. Bishop showed that a very large fraction of conventional mathematics could be added to the store of constructivist truth if one was ingenious enough. Third, Bishop's work was mathematics and only mathematics. It did not contain all the philosophical and mystical accoutrements of Brouwer's thinking and no doubt Bishop was a pleasant fellow into the bargain. Indeed, Bishop developed the constructive approach in order to keep unverifiable philosophical ideas out of mathematics; in the introduction to his book he makes his famous remark that

> Mathematics belongs to man, not to God. We are not interested in properties of the positive integers that have no descriptive meaning for finite man. When a man proves a positive integer to exist, he should show how to find it. If God has mathematics of his own that needs to be done, let him do it himself.

But the last and most interesting possibility is that Bishop's book appeared at a time when the computer was emerging as an important ingredient in the lives of mathematicians and scientists. People were seeing large and difficult problems solved by calculating machines which proceeded by a finite sequence of deductive steps from an input that was written in numbers alone. Understandably, these developments had drawn attention to the nature of computer programming languages in general. These innovations left their mark upon Bishop, who proceeded by regarding mathematics as a 'high-level programming language in which proofs should be written. Algorithms should then naturally accompany existence proofs.' Bishop believed that mathematical objects are completely objective in the sense that every theorem of mathematics can be translated into a verifiable or computable statement about numbers. One can see that when all constructive methods had to be implemented by hand their laboriousness could be a powerful disincentive but with the prospect of rapid electronic implementation of the steps this worry is ameliorated. Or, at the very least, there is an important change of attitude. As more and more mathematicians started to make use of computers, they began unconsciously to take on board the notion that mathematical existence meant 'computable existence'. Thus Bishop was trying to push mathematicians in a direction that the tide of technological progress was already moving them. Brouwer, by contrast, chose to swim against a tide in the affairs of men. In the pre-war era mathematics was focused upon the successful creations in set theory, transfinite number theory, and logic which were entirely non-constructive in nature.* It is tempting to say that Bishop was timely, but of course Bishop did not really 'choose' the optimal time. The emergence of the algorithmic way of thinking influenced Bishop and this we might surmise led him to focus upon the new significance of constructive methods.

As a postscript to this entire episode we offer the following little fable about a frog and a mouse written by the American mathematician John Hays:

The mouse was a German mathematics professor, David Hilbert, who accomplished many feats of creativity and order along the Stream of Mathematics.

On sunny mornings, Hilbert the mouse would sit amidst patterns of twigs and gleaming pebbles and propound, to the gathered assembly of admiring creatures, 'We must use logic to prove the consistency of mathematics, since—*squeak!*—I've already shown...that we can map all of mathematics onto arithmetic.'

And his hearers, bobbing enthusiastically, would scamper off, crying, 'Yes! To work! To work!'

The frog was the Dutch mathematician, Luitzen Egbertus Jan Brouwer. One blue-strewn day, discarding his tadpole tail with a grunt, Brouwer the Frog swam to a lily pad, assumed the lotus position, and delivered, to the shocked passersby on the bank, his doctoral thesis:

'The Lord helps those who help themselves! Brek-kek-kek-kek! Hilbert the Mouse and Russell the Rabbit and the other logicians and mathematicians think they can cop out in their proofs. Brek! When it's too hard to prove a statement *directly*, they assume the *contradictory* of the statement, then seek *a contradictory of the contradiction. Broax-broax!* They suppose that the Good Lord, like a naive, orderly German professor, or an eccentric English aristocrat, has providently bundled all possible into *neat pairs of contradictories*. And, when any sagacious thinker finds an earthly copy of one of a given pair of statements to be defective, he can forward it to heaven in sacrificial smoke, and the Lord sends back its *negation* by return zephyr. Brek-kek-kek! Such nonsense! As every sensible frog knows, the Lord is an industrious ruler, who loves a constructive worker with crafty hands. If you wish a proof of a statement, you must *construct* it, *directly*, of your own initiative, and then the Lord will approve. Henceforth, I shall accept nothing as *rigorously* established in mathematics in absence of its *construction*. We may assume nothing more, except, as Kronecker the Shrew noted, *the intuition of the natural numbers*, passing endlessly by, like the ripples of this cool, sweet stream.' ...

* I have often wondered what Brouwer made of the discoveries of Turing and others regarding computability but my search through Brouwer's works and biographical studies has never revealed any intersection of their lives or their ideas.

When the creatures round about had recovered their wits, they scampered off to tell Hilbert the Mouse of this new heresy. A grasshopper vaulted after them, spitting tobacco juice and declaring: 'There's gonna be a fight! Sputt! Betcha the Mouse and the Frog will get into a scrap as soon as they meet! Sputt!'

And fight they did!—verbally, in loud angry tones that reverberated through the Academic Grove, beyond the Stream of Mathematics.

On one occasion, Hilbert squeaked to the assembly: There have been two great *crises* in *the Foundations of Mathematics*. First, when the Pythagoreans discovered *the irrationality of the square root of two* ... Second, when the philosophical Bishop, Berkeley the Bear, pointed out *the contradiction of dividing by zero* ... But Cantor the Goat, by his *General Set Theory*, has shown us how to resolve these problems, and to dispel crisis from the Foundations of Mathematics! And now we shall go forward to—.'

'Don't listen to that quietist propaganda!' boomed a voice from the rushes. 'Cantor's set theory attacked those problems only to create *a third crisis in the Foundations of Mathematics!* Hilbert is quite familiar with the paradoxes of set theory!'

'But they've been taken care of!' squeaked Hilbert, in shrill protest. 'And mathematics is the richer for Cantor's General Set Theory, and for *the chain of infinite aleph numbers* he derived therein. 'No one,' he snapped, whiskers twiddling, eyes blazing, 'no one shall drive us from the paradise Cantor created for us!'

'Broax! and nonsense! You prate constantly about building a neat little pen of axioms to protect the innocent sheep of your doctrinal theses from the wolves of contradiction. But you build by General Set Theory. And as Poincaré the Otter so aptly asked: *What good is it to build a pen for your sheep if you corral the wolves of contradiction inside with them?*'

At another time Brouwer the Frog sat on a lily pad and dispensed the new reformist gospel to a fascinated but uneasy crowd of creatures on the banks of the Stream of Mathematics.

'Hilbert defends the use—in order to prove any mathematical theorem— of the so-called rule of logic ... *Either a statement or its negation is true with no third possibility*. But logic is not a set of commandments handed down from heaven—or a rock you may skip from one side to another of the Stream of Mathematics. No! Logic is a set of (possibly) useful rules of thought or argument, abstracted from our experience with mathematics— principally from number theory and arithmetic. *As pointed out in my doctoral thesis, we are allowed to abstract a limited form of this [rule of logic] from our experience with* finite collections, *since we know that, should we doubt the conclusion derived by means of this, the conclusions can always be checked out by* finite sequence of trials! *But we have no such experienced assurance of the safety of this procedure with* infinite collections! *So we cannot ... devise* an existence (*or* reductio ad absurdum) proof *to posit properties of infinite collections—of which we know so little!*'

But Hilbert the Mouse was now jumping up and down on the bank in anguished distress. 'To deny ... *the law of the excluded middle*, to the mathematician is like taking the telescope from the astronomer, or the microscope from the biologist! Don't listen to Brouwer! You've seen what a shambles there would remain of mathematics if you try to follow his *constructivist methods!*'

Croaked the Frog: 'It wouldn't be a shambles! But it also wouldn't be the Cloud-Cuckoo-Land you create with your *Axiom of Choice* ... Do you know what delusions you can arrive at by means of the Axiom of Choice?'

'No,' they cried. 'What?'

'Behold! By appealing to the Axiom of Choice, I can claim that I can dissect the *very moon* into but *five parts,* then put it back together and pop it into my mouth—like this fly! Gulp! ... This axiom assures us that every collection—be it finite or infinite ... is as countable as the ripples of this cool, sweet stream!

'Shazammm!' gasped the creatures. 'And how do we do that?'

'Hump! Don't ask Hilbert the Mouse, or Russell the Rabbit, or Zorn the Muskrat, for they don't know how ... They claim that the Lord is their *Thesis Adviser* and has neatly filed away all finite and infinite collections—into file drawers. Whenever an example, or counterexample, is required in a proof, they've only to appeal to the Lord, their Thesis Adviser, and He will beneficently retrieve it from a neat file of *ordered* manila folders and dispatch it to earth—in the latest mathematical journal.'

'Shazamm!' gasped the creatures.

'Shazamm, nothing! Hoakum!' At this moment, Brouwer filled a great bubble with air in his throat, and boomed it forth in angry explosion.

'If a criminal, in committing his crime, manages to destroy the only evidence that could be used to convict him—must I, therefore, accept the verdict that he's innocent? No! Either he can be proven *guilty as charged,* or *innocent of charge,* or we rest with the verdict of *unproven by existing evidence.* Similarly, in mathematics, a thesis can be proven or disproven *constructively, or we must declare it unproven.* Brek-kek-kek!'

Other proponents of differing schools of thought would quarrel for hours on the banks of the Stream of Mathematics ...

Weyl the Badger, long considered a disciple of Hilbert the Mouse, was so persuaded by the arguments of Brouwer the Frog that he ventured the following pessimistic opinion: 'We must learn a new modesty. We have stormed the heavens but we have succeeded only in building fog upon fog, a mist which will not support anybody who earnestly desires to stand upon it.' ...

Eventually, the different facets of the controversy factioned the creatures of the region into many little bands of quarrelling doctrinaires: Formalists, Logicists, Intuitionists, Neo-intuitionists, Finitists, etc. It just wasn't possible to keep up with the issues and actions percolating out of so many storm centers. So, the Grasshopper decided it was time to bring the controversy to focus by staging a great *debate* between the two principal (and original) adversaries, Hilbert the Mouse and Brouwer the Frog.

On the morning of the great event, all the creatures who lived along the Stream of Mathematics, and many from the neighboring Forest of Logic, were crowded under the great oak tree, in the Academic Grove, to hear the opening arguments.

But the discussion was chaired by the Ant, who announced that his neighbor, the grasshopper, unexpectedly called away, had asked him to substitute. So they began.

Within an hour, the fervour of the two speakers, and the intellectual passion surging through the crowd, set off an angry antiphonal exchange between the two proponents of totally different philosophies of mathematics.

Suddenly, into the midst of the shouting, there sprang the giddy Grasshopper, ablaze with excitement: 'Wait! Hold it! This is even better than a debate! Gödel the Fox has achieved a completely unexpected result: He has proven that *you cannot prove the consistency of arithmetic!*

'What?' exclaimed a stunned Hilbert the Mouse.

'Oh, did he really?' boomed a doubtful Brouwer the Frog.

'Yes, he did, Brouwer. Indeed, indeed, Gödel the Fox used methods you cannot quarrel with. You see, the tools of his proof are constructed according to a model he abstracted from the structure of the natural numbers, your paradigm for all of mathematics. And—Hilbert, my friend—Gödel proved that any *axiomatic system*, such as you favor (and use) must *either fail to capture all essentials of arithmetic, or else it must already have penned into it at least one of the wolves of contradiction!*'

The pervasive silence that smothered the assembly under the oak tree was broken by the voice of the practical Ant: 'Then—the debate is over. So, let us get back to our labors.' ...

And in the Meadow, led by a (classical mathematical) shepherd, the sheep bleated as they ran:

The Lord is my thesis adviser; I shall not err.
He arranges for me to be published in the
respectable journals; he teaches me how
to use the *reductio* argument.
He enshores my validity; he leads me by
the classical logic, for the truth's sake.
Yea, though I walk through the valley of
the existence proofs, I will fear no contradiction;
for he edits my work.
The Axiom of Choice and Zorn's Lemma, they comfort me.

He invites a colloquium on classical analysis,
for my participation, in the absence of the constructivists;
He frequently and approvingly abstracts me
in *Mathematical Reviews*;
my reputation flourishes internationally.
Surely, honors and grants shall follow me
all the days of my career,
And I shall rise in the ranks of the Department,
to Emeritus.

Amen.

## THE FOUR-COLOUR CONJECTURE

*You say you got a real solution*
*Well, you know,*
*We'd all love to see the plan.*

JOHN LENNON AND PAUL MCCARTNEY

In our earlier revelations about the extraterrestrial mathematicians, you recall that the aliens' notion of proof was such an efficient one because they had powerful computers at their beck and call. Many people see the work of the mathematician to be proving that one thing is equal to another by 'pure thought' aided only by a pencil and paper. Yes, computers are useful, but only for generating lists of numbers or doing repetitious tasks like evaluating electricity bills. No one imagines that they actually do what human mathematicians are supposed to do—think. This used to be true; but things are changing. In 1976 a computer was employed to establish the truth of one of the classic unsolved problems of mathematics. The fact that it was used, how it was used, and what our reactions should be to the results, have extended our notion of proof a little in the direction of the extraterrestrials. This is the story of how the 'Four-Colour Problem' became first the 'Four-Colour Conjecture' and then the 'Four-Colour Theorem'.

Suppose you need to colour a map of the counties of England so that each region is to be in a different colour from its neighbours (see Fig. 5.8). What is the least number of colours that can do the job on a flat sheet of paper if neighbouring regions don't ever meet just at a point? This problem first occurred to an English student working in London in 1852. Francis Guthrie had been trying to colour a map of England subdivided into counties. After much trial and error he wrote to his younger brother asking if 'every map drawn on the plane [can] be coloured with four (or fewer) colours so that no two regions having a common border have the same colour'. Guthrie's brother failed to come up with a solution but quickly passed the problem on to other more able mathematicians and it became known among a small group of mathematicians as an entertaining unsolved problem, albeit one with little obvious consequence for any other parts of mathematics. It only became widely known after 1878 when the prolific mathematician Arthur Cayley brought it to the attention of mathematicians at the London Mathematical Society and described it in the Society's journal. The following year a London barrister and amateur mathematician, Arthur Kempe, published what he believed to be a proof that four colours suffice to colour the map. Although Kempe's proof was extremely novel, it contained a subtle gap which was

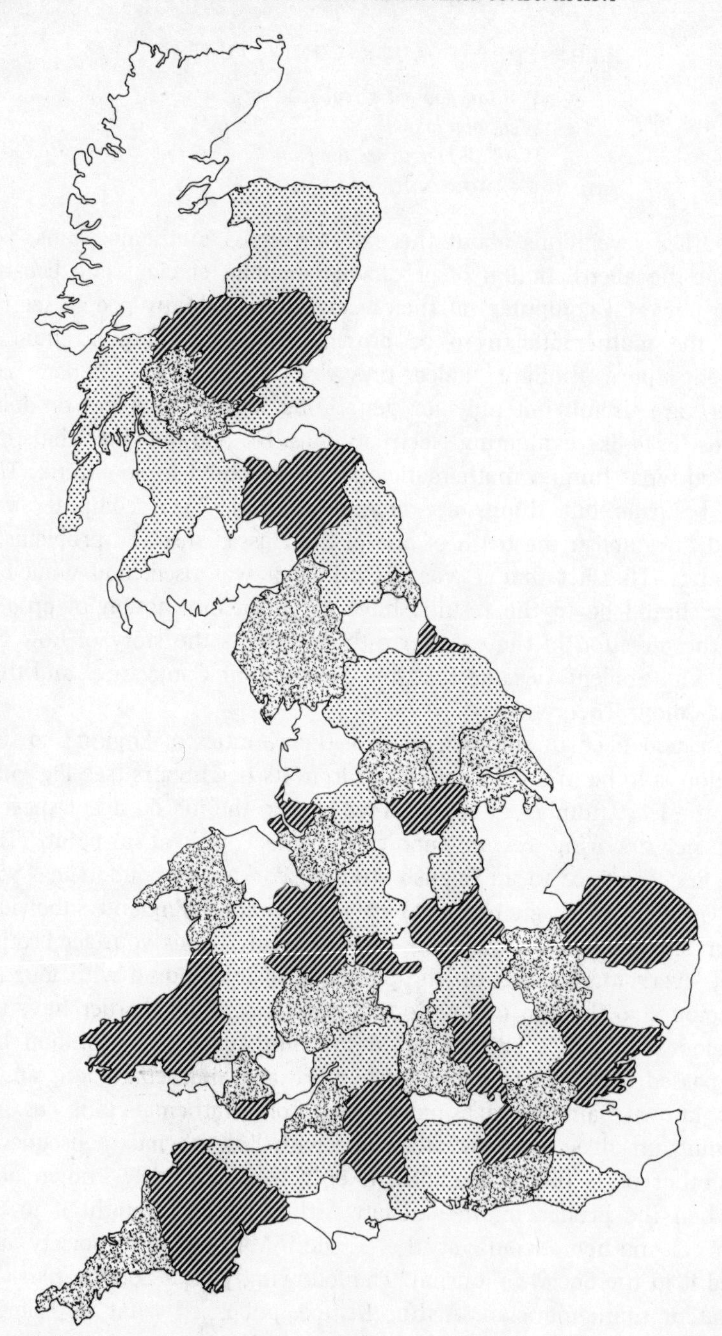

**Figure 5.8** A map of mainland Britain in which four varieties of shading suffice to 'colour' all the counties.

only noticed eleven years later by Percy Heawood. This rejuvenated interest in what was thought to be a solved problem and more flawed proofs and their subsequent critiques were published in the next few years. Heawood did something else characteristic of mathematicians. He generalized the problem and asked how many colours would be required to colour a map drawn on other types of surface, like a sphere, a ring doughnut (what mathematicians call a torus), or a sphere with several handles attached (see Fig. 5.9). Progress then came to a halt until 1968 when the problem was solved for the unusual shapes like the torus and the Klein bottle—where 7 colours suffice—but *not* for the map on a flat plane or a sphere although it was appreciated that they require the same number of colours. Unfortunately, the solutions to the other problems didn't seem to help with the original problem.

Then, in 1970 Wolfgang Haken followed up some earlier suggestions of Heinrich Heesch dating from the 1950s by reducing the problem to a list of the possible types of map in which the conjecture of four colours could conceivably fail. This list is so extensive that there is no possibility of working through it case by case using pencil and paper and pure thought. So, he considered what impact the computers of the time might have upon the process. Unfortunately, given the way he had enumerated things it

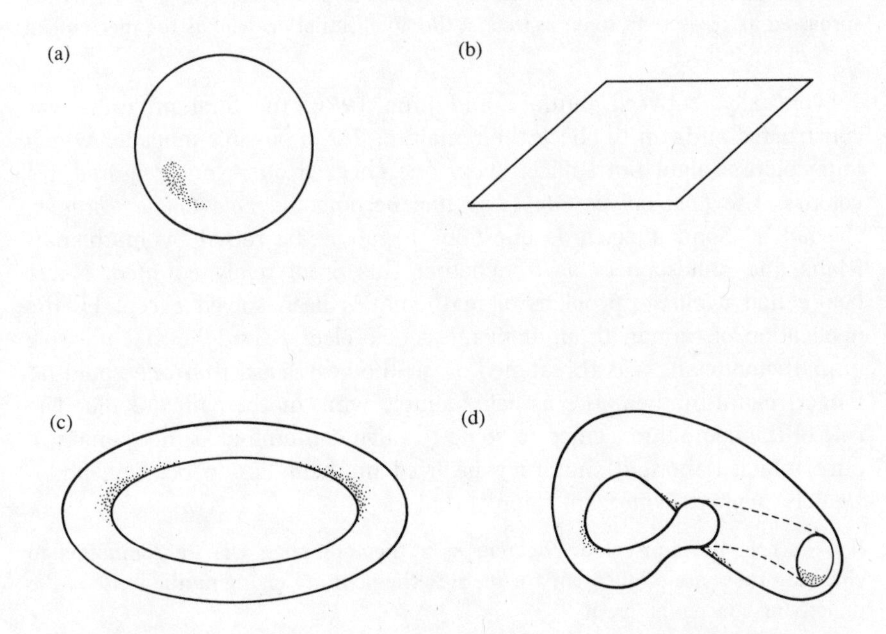


**Figure 5.9** (a) A sphere; (b) a plane; (c) a torus; (d) a Klein bottle.

looked as if more than a hundred years of computer time would be necessary to examine every possibility. A little more human thought was necessary, and between 1972 and 1974 Haken, with the help of Kenneth Appel, began to streamline a computer program which could carry out this series of checks in a sensible period of time. It took over a year of trial and error, learning from the way the computer program followed instructions and testing for possible maps requiring more than four colours, before the whole venture began to appear possible. Appel and Haken's description of this period is very interesting because it reveals how their interaction with the computer program unexpectedly taught them things about the problem which they did not previously know. It assumed the role of an almost human collaboratcr rather than merely a calculational tool; in effect, it was an extension of their intuition to encompass many more examples than they would normally be able to hold in their minds for examination at any one moment. They reveal that

> the program began to surprise us. At the beginning we would check its arguments by hand so we could always predict the course it would follow in any situation; but now it suddenly started to act like a chess-playing machine. It would work out compound strategies based on all the tricks it had been 'taught', and often these approaches were far more clever than those we would have tried. Thus it began to teach us things about how to proceed that we never expected. In a sense it had surpassed its creators in some aspects of the 'intellectual' as well as the mechanical parts of the task.

Eventually, between January and June 1976, the final program was constructed and run to check the remaining list of possible maps for which four colours might not suffice. Every one checked out. None required five colours. The *Four-Colour Conjecture* had become the *Four-Colour Theorem*. Or had it? Soon afterwards questions began to be raised by mathematicians and philosophers as to whether this proof really counted. Never before had a classic problem of mathematics been solved except by the application of human thought. Some people clearly thought that the role of mathematicians was threatened or, at the very least, that one could no longer regard mathematics as being simply what mathematicians did. The role of the computer seemed to some to have contaminated the domain of pure unaided thought that they believed mathematics to be. One philosopher, Stephen Tymoczko, claimed that

> If we accept the four-colour theorem as a theorem, then we are committed to changing the sense of 'theorem', or more to the point, to changing the sense of the underlying concept of 'proof'.

Why would he claim such a thing? Because the proof has not all passed through a single human mind, or a group of human minds, it is being regarded as inadmissible. There are a number of reasons one might speculate to be the source of such scepticism. There is the deep-seated feeling that we often see surface in opposition to the goals of the quest for 'Artificial Intelligence', that the human mind does 'something'—we don't quite know what; but it has something to do with that quality of self-reflection that we call consciousness—that no machine can emulate. There is also the more practical concern that the computer may contain some programming error or carry out some operation which rounds-off a number in some way that has disastrous consequences for the truth of what it ultimately concludes. When human mathematicians carry out a calculation they may make some approximation at stage one but they have some feeling for what is being done and an overview of the structure of the argument as a whole. Appel and Haken say that

> A person could carefully check the part of the discharging procedure that did not involve reducibility computations in a month or two, but it does not seem possible to check the reducibility computations themselves by hand.

And, in fact, when their article was sent to a mathematics research journal and refereed anonymously by other mathematicians to see if it was correct in so far as they could judge, they checked part of the paper by using a separate computer program of their own as an independent check. This is rather like a team of scientists checking the claim of another by repeating their experiment in a different way and obtaining the same answer. These checks can never prove that the answer is correct, of course; the checking program, just like the follow-up experiment, might contain the same error, or even a new one, buried within it. Actually, these concerns are hard to sustain if we look at them closely. The risk of error is always present. Just because a proof is carried out by an unaided human mathematician is no guarantee that it is correct. Mathematicians, like everyone else, make mistakes and a number of papers published by mathematicians are corrections to other ones, just as had been the case in the early history of the Four-Colour Problem. There exist proofs of other mathematical theorems that are many hundreds of pages in length and it is very difficult for one to be totally sure that there is no flaw or gap in the proof. Cross-checking of the consequences of the result with other results and facts is of course necessary and would often cast suspicion upon a falsity that had been mistakenly 'proved' true. But even this is no defence against the true result that is established by an incorrect proof. The length of the proof is too long to be checked manually. There may be a flaw in the program but one does not hear mathematicians raising this concern with any great conviction. This may be because they feel happy with the result. It was expected that the Four-Colour Conjecture was correct. Of course, for it to have been proved incorrect—even by computer—we might have had to have been shown a counterexample in the form of a particular map which required more than four colours and there could have been no possible dispute over that. The only alternative would have been some demonstration that its truth would create a logical contradiction (this is the sort of non-constructive proof that Brouwer would have regarded as inadmissible). But if the latter demonstration had been carried out, it would have provoked a thorough computer search for counterexamples which must exist. Hence, if the Four-Colour Conjecture had been false and the computer searched through a collection of maps which might contain a counterexample before presenting us with one, then no objections of fallibility could be raised against the computer because we could check the counterexample for ourselves as well.

In fact, one feels that the psychological reaction to computer disproof might be rather different to that of computer proof. What seems of the greatest weight in mathematicians' reactions to the Four-Colour Theorem is something about the type of proof used. It teaches us something very important about mathematics and mathematicians. Contrary to popular mythology, and the impression created by the more formalistic philosophies of mathematics, mathematicians are not primarily interested in the collection of all truths, or theorems, that can be established from starting axioms. There are millions of possible extensions that could be made to the collection of known mathematical truths, but most are uninteresting to the mathematician and might not be accepted for publication in a research journal. Newness is necessary but not sufficient to attract the interest of mathematicians. And for this reason there were some mathematicians who were deeply disappointed by the computer-aided proof of the Four-Colour Conjecture. The mathematicians Martin Davis and Reuben Hersch wrote about their sense of bathos in the following graphic terms:

> When I heard that the four-colour theorem had been proved, my first reaction was, 'Wonderful! How did they do it?' I expected some brilliant new insight, a proof which had in its kernel an idea whose beauty would transform my day. But when I received the answer, 'They did it by breaking it down into thousands of cases, and then running them all on the computer, one after the other,' I felt disheartened. My reaction then was, 'So it just goes to show, it wasn't a good problem after all.'

What these commentators are getting at is that mathematicians are usually more interested in the chase than the kill. They are keen to find some novel type of argument in a proof that breaks new ground or recognizes some unsuspected link between different problems that reveals their solutions to be linked in some deep and subtle way. The Haken–Appel proof possessed none of those inspiring ingredients. It was a demonstration of technique rather than insight, of patience rather than inspiration. Indeed, one has often encountered the case of the 'heroic failure' in mathematics where an incorrect proof has contained some vital spark of insight that attracts the interest of mathematicians and forms a springboard for others to find a correct proof, not just of the problem in question, but of other problems as well.

This focus upon the novelty and content of a proof places many philosophies of mathematics in an awkward position. When one looks to see what mathematicians actually *do*, rather than what others say they do, one discovers that they do not spend very much of their time either individually or collectively doing the sorts of things that commentators say they do. They don't sit around thinking up new systems of axioms, or deriving all the theorems they can from a given set of axioms. They don't very often prove new theorems. Most of the literature of modern mathematics does none of these things. Instead, it is populated by works of refinement and distillation which find simpler, shorter, and clearer ways of proving results which are already known. The essential interest of such work is in the structure of the proof being used, whether certain results are within the scope of particular simple methods of attack, or whether they really require the advanced and novel methods that were used to prove them for the first time. If one focuses upon any great result of modern mathematics, one will usually find that it has a history which begins with its proof via some lengthy and difficult process that only specialist mathematicians working in that branch of the subject would be able to follow. Then other simpler proofs will appear showing that some of the pieces of the original proof were unnecessary or long-winded. This process will continue for some time with the agreed proof becoming shorter, and harder, as more sophisticated techniques are employed to obtain the result in other ways. Then someone will produce an 'elementary' proof. This will not necessarily be an easier one from the point of view of the casual mathematical reader and it will certainly be much longer, but it will use much simpler mathematical ideas, ideally, only those that university students of mathematics would be familiar with so that it can enter their curriculum. This process of refinement has as its goal a sort of 'trivialization' of all the results of mathematics—a reduction of their proof to the application of simple well-tried ideas. Indeed, one could characterize the whole quest for a human understanding of the Universe in such a way. We seek to reduce the inexplicable and the novel to the known and the familiar. Of course, this sometimes creates conflict with other ways of viewing the world. The poet or the writer often seeks to do just the opposite—to take the mundane and familiar and expand its meaning in ways that render it mysterious or unusual and worthy of continued study or contemplation.

The evaluation of mathematical proofs is not dissimiliar to someone making a journey through the heart of England. There is a fastest route but many motorists might not want to follow it, choosing instead some more interesting or scenic route. Mathematicians are not merely interested in mathematical truths, they are also interested in the ideas and procedures that must be used to establish them. Returning to the Four-Colour Theorem, we can be sure that mathematicians will continue to seek proofs of this result which make no use of the computer. They will still hold out hope for the sort of proof they wanted to see in the first place, a bold new idea that reduces the complexity of the problem to a much simpler set of obvious ideas.

## TRANSHUMAN MATHEMATICS

> *Yes, yes, I know that, Sidney ... everybody knows that! ... But look: Four wrongs squared, minus two wrongs to the fourth power, divided by this formula, do make a right.*
>
> GARY LARSON

The use of computers as an aid to human thought may one day completely alter the principle of 'proof' that is adhered to; at present it merely alters its practice and its speed. Human mathematicians map out a pattern of proof that they want to see carried through. If the process is going to take an enormous length of time, then they can call on the computer to do it faster so long as the steps involved have a certain type of predictable repetitiveness. This has interesting consequences. Suppose we imagine the great ocean of mathematical truth stretched out in front of us. Then Gödel has taught us that there is only some part of it that we can discover to be true or false by use of axioms and logic. We do not know the shape of that boundary or how much of the ocean it encompasses. We could also divide the ocean in another way. There will exist some truths which are so deep that they require proofs of such enormous length from our starting axioms that no human mathematician could reach them unaided (Fig. 5.10). Truths in this part of the ocean require computers to reach them, but if there exists a fundamental limit to the accuracy and speed of computation then there will exist another

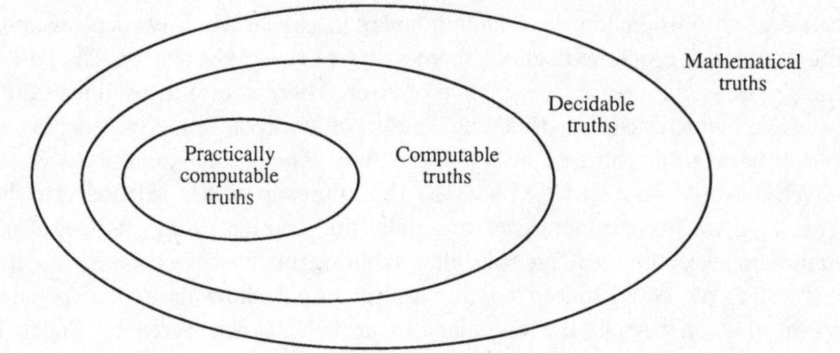


**Figure 5.10** The realm of mathematical truths with its division into decidable and undecidable statements and a further subdivision of decidable truths into those whose shortest demonstration places them beyond human reach during the age of the Universe.

boundary which separates the realm of machine-attainable truth from what lies beyond.*

If one regards the ocean of mathematical truth as a pre-existing thing which we gradually discover, then this picture puts human mathematical activities firmly in their place. We are confined to discovering what might be the shallowest of results: the short links in the network of logical connections. All the greatest and deepest truths will require more than a human lifetime,† maybe more than the lifetime of our planet or even of the Universe itself to discover. Formalists don't see the ocean as sitting out there to be charted. Rather, we set up a few little tributaries here and there and then follow the waters as they flow ever wider and deeper. There is no end to the process of extension, and in our lifetime, or that of our fastest computers, there would only be time to make a small fraction of the explorations that are possible in principle.

It is apparent that as we start to prove truths that lie close to the boundary of what is practically attainable the criteria of proof must weaken. Just as with the Four-Colour Problem, there are difficulties in carrying out independent checks of what has been done; there are limits to the length and volume of logical argument that human minds

* In fact, there are also many truths which can be established by short proofs but which could not be reached by a computer. There are certain operations, called 'computable' which any algorithmic device can carry out, but not all mathematical operations are computable, as was first shown by Alan Turing in the 1930s.

† With regard to logical argument that lies beyond the comprehension of single or small groups of individuals, it is interesting to draw attention to several lengthy and extremely complex business fraud trials which have created a network of fact and deduction that is beyond that which jurors or judges can apprehend as a coherent whole. Only a relatively small increase in initial complexity might render human adjudication rather unreliable.

can deal with; the chances of finding errors in purported proofs depend upon the number of people who check them or try to re-establish them. The longer the proofs so the greater the chance of error. There is a form of 'uncertainty principle' which tells us that the length of a proof times our degree of certainty about its truth is always bigger than some positive quantity.

What would happen if we adopted the approach of the extraterrestrials? That is, if we just explored experimentally and counted things as true if we found no exceptions to the rule after running billions of examples on the computer. We could indeed do this and it would allow us to find possible truths that lie beyond the boundary of provability discovered by Gödel. It would also enable us to expand the boundary of attainable results considerably because we would be limited by the longest computer search that we could carry out through examples rather than by the longest logical argument that could be run through. However, it is not quite as simple as that. Mathematical truths have a habit of being annoyingly subtle. One might check a huge number of cases and suspect a general truth to hold only for there to exist a counterexample waiting far upstream.*

But more worrying is the problem the computer has in discovering what examples to test out in its search for generalizations. Unless one has other ways of studying mathematical structures, we will not have a way of generating lots of possible truths for the computer to check. Of course, our extraterrestrials were not imagined to be stupid; they were well aware of this and knew all about the type of proof that we demand before accepting something as true. They just found this a very restrictive way of exploring the realm of true statements. It was convenient for the deduction of simple truths. But they didn't see why they should wait around for the same methods to prove the ultimate validity of all those conjectures which were far too elaborate for checking by hand (or tentacle?).

## NEW-AGE MATHEMATICS

> *On the basis of my historical experience, I fully believe that mathematics of the twenty-fifth century will be as different from that of today as the latter is from that of the sixteenth century.*
>
> GEORGE SARTON

Mathematics seems the ultimate bastion of conservatism to the outsider. The methods of proof go back thousands of years and everyone seems happy about what is and what isn't mathematics. Or do they? We have

* Interesting examples of this sort are provided by the attempts to find a magic formula which will generate all the prime numbers. One such example is $n^2 - 79n + 1601$. If any natural number from 1 to 80 is input for $n$, then this formula generates a prime. But it fails at 81.

already seen something of the fuss that grew up over attempts to redefine mathematics in very restrictive ways. The first of these, formalism, was really just an attempt to tidy things up after the fact and to put the entire enterprise on a firm foundation. The other, intuitionism, was something else entirely: it sought to alter the character of mathematics, change what mathematicians do and evict a large fraction of what they had always been doing from the collection of statements and proofs called 'mathematics'. But what about the future; could there be any future crisis as to the meaning and nature of mathematics? Could the subject split into factions pursuing different methods for different ends? Is the traditional method of proving statements to be true by a stepwise process of logical deduction simply unquestionable?

It is quite possible, probable even, that mathematics is in the process of evolution into something rather different to what it has been and what it is fondly imagined to be by many of its users. At the beginning of the chapter we told a story about contact with an imaginary alien civilization whose way of doing mathematics was rather different from ours. Or, to be more precise, their notion of what constituted a convincing proof was different to ours. They were happy to do *experimental mathematics*. That is, they exploited their sophisticated silicon technologies to build machines that were able to search for possible relationships between numbers and other more abstract mathematical quantities. This procedure has greater scope than logical deduction because it can find evidence of those true but undecidable propositions which Gödel taught us must exist.

The recent growth in the capability of computers makes it very likely that something like this will occur in the realm of terrestrial mathematics. Maybe one day mathematicians will build truth-seeking super-computers in the way that particle physicists now build accelerators. Indeed, it is already beginning to happen. Computers have started to play a role in the process of exploring mathematical possibilities and in proving their truth. Some of these applications are predictable. They are mundane in the sense that they simply search through very long lists of possible exceptions to the rule being asserted by the theorem that we want to prove. In this sense it might be said that the computer is playing the same role as a pocket calculator, a slide-rule, or a set of log tables; it is a labour-saving aid rather than an extension of the conceptual intelligence of the brain. This is the type of computer-assisted proof that was used to transform the the Four-Colour Conjecture into the Four-Colour Theorem. One wonders what an intuitionist like Brouwer would have made of it. It is not easy to say. On the one hand he would have been pleased by the manner in which programming the operation of the computer guarantees that all its 'proofs' are step-by-step demonstrations starting from the natural numbers. No idealistic notions can seemingly arise: no actual infinities. But maybe this is too simplistic a verdict. First, there is the problem that human intuition, which Brouwer regarded as the vital spark that leads to the creation of mathematics, is not present in the computational process. Could the computer have some other form of intuition that is not human? This might appear quite plausible, but surely the computer is programmed to carry out those arithmetic processes that humans are familiar with. Thus one might claim that computers are merely being used as implementers of human intuition. They could only differ in substantive ways if the speed and quantity of information that can be coordinated and manipulated in one step could lead to qualitatively new levels of complexity which we could interpret as a higher level of 'intuition'. However, computers exist which can arguably create things: they can write music and produce patterns and high-level organization. They are no longer merely glorified pocket calculators. Yet, Brouwer would at root have been most disturbed by the fact that some computers are analog machines. That is, they model the process of arithmetic by some physical process like the flow of electricity. There is nothing unique to electronic computers about this; a pendulum clock counts the swings of a pendulum hanging in the Earth's gravitational field and calls this a measure of time. What this implies is that the physical world is so constituted that by manipulating it in appropriate ways it can be coaxed into producing sequences of events which simulate the counting process. In fact, it can be made to simulate the operation of much more complicated mathematical operations. Such processes are called *computable functions*. It is a famous result, first established by Alan Turing, Alonzo Church, and Emil Post, that not all mathematical operations that we can conceive of are computable in this sense. There exist *non-computable functions*. This would have been a problem for Brouwer because he believed that the laws of Nature are imposed by the mind upon reality. Thus he would have found himself in the very awkward position of having to believe that the process of arithmetic, a fruit of human psychological evolution, was being performed by the evolution of a particular physical system in accord with laws of Nature which are merely imposed upon that part of reality by the mind. There is a possible escape from this impasse. If Nature is intrinsically mathematical, then one might expect successful evolution to select for mental processing that was closely tuned to the mathematical aspects of Nature. This would lead to the 'intuition' that Brouwer appeals to and its source would be the intrinsic mathematical nature of the universe. But mathematical nature is also the source of the fact that bits of the world can be arranged so that they simulate the operation of mathematical functions and hence of the computer. Whilst there are indeed uncomputable functions which the machine could not mimic, such operations transcend its capabilities by the very fact that they cannot be carried out in a finite sequence of step-by-step procedures. So, these would not be mathematical operations which Brouwer would be willing to admit as part of mathematics because they could not be established from the human intuition for the natural numbers in a finite number of counting steps.

In recent years the concept of experimental mathematics has begun to take on a new and more adventurous complexion than that revealed by the proof of the Four-Colour Theorem. Computers have been used to explore the realm of mathematical truth in systematic and haphazard ways. New types of computer program now exist which do not merely manipulate numbers. They do algebra, manipulating strings of symbols, transforming them into other strings, gathering up all the terms of the same type into a single collection, dividing them up into neat forms which cancel out other identical collections elsewhere in the equation. As a result there are areas of mathematics whose development is being directed through the use of such programs by mathematicians. These areas usually focus upon very intricate and complicated structures which run through an evolution in time according to definite rules of transformation. The result is rather like a movie film in which each frame of the movie has been created from the previous frame by the operation of some particular rule. Whilst one could calculate these changes by hand, it would be very slow and laborious and one would fail to appreciate the deep structures or repetitions that can arise over long periods of time. An example is the 'Game of Life' invented by the English mathematician John Conway shown in Figure 5.11.

An understanding of the consequences of Conway's simple rules were built up through hundreds of hours of experience just playing the game, or alternative versions, on a 'Go' board at tea and coffee breaks in the Cambridge University mathematics department over a long period of time. Subsequently, some proofs were found of properties of the game but on their own they fail to reveal all but a tiny fraction of the game's deep and fascinating properties. With the aid of a simple computer program, one can explore the consequences of many generations of the evolution of the 'Game of Life' in a way that would not be possible in a human lifetime of shifting counters on a 'Go' board.

The Game of Life was a prototype for a whole panoply of similar algorithmic 'games' that evolved from one generation to the next on the computer screen according to some rule. These have become known as

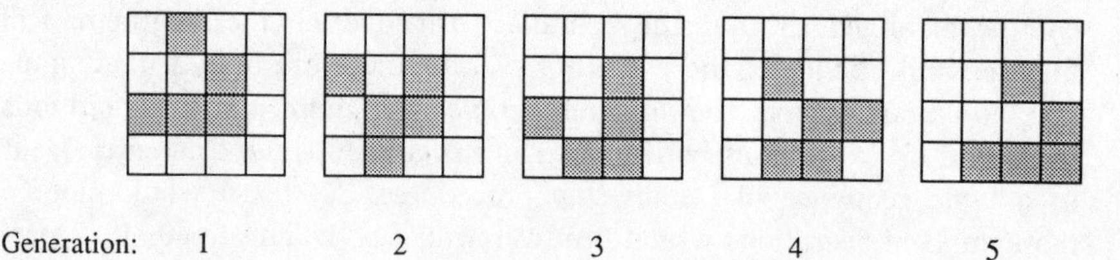


**Figure 5.11** John Conway's 'Game of Life': a generation of live cells consists of a distribution of points on the board. The next generation of live cells is determined from it by the application of the following two rules: (1) A cell remains alive if it is neither 'overcrowded' nor 'lonely'. This requires that it possess either two or three living neighbours. (2) A dead cell remains dead unless it has three living neighbours in which case it is rejuvenated. We have shown the evolution of a simple starting configuration through a number of generations according to these rules.

'cellular automata' and are a subject of considerable interest amongst mathematicians and scientists as they seem to mirror many fundamental processes in Nature. Their most persistent common feature is that which defines the phenomenon of 'chaos'. This means that a very small change in the starting state of the game would lead to very, very different configurations after a number of moves. This sensitivity to starting configurations betrays an underlying aspect of the transformation process for creating one generation from the last one: it takes points that are very close together and moves them very far apart in the next generation. This is rather similar to what goes on when you knead dough. If you marked two points in the dough which were very close by before you started kneading then, after just a few turns of stretching, squashing, and dividing the dough you would find the two marked points widely separated. The effect on two adjacent playing cards after shuffling the pack is very similar. Because of the complexity that can be created after just a few evolutionary moves in a cellular automaton, the ability to represent what is happening by good computer graphics, perhaps colour-coding different aspects of the pattern from generation to generation, is a vital aspect in the apprehension of what is happening. It is no coincidence that the entire subject of 'chaos' and the migration of applied mathematicians and scientists away from the study of orderly and simply predictable processes to the detailed scrutiny of disorderly and unpredictable ones has accompanied the revolution in powerful small computers. The development of this branch of mathematics is just not possible without such aids. They are used to explore the outcomes of whole ranges of possible algorithms which display sensitivity to their starting state. This subject has gone so far in such a short time that one might hazard the prediction that it will always be dominated by the habit of experimental investigation. Whilst its practitioners are not uninterested in conventional proof or theorems about all possible cellular automata, they realize that the sheer diversity of these systems ensures that any general theorems which tell us about the properties of all of them must necessarily tell us something rather weak and unspecific in order that it be a shared property of all of them.

The intensive study of chaotically unpredictable systems has recently brought to light a fascinating connection between the inability to predict *in practice* which chaos creates, and the inability to determine the solution of a problem *in principle* which Gödel undecidability forces us to contemplate. For it can be shown that there is no general algorithmic criterion which would enable us to determine whether any given system is chaotically random or not. Moreover, a whole host of related questions about the evolution of very complicated changing systems (like the motions of bodies in the solar system, for example) are also, in general, undecidable. This is not the case for all chaotic systems, just as there are a host of statements about arithmetic that are decidable; but one cannot generate a catalogue of the ones about which there is undecidability. Another tantalizing curiosity is the fact that we can always write down an elaborate set of equations which will predict how any given Turing machine, which might just be a variety of electronic computer, works. At the most basic level these equations might describe how electric current flows through the components of its circuits. In this way one has created a 'model', or an 'analogue', of the Turing machine in the form of a mathematical system of equations. One can now solve these equations by means of pencil and paper or even by running them on a different computer. But one is at liberty to alter the rate at which time flows in the paper version of the Turing machine's operation (for example one might change the time to the tangent function of the time). This allows us to compress an infinite amount of the real time that would be required to run a particular program on the Turing machine into a finite amount of time in the evolution of our equations on paper or the analog machine. This would enable us to render finitely computable operations that would take an infinite amount of time on the original Turing machine. It is possible to provide a recipe for constructing an analog computer which can, in a finite amount of running time, decide the truth of a proposition that requires systematic search through an infinite number of cases by the original computer. If such a device could be fabricated then one could construct analogue simulations that have very different sorts of behaviour depending upon whether the original computational problem was decidable or not, and by making those behaviours sufficiently distinct one could settle the decidability of the original problem by observation.

There are some who fear that this runaway use of experimental mathematics will just result in the eventual accumulation of a rag-bag of peculiar discoveries that do not dovetail together to create any coherent form of general knowledge. Worse still, it may give rise to a corpus of unreliable knowledge, founded upon trends and upon factors that happen to have been shared amongst the particular collection of examples that we happen to have studied, but which would be found false if only one looked in the right place for the counterexamples.

There is another type of experimental mathematics which is the most unusual. It is very new but conceptually the most powerful and the most likely to change the face of mathematics over a range that is not just confined to the traditional domain of 'applied' mathematics. It is most easily characterized as the mathematicians' version of the quest for virtual realities which has begun to dominate many aspects of popular culture. One recalls the early popularity amongst children of 'role-playing' games where once they might merely have indulged in charades or amateur dramatics. The computer revolution and the creation of very-high-resolution graphics has enabled this to become even more realistic or bizarre with the creation of simulated environments and realities. These have many legitimate purposes, from training airline pilots to providing fairground entertainment, but one senses that they may become a new paradigm for our picture of Nature, taking over from the simple modern image of the Universe as a great abstract computer whose software rules are called the laws of Nature which run on a hardware composed of the elementary particles of matter. Now mathematicians have begun to exploit this possibility of creating alternative 'virtual' realities as a way of demonstrating the truth of mathematical conjectures and exploring the possible truth of others. The way one proceeds is to create a video or a movie of a world which is governed by the particular axiomatic system of mathematics that one is interested in. For example, it might be a very unusual non-Euclidean geometry. The geometrical structure of this world can be simulated on the computer using the latest graphics techniques and one can just explore it to see what sort of things are true within it. It will be a world of unusual patterns which one can situate oneself within, or orbit around to inspect its intricate substructure. This new approach to mathematics—the simulation of mathematical structures—has played an important role for many years in the physical sciences. Astronomers program computers to simulate the clustering of galaxies in model universes and then position imaginary observers within the simulation who carry out observational surveys of their 'sky' in a manner analogous to that used by astronomers with real telescopes with the same biases and limitations built in. The resulting catalogues of observations can be compared with those gathered by real observers to test whether any of the model universes capture the essential pattern of galaxy-clustering found in the real sky. Mathematicians have recently 'proved' a long-standing conjecture about the properties of a non-Euclidean geometry by making a video of the world that it would create and then observing that the required property does indeed hold.

### PARADIGMS

*Brother, can you spare a paradigm?* ANONYMOUS GRAFFITO

The predictions that we have just been making about the future course of mathematics have been linked to trends in contemporary culture. This is by no means a baseless speculation. History teaches important lessons concerning the way in which our image of Nature and the mode in which it is best represented take their cues from cultural and sociological influences. From the times of the Pythagoreans to the present we can pick upon a series of paradigms which have temporarily held sway as the dominant image of the Universe around us. For the early Greeks, caught up with their first discoveries in geometry and the systematic motions of the heavens, the world was a geometrical harmony. Later, the school of Aristotle, with its keen interest in the study of living things and their apparently purposeful actions, would regard the Universe as a great organism—a living thing with a goal and purpose all of its own. Thousands of years later, for Newton and Huygens, living in an age of new mechanical devices, the Universe was itself envisaged as a great determined mechanism in which the future followed, predetermined like clockwork, from the present. For the Victorians of the Industrial Revolution, fascinated by the power and potential of machines and steam engines, the Universe was seen primarily as a great heat engine governed by the laws of thermodynamics which condemned it to a lingering 'heat death' of overwhelming thermal equilibrium. So, today, it not unexpected to find the 'computer' or the 'program' as central paradigms in our attempts to interpret the Universe in terms of lesser concepts. The growth of the personal computer industry and the ubiquity of the computer on the desks and in the minds of modern scientists has influenced our outlook on everything from the nature of human intelligence to the laws of Nature. But, by the same token, we must expect this cultural resonance to pass. In the future there will be new images seeking a place on the roster of paradigms. The three-dimensional video, the hologram, alternative realities are the new emergent concepts. What will they evolve into?

The march of paradigms we have just seen reveals the way so much of our thinking about the Universe is incubated by the ambient cultural medium in which it finds itself immersed. The same might be said of mathematical concepts. Let us explore what could be made of the notion that mathematics is a culturally derived activity.

If one treats mathematics as a cultural artefact then one is committed to recognizing it as a changing entity. It is more akin to a living species than a static subject. If one takes this starting-point, then one can proceed by applying the techniques of the sociologist to the development of mathematics. This position is easier to sustain, the earlier one looks at the growth of mathematics. It cannot be denied that the early intuition about geometry grew out of attempts to measure out plots of land, that counting derives its sophisticated aspects as a result of monetary transactions and trade. The Egyptians, the Greeks, and the Babylonians all display strong evidence of the growth of aspects of mathematics out of the activities of everyday life which were necessary to sustain and enhance the quality of life. One attractive aspect of this thesis is that it appears to offer a reasonable explanation for the utility of mathematics in describing the workings of the world about us. Because the problems which mathematicians work on, and the solutions that they arrive at, are imposed by the cultural influences around them, it is not surprising that mathematics will be found applicable to the sciences if there are enough contacts between the two. Moreover, the evolving nature of mathematics on this picture of things means that the applicability of mathematics is not something we can assume will remain for ever the same. Mathematics may begin to evolve in directions that do not have many useful applications. Suppose we look at the evolution of mathematics in a manner akin to the biological evolution of intelligent organisms. At first there is a phase where evolution proceeds slowly because organisms do not have the power to imagine or simulate the effects of their environment upon them. They can only learn the hard way: by experience. But a critical stage was reached in human evolution when the brain was able to support information-processing and organization at such a level of complexity that a new phenomenon was manifested which we call 'consciousness'. After this stage is reached evolution becomes a much more complicated and unusual process. Humans are able to pass on information to one another in many new ways and ideas can be preserved within the species. The zoologist Richard Dawkins has highlighted the significance of this new form of intellectual inheritance by dubbing the entities in which it deals 'memes' in contrast with the purely biochemical information that is coded within genes during the evolutionary process. The ideas that are found in the world of memes are able to propagate and change. If they are interesting or fruitful, they may give rise to whole philosophies of living or of human practice. Such influences are at present far more powerful factors in shaping the direction in which humanity is evolving than any genetic influence. One reason for this is that the influence of memes can be so rapid. One does not have to wait for a human lifetime before its effects can multiply and propagate. Dawkins writes:

> Examples of memes are tunes, ideas, catch-phrases, clothes fashions, ways of making pots or building arches. Just as genes propagate themselves in the gene pool by leaping from body to body via sperms or eggs, so memes propagate themselves in the meme pool by leaping from brain to brain via a process which, in the broad sense, can be called imitation. If a scientist hears, or reads about, a good idea, he passes it on to his colleagues and students. He mentions it in his articles and his lectures. If the idea catches on, it can be said to propagate itself, spreading from brain to brain ... memes should be regarded as living structures, not just metaphorically but technically. When you plant a fertile meme in my mind you literally parasitize my brain, turning it into a vehicle for the meme's propagation in just the way that a virus may parasitize the genetic mechanism of a host cell. And this isn't just a way of talking—the meme for, say, 'belief in life after death' is actually realized physically, millions of times over, as a structure in the nervous systems of individual men the world over.

Likewise, with respect to those memes which carry information about our environment which we use to develop lines of mathematical enquiry. Having reached a critical stage, mathematics can evolve much more rapidly by the medium of human cultural exchange than by simply waiting for the human condition or surroundings to suggest new problems that require solution.

## COMPUTABILITY, COMPRESSIBILITY, AND UTILITY

*The construction itself is an art, its application to the world an evil parasite.*

LUITZEN BROUWER

The mechanization of the step-by-step deductive process in the form of computer has important lessons to teach us about the meaning of the effectiveness of mathematics as a description of the physical world. We are used to thinking of computers as the products of rather sophisticated technological developments, but their essential nature is really rather simple. Indeed, the term 'computer' arose as an anthropomorphism, describing the activities of a human calculator whose activities a machine was then devised to simulate. A computer is an arrangement of some of the material constituents of the Universe into a configuration whose natural evolution in time according to the laws of Nature simulates some mathematical process. There are many simple examples. The swing of a pendulum in the Earth's gravitational field can be used to make a 'computer' that counts in a regular way. We call this a 'clock'. The decay of a radioactive atom depletes its source in a way that simulates a particular mathematical operation that we call 'the exponential function'. Thus we see that the fact that any computers can exist witnesses to the fact that the world *is* mathematical in some sense. In the 1930s Alan Turing developed the conception of an idealized computer whose basic capabilities are still those at the heart of almost every modern computer. They are called Turing machines and their capabilities are illustrated in Fig. 5.12. In essence they read a list of natural numbers and transform them into another list of numbers. Turing showed that there exist mathematical operations, the non-computable functions we mentioned earlier, which cannot be mimicked by the operation of this generic device. This famous result enables us to recast the problem of the applicability of mathematics into a new and sharper form. If a mathematical operation is computable, it means that it is possible to simulate that function by some natural fabrication or arrangement of matter and energy in the Universe. Conversely, the fact that the structure and arrangement of material and energy in the Universe turns out to be well described by mathematics can

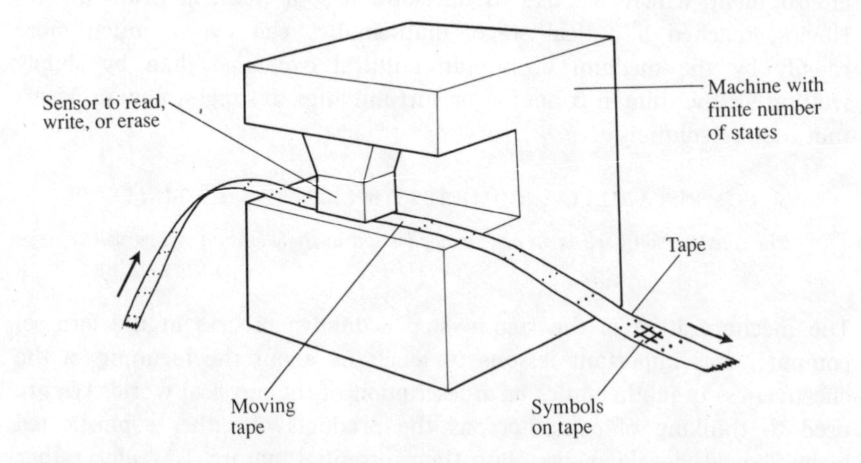


**Figure 5.12** Alan Turing's all-purpose machine which he conceived of in 1936. It consists of a finite collection of symbols, a finite set of states in which it can reside, an endless tape marked off in bits, each of which carries a single symbol, a sensor head which can scan the tape one square at a time and read it or write on it, and a set of instructions giving the rules that cause changes (or no change) to occur to the tape when each square is read.

be ascribed to the existence of computable functions in Turing's sense. If virtually all the functions of mathematics beyond those of addition, subtraction, multiplication, and division were *non-computable* functions, then, although the world could still be intrinsically mathematical, we would not find mathematics a terribly useful tool for describing its workings. The fact that we do find mathematics to be extremely useful in practice is equivalent to the fact that there are so many simple mathematical functions which are also computable functions.

This state of affairs provokes us to ask whether any use has yet been found for non-computable functions in the description of the physical Universe. The answer is not yet clear. But it is clear that were non-computable functions to play an essential role in Nature then it would have considerable ramifications. If the human brain conducted its operations by the employment of any non-computable operations, then it would not be possible for all the goals of the 'Artificial Intelligentsia' to be achieved. One could not replace the brain's behaviour by some computer algorithm. More generally, the appearance of non-computable operations signals the breakdown of traditional assumptions about a problem. In order to get a better and better approximation to, or description of, a computable function, one can proceed by a sequence of computational steps that are repetitious. All one needs is persistence to make the description just a little bit better after the next iteration. Non-computable problems are not like this. They do not respond to the continued application of a formula. Each step in the search for a better and better description requires something novel and qualitatively new to be injected. Something more than technique is required.

If we pursue our computer picture of the world a little further, then we can recast the dilemma of the utility of mathematics into the language of algorithmic compression that we introduced earlier. We saw that if some string of pieces of information can be compressed, or abbreviated, into a 'formula' which is shorter than the length of the string then it must contain some pattern or order which renders it non-random. Science is, at root, just the search for compressions of the world. And the fact is that science has been successful in making sense of a very large part of the physical world, setting up systems of laws and symmetries which we can use to summarize the behavioural properties of the world far more succinctly than we could if we had to keep a record of all the events that ever happened. In short, the world is surprisingly compressible and the success of mathematics in describing its workings is a manifestation of that compressibility.

Finally, to close our discussion of the computational picture of the world we should return full circle to the Intuitionists. They appear, like Brouwer, to have no interest in explaining the applicability of mathematics to the physical world, and are happy to effect a divorce between mathematics and science by the adoption of a three-valued logic and the banishing of actual infinities from the language of mathematics. Traditionally, it has been felt that intuitionistic mathematics does not describe the physical world; but we may be in for a surprise. The growing use of computer concepts like information, programs, software, hardware, computability, and so forth in the foundations of physics signals the emergence of a new paradigm which vies for a place in the minds of physicists. For twenty years they have been persuaded that at root the Universe is a great static symmetry whose determination would reveal the true 'Theory of Everything'. But the picture of the Universe as a great kaleidoscopic symmetry is being challenged by the image of the Universe as a great computer program, whose software consists of the laws of Nature which run on a hardware composed of the elementary particles of Nature. If this conception is more basic than the physicists' belief in the primacy of symmetry then, not only would the appearance of symmetry be a consequence of some primitive rules of computation, but at root the world would not be a continuum: it would be discrete, changing by definite information jumps. This would amount to a vast deepening of its intrinsic structure. Continuous changes are very special types of change. If we appeal to Cantor's hierarchy of the infinite, then the collection of all possible discontinuous changes is a higher order of infinity than the number of continuous changes. By building our study of the physical world ultimately upon the assumption of the ubiquity of continuous changes at the most macroscopic level, we may be making not merely a gross simplification but an infinite simplification.

# CHAPTER SIX

# *Platonic heavens above and within*

*A mathematician is a blind man in a dark room looking for a black cat which isn't there.* CHARLES DARWIN

## THE GROWTH OF ABSTRACTION

*I knew a mathematician who said, 'I do not know as much as God, but I know as much as God did at my age.'* MILTON SHULMAN

The giant leap that was needed for Mankind to pass from the mundane ancient world of counting into the realm of modern mathematics is a strange but simple one. It is the ability to pass from seeing numbers as mere properties of collections of things, of cabbages or kings, to an appreciation that they have an identity that is larger than these examples; that amid all the collections of 'threes' in our experience there exists a common factor that has nothing to do with any intrinsic property of the individual members of those collections—the abstract notion of 'threeness'. Once seized, this master key opens many doors. For now one can reason and speculate about the attributes of all possible collections of threes without necessarily having any one of them in mind. And any conclusions one draws will hold good for all the threes in Heaven and Earth. We have taken the first steps along the path of generalization. In this way mathematics starts to be a mental science in which we reflect upon the results of certain specific thought processes. The fact that the results have widespread application to other specific examples is a wonderful but not unexpected bonus. Yet, whereas other sciences have a domain of attention which is restricted to certain varieties of things, mathematics is far more general: its conclusions apply regardless of the specific identities of the objects being counted.

These steps sound like a triviality, but no ancient cultures took them before the Greeks. Evidently, it requires something more than lots of examples of threeness to provoke the final step which considers disembodied threeness as a concept as valid and useful as that of three apples or three pears. Such a development seems bound up with the conception of some form of 'idealism' which is willing to attribute the notion of 'existence' to immaterial things. At first, one might think this would have been far easier for primitive peoples than for modern ones. Did they not imagine a world inhabited by countless disembodied spirits and deities who controlled every facet of their lives, and did their leaders not seek to perpetuate the *status quo* by reinforcing and exploiting any such beliefs? Clearly they did; but these beliefs are not as abstract as the notion of 'threeness'. Almost all ancient deities were represented in some physical form, perhaps as idols of wood or stone, or personified by witch-doctors or priests clad in a garb which transmogrified any vestiges of their own character. Moreover, the activities of these deities were interpreted as definite acts of intervention in the physical world and their interaction with their colleagues in the spiritual realm was characterized by just the same motives as those which appear in human life: anger, jealousy, boldness, love, or indifference. They were really just larger-than-life people. The well-known exception, which thus seems to prove the rule elsewhere, is the Hebrew tradition in which the notion of a single omnipotent deity was extremely abstract. All representations of Yahweh were forbidden, but this was clearly a difficult rule to adhere to since we read of the continual temptation for the Israelites to create representations, or 'graven images' of other gods.

Other, more mundane, explanations of the growth of an abstract notion of number are possible. Trade and commerce, albeit sometimes in rudimentary forms, are wellnigh universal. In the sophisticated ancient cultures they were highly developed and both underpinned their prosperity and aided the diffusion of improved methods of counting and reckoning. These commercial activities played a key role in the development of tokens and tallies which kept track of the relative values of things traded. Ultimately, this constituted what we call 'money'. At root, money is a means for comparing qualitatively different things in a common way. As such it might well have played a key role in the gestation of an abstract view of number. It induces people to idealize things in a particular sense: in the commercial context they are seen as devoid of all their particular qualities save for one only—that which we might call its value. Moreover, this numerical value, expressed in some units of the adopted currency, is an abstract use of number because it can be applied, without any particularization, to all varieties of things. We see here a possible beginning of the notion that there might be a universal aspect of quantity which can be conceived independently of our knowledge of all the specifics to which it will ultimately be applied. Once numbers were only nouns;

now they are free to become adjectives as well. Where once we could speak only number words, now we are free to speak *about* numbers with words.

## FOOTSTEPS THROUGH PLATO'S FOOTNOTES

*Had Diotallevi turned arithmetic into a religion, or religion into arithmetic? Perhaps both. Or maybe he was just an atheist flirting with the rapture of some superior heaven.* UMBERTO ECO

Only in one place in the ancient world do we find a highly developed notion of number in the abstract. That place is Greece and with it comes the first philosophical recognition of the very problems we are discussing. But its beginnings are very strange. They take us back to reconsider the events of the sixth and fifth centuries BC when the mystical Pythagorean Brotherhood emerged and thrived. They were fascinated by number and the means it gave them to endow their religious beliefs with a rational basis. From the Babylonian traditions in astronomy and mathematics, they had learnt that numbers were associated with the constellations; their interests in music and rhythm revealed the numerical relations linking the harmonics of a plucked string. The ubiquity of number in these important aspects of the Pythagoreans' world, together with the commercial aspects of everyday life, led them to see number as a cosmic generality that bound all things together. Perhaps number was the common factor unifying all mankind's great interests from ethics to the nature of the Universe?

It appears that the Pythagoreans began to see things solely as numbers, stripping away all other defining characteristics as inessential trappings: they maintained 'that things themselves are numbers' and these numbers were the most basic constituents of reality. All things derived from numbers but numbers themselves could not be derived from any more fundamental entity. What is peculiar about this view is that it regards numbers as being an *immanent* property of things; that is, numbers are 'in' things and cannot be separated or distinguished from them in any way. It is not that objects merely possess certain properties which can be described by mathematical formulae. Everything, from the Universe as a whole, to each and every one of its parts, was number through and through. It is easy to collect evidence to fit such a view—everything one sees has attributes that can be counted. But the Pythagoreans went one stage further and associated numbers with abstract things like 'justice' or 'opportunity' as well. Gradually they saw experience to consist only of number. Nothing was merely *like* number, or possessed of some numerical properties, it *was* number. Philolaus of Croton, where Pythagoras founded his sect, wrote the first account of their teachings in the fifth century BC and claimed their doctrine then to be that

> all things which can be known have number; for it is impossible for a thing to be conceived or known without number.

Beset by such beliefs, science as we know it could not develop. If all things were made of number, then the true nature of reality could only be discerned by studying numbers themselves, their interrelationships, their properties, divorced from any utilitarian motive. Things, experiences, causes, and effects were only of secondary interest. Weird as this numerical religion seems to be, it was the first extensive and influential focus of interest upon numbers which was divorced from any practical applications. It was what is now called 'pure' mathematics to emphasize that its focus is not practical applications. But the Pythagoreans' programme did not become abstract and metaphysical. Rather, we find that it led them to treat numbers as if they were living things.

The influence of the Pythagoreans reached Plato when he travelled in southern Italy (then part of Greece) and as a result he became persuaded of the importance of mathematics. He was impressed by the way in which the Pythagoreans had associated mathematics with other forms of abstract knowledge. From these seeds there was to grow up a philosophy of the nature of mathematics, and of things in general, that now goes by the name of 'Platonism'.

Plato's philosophy of mathematics grew out of his attempts to understand the relationship between particular things and universal concepts. What we see around us in the world are particular things—this chair, that chair, big chairs, little chairs, and so on. But that quality which they share—let's call it 'chairness'—presents a dilemma. It is not itself a chair and unlike all the chairs we know it cannot be located in some place or at some time. But that lack of a place in space and time does not mean that 'chairness' is an imaginary concept. We can locate it in two different pieces of wood and upholstery. These things have something in common. Plato's approach to these universals was to regard them as real. In some sense they really exist 'out there'. The totality of his reality consisted of all the particular instances of things together with the universals of which they were examples. Thus the particulars that we witness in the world are each imperfect reflections of a perfect exemplar or 'form'.

Plato's interest in the status of universals had two particular applications that have left their mark upon human thinking ever since. The first was in the realm of ethics and morals where he sought to reconcile the ideal notions of virtue and goodness with the imperfect approximations that humanity strove to demonstrate in practice. But the second application that preoccupied him was the question of the meaning of mathematics. We seem able to conceive of perfect mathematical entities—like straight lines, triangles, right angles, numbers, or parallel lines—yet all the examples we see of them are imperfect in some respect: we cannot draw perfectly straight lines or exact right angles. We need a conception of the ideal examples in order to be able to judge all the particulars to be mere approximations to them. So, 'somewhere', Plato maintained, there must exist perfect straight lines, perfect circles and triangles, or exactly parallel lines; and these perfect forms would exist even if there were no particular examples of them for us to see. This 'somewhere' was not simply the human mind. Strikingly, Plato maintained that we discover the truths and theorems of mathematics; we do not simply invent them. And lest one should be tempted to think that the truths of mathematics are nothing more than particular thoughts, he points out that we can readily distinguish between the discovery of a new theorem of mathematics and our thoughts about it, and even our thoughts about our thoughts about it. He argues that

> no one who has even a slight acquaintance with geometry will deny that the nature of this science is in flat contradiction with the absurd language used by mathematicians, for want of better terms. They constantly talk of operations like 'squaring', 'applying', 'adding', and so on, as if the object were to *do* something, whereas the true purpose of the whole subject is knowledge—knowledge, moreover, of what eternally exists, not of anything that comes to be this or that at some time and ceases to be.

For Plato, who was a seeker after the truth behind the superficial appearances of human discourse, mathematics seemed to offer a wonderful example of a particular form of knowledge that owed nothing to the process of human recognition. It was our truest experience of eternal and absolute truths about the Universe. In the *Meno* he gives an account of Socrates questioning a slave boy, who knows nothing of mathematics, in a way that leads the boy to 'see as inevitable the fact that if one draws a smaller square inside another square in such a way that the corners of the enclosed square are the mid-points of the sides of the larger square then the area of the smaller square will always be half that of the larger one.' Socrates argues that this knowledge must be a memory of some previous life because the boy had not learnt this mathematical fact in his earthly life. But Plato argues that this knowledge of the truths of geometry is not a memory or the result of education: it is an example of our ability to perceive the contents of a changeless world of universal truths.

Another telling commentary can be found in Plutarch's writings, where he reveals how two mathematicians had invented what we would now call very simple 'computers'—devices whose operation demonstrated the truth of some mathematical conjecture or another. This is an early example of the experimental approach to mathematics which we prognosticated about in the last chapter. Plato, we are told, reacts vehemently against rendering temporally concrete what should be eternally abstract:

Eudoxus and Archytas had been the first originators of this far-famed and highly prized art of mechanics, which they employed as an elegant illustration of geometrical truths, and as a means of sustaining experimentally, to the satisfaction of the senses, conclusions too intricate for proof by words and diagrams. In the solution of the problem ... both these mathematicians had recourse to the aid of an instrument, adapting to their purpose certain curves and sections of lines. But what of Plato's indignation at it, and his invectives against it as mere corruption and annihilation of the one good in geometry, which was thus shamefully turning its back upon the unembodied objects of pure intelligence to recur to sensation, and to ask help (not to be obtained without base supervisions and depravation) from matter; so it was that mechanics came to be separated from geometry, and, repudiated and neglected by philosophers, took its place as a military art.

The first problem Plato's picture of reality presents is to clarify the relationship between universals and particular examples of them. Plato's idea that there exist perfect universal blueprints of which the particulars are imperfect approximations does not seem very helpful when one gives it a second thought; for, as far as our minds are concerned, the universal blueprint is just another particular. So whilst we could say that Plato maintained that universals would exist even in the absence of particulars, this statement does not really have any clear meaning. If all the particulars vanished, so would all those mental images of concrete perfect blueprints together with all the blueprints themselves. Aristotle, who maintained that the Forms which gave things identity were 'in' things rather than abstract Ideas in another realm, also picked upon this weakness in Plato's doctrine of Ideas, arguing that it leads to an endless regress:

If all men are alike because they have something in common with Man, the ideal and eternal archetype, how can we explain the fact that one man and Man are alike without assuming another archetype? And will not the same reasoning demand a third, fourth, and fifth archetype, and so on into the regress of more and more ideal worlds?

The attempt to create a heavenly realm of universal blueprints that are truly different from the particulars founders under the weight of another simple consideration that was not noticed in early times. Plato wants to relate the universal abstract blueprint of a perfect circle to the approximate circles that we see in the world. But why should we regard 'approximate' circles, or 'almost parallel lines', or 'nearly triangles' as imperfect examples of perfect blueprints. Why not regard them as perfect exhibits of universals of 'approximate circles', 'almost parallel lines', and 'nearly triangles'? When viewed in this light the distinction, and the basis for any distinction, between universals and particulars seems to be eroded.

Plato's emphasis upon the timeless and unchangeable blueprints behind the changing secondary appearances of the world about us began a philosophical tradition that is usually called 'idealism' in order to stress its foundation upon the assumption that 'ideal' perfect exemplars of the objects of experience exist. It is often regarded as a 'realist' philosophy of the nature of knowledge because it maintains that we discover examples of this world of universals without distorting their true nature by any subjective mental processing. The world is 'out there' not 'all in the mind'.

The Platonic picture is one of those ideas which at first seem to be eminently simple and unencumbered by abstraction, but which, as we shall see, is infested with all manner of problems. It is a philosophy of mathematics that is popular amongst physicists but less so amongst mathematicians, and is almost universally regarded as an irrelevance by consumers of mathematics like computer scientists, psychologists, or economists. One immediate consequence of the abstract world that it introduces is the downgrading of the meaning of traditional mathematical objects like numbers. Whereas the Pythagoreans saw them as symbols imbued with irreducible meaning, Plato sees them as empty vessels whose significance lies entirely in the relationships they have with other symbols. Because the Platonic view points to innumerable particular representations of the same ideal formal concept, it places no significance upon those particular examples.

The Platonic realm of perfect ideas and concepts held great attraction for many theologians and there is a long tradition of interaction between Christian theology and Platonic philosophy as a rationale for the concept of absolute moral principles. At first, from the first century BC until about AD 300, God, the Ideas from which the particulars of the world were cast, and the matter out of which they were made, were all regarded as eternal. The abstract ideas eventually found themselves lodging in the mind of God. Attractive as this might at first seem, it takes us no nearer a resolution of the subtle problem of universals and particulars because such 'universals' are again merely particulars. Moreover, attractive as the general tenor of this scheme might appear at first sight to those seeking a place for God as the source and inspiration of all that is perfect, most Christian theologians regard it as contrary to traditional Christian doctrine because it regards the material world and the Incarnation as in some way inferior to the perfect immaterial world of the perfect forms.

St Augustine reveals something of the tension that must have existed in relating the omnipotence of God to the truths that lay in the Platonic world of perfect forms. Whilst one might regard those truths as simply the thoughts of God, it is difficult to conceive of how they might be other than how they are if God chose to think differently. Some principles of logic and arithmetic seemed to be irrevocable and hence appear to place some constraint upon God's freedom of action. A representation of the possible viewpoints is given in Fig. 6.1. In this vein we find Augustine transposing the usual deduction that six was a special number because God chose to create the world in six days into a claim that *because* six is a 'perfect' number (i.e. its divisors one, two, and three sum to it) God chose to complete his creative work in a six-day period:

> Six is a number perfect in itself, and not because God created all things in six days; rather the inverse is true, that God created all things in six days because this number is perfect, and it would remain perfect, even if the work of the six days did not exist.

This is one of the reasons for the status assumed by Euclidean geometry for so long, which we discussed in the opening chapter. It just could not

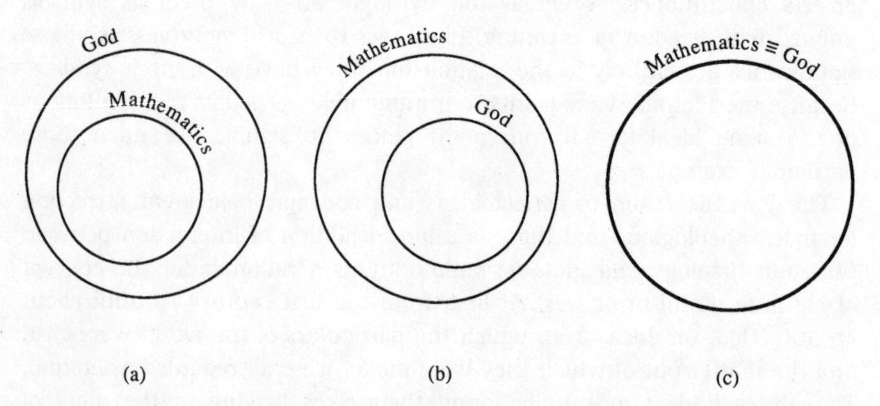


**Figure 6.1** A simple set of three alternative viewpoints concerning the relationship of God and Mathematics: (a) a traditional Western position in which mathematics is part of the created order; (b) the position that mathematics is a necessary truth that is larger than God in the sense that He cannot suspend or change the rules of mathematics; and (c) the view that God and mathematics are the same.

be envisaged how the laws of geometry could be violated any more than could two and two make five. Both scientists and theologians could, and often did, contemplate the possibility that the laws of Nature might be violated or suspended. But a violation of the laws of logic was something else entirely. Different laws of Nature merely permitted other, easily visualized, chains of events to occur. Different laws of logic were inconceivable.

The attraction for these interpretations of the nature of reality for the early (and many contemporary) Platonists witnesses to an interesting natural prejudice that we seem to have inherited, regarding permanence as superior to change. If we see things that are old, they are regarded as somehow more valuable, having 'stood the test of time', than the manifestations of 'here today, gone tomorrow' fashions. Platonic philosophy had negative effects upon the early development of science because it regarded the things we see around us, the very things that an observational or experimental science would have to be based upon, as secondary compared with the perfect timeless Ideas behind the appearances within which were encoded the true nature of things. The American philosopher Robert Nozick protests about this prejudice towards unchanging notions in mathematics, and elsewhere, which seeks to endow them with an elevated status:

> Some mathematicians have this attitude towards the permanent and unchanging mathematical objects and structures they study, investigate and explore ... Despite the pedigree of this tradition, it is difficult to discover why the more permanent is the more valuable or meaningful, why permanence or long-lastingness, why duration in itself, should be important. Consider these things people speak of as permanent or eternal. These include (apart from God) numbers, sets, abstract ideas, space-time itself. Would it be better to be one of these things? The question is bizarre; how could a concrete person become an abstract object? Still, would anyone wish they *could* become the number 14 or the Form of Justice or the null set? Is anyone pining to lead a setly existence?

The existence of mathematical entities inhabiting some realm of abstract ideas is a lot for many modern mathematicians to swallow, but three hundred years ago a Newton or a Leibniz would have taken for granted the existence of mathematical truths independent of the human mind. They had faith in the existence of the Divine Mind in which perfection lived and so they saw no problem at all with the concept of perfect forms. Their problem was to reconcile them with the existence of the imperfect, material objects they saw around them. In the following centuries it was the gradual emergence and predominance of rationalism and empiricism, with their primary focus upon the objects of the real world that tipped the scales so that the non-ideal material objects were accepted but the ideal abstractions were questioned and discarded. Because of this reversal of emphasis the question of 'how' mathematical knowledge came to be obtained was just ignored. The only exception within the ranks of the empiricist philosophers was John Stuart Mill who proposed that mathematics was just another science and mathematics the record of its observations and predictions. It was simply the record of our habitual experience that two beans plus two beans make a number of beans that we call four: a view that was savaged during the nineteenth century with the emergence of non-constructive existence proofs, infinities, and formal systems which you cannot 'see'. There was simply too much mathematics for the Millian.

## THE PLATONIC WORLD OF MATHEMATICS

*O world invisible, we view thee,*
*O world intangible, we touch thee,*
*O world unknowable, we know thee,*
*Inapprehensible, we clutch thee!*

FRANCIS THOMPSON

The Platonic view of reality has crept unseen upon many modern scientists and mathematicians. It seems simple, straightforward, and inspiring. There is an ocean of mathematical truth lying undiscovered around us; we explore it, discovering new parts of its limitless territory. This expanse of mathematical truth exists independently of mathematicians. It would exist even if there were no mathematicians at all—and indeed, once it did, and one day perhaps it will do so again. Mathematics consists of a body of discoveries about an independent reality made up of things like numbers, sets, shapes, and so forth. 'Pi' really is in the sky.

Despite its obvious correspondence with Plato's more general idealistic philosophy of our knowledge of the world, this philosophy of mathematics only became generally known as 'mathematical Platonism' after 1934, when it was so described by Paul Bernays, Hilbert's close collaborator on the development of consistency proofs for formal mathematical systems. He delivered a lecture at the University of Geneva on 18 June of that year entitled 'On Platonism in mathematics' which he characterizes as a view of mathematical objects 'as cut off from all links with the reflecting subject'—that is, cut off from the personal influence of the mathematician. He continues,

> Since this tendency asserted itself especially in the philosophy of Plato, allow me to call it 'Platonism'.

Before we explore the range of interpretations that have been placed upon a Platonic interpretation of mathematics, it is instructive to look at the enthusiastic adherence that is found to this doctrine amongst mathematicians. Even Hermann Weyl who, as we saw in the last chapter, was an early subscriber to a moderate version of Brouwer's intuitionism, felt a natural leaning towards the Platonic attitude of regarding mathematical entities as real and transcending the human creative process. He found that his tendency to veer from this dogma towards the more conservative constructive philosophy of the intuitionists felt unnatural and even acted as a restraining influence upon his mathematical creativity because although

> outwardly it does not seem to hamper our daily work, yet I for one confess that it has had a considerable practical influence on my mathematical life. It directed my interests to fields I considered relatively 'safe', and has been a constant drain on the enthusiasm and determination with which I pursued my research work.

Here we see a viewpoint that is almost religious in the sense that it provides an underpinning necessary to give a meaning to life and human activity. It hints that, even if Platonism is not true, it is most effective for the working mathematician to act as if it were true. The French mathematician Emile Borel, writing a century ago, leans a little closer to this attitude:

> Many do, however, have a vague feeling that mathematics exists somewhere, even though, when they think about it, they cannot escape the conclusion that mathematics is exclusively a human creation. Such questions can be asked of many other concepts such as state, moral values, religion, ... we tend to posit existence on all those things which belong to a civilization or culture in that we share them with other people and can exchange thoughts about them. Something becomes objective (as opposed to 'subjective') as soon as we are convinced that it exists in the minds of others in the same form as it does in ours, and that we can think about it and discuss it together. Because the language of mathematics is so precise, it is ideally suited to defining concepts for which such a consensus exists. In my opinion, that is sufficient to provide us with a *feeling* of an objective existence regardless of whether it has another origin.

If we look back to the end of the nineteenth century when Borel was writing then we find some of the most emphatic Platonists. Charles Hermite regarded mathematics as an experimental science:

> I believe that the numbers and functions of analysis are not the arbitrary product of our spirits: I believe that they exist outside of us with the same character of necessity as the objects of objective reality; and we find or discover them and study them as do the physicists, chemists and zoologists.

Henri Poincaré highlighted Hermite's approach as especially remarkable in that it was not just a philosophy of mathematics, but more a psychological attitude. Poincaré had been Hermite's student and was also much taken to careful introspective analysis of his mathematical thought processes and creativity.*

Unlike many subscribers to the Platonic philosophy, Hermite did not regard all mathematical discovery as the unveiling of the true Platonic world of mathematical forms; he seems to have drawn a distinction between mathematical discoveries of this pristine sort and others, like Cantor's development of different orders of infinity which we met earlier, that he regarded as mere human inventions. Poincaré writes, in 1913, twelve years after Hermite's death, that

> I have never known a more realistic mathematician in the Platonist sense than Hermite ... He accused Cantor of creating objects instead of merely discovering them. Doubtless because of his religious convictions he considered it a kind of impiety to wish to penetrate a domain which God alone can encompass [i.e. the infinite], without waiting for Him to reveal its mysteries one by one. He compared the mathematical sciences with the natural sciences. A natural scientist who sought to divine the secret of God, instead of studying experience, would have seemed to him not only presumptuous but also lacking in respect for the divine majesty: the Cantorians seemed to him to want to act in the same way in mathematics. And this is why, a realist in theory, he was idealist in practice. There is a reality to be known, and it is external to and independent of us; but all we can know of it depends on us, and is no more than a gradual development, a sort of stratification of successive conquests. The rest is real but eternally unknowable.

Poincaré was most struck by the way in which 'abstract entities were like living things for him. He did not see them, but he felt that they were not an artificial collection [and] had an indescribable principle of internal unity'. The same conviction can be found in the writings of famous mathematicians throughout the twentieth century. G.H. Hardy, the most famous British mathematician of the first half of the century, whom we met when discussing his role in inducting the young Ramanujan into modern mathematics, maintained a solid Platonic realism about the nature of mathematics:

* Poincaré was actually the subject of a very detailed psychological study, including a large number of practical tests, carried out by E. Toulouse in 1897 and published with Poincaré's blessing in 1910. Toulouse was the Director of the Laboratory of Psychology at L'École des Hautes Études. Poincaré laid great stress upon his unconscious and subconscious reasoning processes, which he called 'the play of associations', yet totally avoided the use of any visual image of a structure, preferring instead to remember the listing of its component parts, seeking patterns to help the memorization process. Whilst not a full-blooded intuitionist in the Brouwer mode, Poincaré was sympathetic to the picture of mathematics as a human construction (but was happy to include the *reductio ad absurdum* as an allowed form of reasoning) rather than a discovery; hence the strong counter-impression that Hermite's views made upon him.

I believe that mathematical reality lies outside us, and that our function is to discover or *observe* it, and that the theorems which we prove, and which we describe grandiloquently as our 'creations' are simply our notes of our observations ... 317 is a prime number, not because we think it so, or because our minds are shaped in one way rather than another, but *because it is so*, because mathematical reality is built that way.

The most famous exponent of Platonism was undoubtedly Kurt Gödel, who maintained a striking attachment to the objective reality of the entities that are the day-to-day concern of the logician and the set theorist because

despite their remoteness from sense experience, we do have something like a perception also of the objects of set theory, as is seen from the fact that the axioms force themselves upon us as being true. I don't see any reason why we should have less confidence in this kind of perception, i.e. in mathematical intuition, than in sense perception, which induces us to build up physical theories and to expect that future sense perceptions will agree with them and, moreover, to believe that a question not decidable now has meaning and may be decided in the future. The set-theoretical paradoxes are hardly any more troublesome for mathematics than deceptions of the senses are for physics.

Most recently, the belief in 'pi in the sky' Platonism in mathematics has been forcefully restated by Roger Penrose who uses the example of the intricacy of structure displayed by fractals like the Mandelbrot set, which he claims 'is not an invention of the human mind: it was a discovery. Like Mount Everest, ... it is just there',* to argue that this bottomless structure is not invented by the mind, rather,

though defined in an entirely abstract mathematical way, nevertheless [it has] a reality about it that seems to go beyond any particular mathematician's conceptions and beyond the technology of any particular computer ... it seems

* Anyone unsympathetic to the Platonic viewpoint would point to the fact that the Mandelbrot set, like other spectacular fractal structures, is *constructed* explicitly from an algorithm. Penrose claims that 'the computer is being used in essentially the same way that the experimental physicist uses a piece of experimental apparatus to explore the structure of the physical world'. This seems to be a peculiar analogy; the computer is used to generate the structure not to explore it. Its function is quite different to that of a piece of apparatus like a telescope or a Geiger counter. The impressiveness of fractal structures is, I suspect, a subtle consequence of natural selection. We see that the natural world makes abundant use of fractal algorithms; that is, a basic pattern is copied again and again on a smaller and smaller scale. This is a very simple copying plan to implement. We see it in the branchiness of trees as we go from the spreading of the stoutest limbs to that of the tiniest twigs and in the microscopic patterns of snowflakes and flowers. The reason why the fractal patterns that mathematicians have generated seem so aesthetically attractive, to the extent that they have been exhibited at major art galleries around the world, is that they capture the basic patterns that Nature uses to develop the complexity that surrounds us and to which we have become pleasantly accustomed by the evolutionary process.

clearly to be 'there', somewhere, quite independently of us or our machines. Its existence is not material, in any ordinary sense, and it has no spatial or temporal location. It exists instead in Plato's world of mathematical entities ... In general the case for Platonic existence is strongest for entities which give us a great deal more than we originally bargained for.

The sources of this attachment to the Platonic viewpoint are easily found. For the practising mathematician or scientist it seems akin to the 'common-sense' view that things are what they seem: 'what you see is what you get'. When we do mathematics we talk of 'finding' the answer to the problem, of 'discovering' a new property of numbers, or proving the 'existence' of a particular type of number. We feel that we are uncovering things that would be true even if we had not chosen to unveil them. The only alternative seems to be a human-centred mathematics. Moreover, the immediate alternative, that the mind is somehow conjuring these structures out of nothing more than mundane experience, seems deflating. If they were entirely human inventions, why do they so often defeat us in the quest to unravel their subtleties and why do they have so many unsuspected applications to the esoteric workings of the natural world far removed from natural human experience?

Nevertheless, as we have seen in earlier chapters, there are many dissenters to such a confident Platonic view and they are often to be found amongst the consumers of mathematics, like economists and social scientists for whom the cultural and human contribution to human knowledge is of primary interest. In Chapter 4 we saw something of the subtle change of direction in the titles of applied mathematics books which highlight changing attitudes to the status of mathematical descriptions of the world. They emphasize the use of mathematics as a tool for deriving approximate descriptions ('models') of the real thing. There is no implication that the mathematics being presented *is* the reality. Here we see the influence of the inventionist and constructivist philosophies which we discussed in earlier chapters. They lead rather naturally to a perspective that denies the status of absolute truth to mathematical descriptions of natural phenomena. They see the work of the mathematician as creative rather than exploratory, artistic rather than explanatory. Forceful philosophers of science in the operationalist mould, like Percy Bridgman, regarded a blind adherence to belief in the Platonic world of mathematics as a source of understanding of the physical world to be an unproven and dangerous religious belief that could drag science into error. Sixty years ago Bridgman picked upon cosmologists' reverence for the mathematical beauty of Einstein's general theory of relativity in the face of its almost impenetrable difficulty as a classic example of undue reliance upon aesthetic qualities which cannot be tested by experience:

The metaphysical element I feel to be active in the attitude of many cosmologists to mathematics. By metaphysical I mean the assumption of the 'existence' of validities for which there can be no operational control ... At any rate, I should call metaphysical the conviction that the universe is run on exact mathematical principles, and its corollary that it is possible for human beings by a fortunate *tour de force* to formulate these principles ... when, for example, I ask an eminent cosmologist in conversation why he does not give up the Einstein equations if they make him so much trouble ... he replies that such a thing is unthinkable, that these are the only things that we are really sure of..

Of course, even the most ardent Platonist recognizes that some parts of mathematical activity are merely outlines of very complicated physical processes whose exact description is beyond the scope of our mental abilities. Moreover, it is undeniable that many aspects of mathematics have been abstracted from Nature. But he would argue that this strengthens, rather than undermines, our confidence in the existence of an intrinsic mathematical facet of the real world. If our minds have derived a special mathematical facility from the real world, it is likely that they have done so as a result of an evolutionary process which has selected for those mental images and representations of the world because they most faithfully represent how the world truly is.

Another consideration that weighs strongly for the Platonist is the nature of mathematical discovery. The fact that the same mathematical ideas and theorems seem to be found by independent mathematicians who are widely separated in space and time, race and religion, cultural upbringing and political viewpoint strikes many as a dramatic endorsement of the suspicion that all these mathematicians are discoverers of an objective reality that 'exists'. They may use different symbols to describe what they find, but '7' means to us what 'VII' meant to a Roman centurion. The cultural aspects are believed to be tied solely to the mode of representation of the mathematical concepts and to the order in which they are discovered. Thus, if one lives in an seafaring community, one will be motivated to develop mathematics that is related to the needs of navigation and astronomy. If one lives in a community that places onus on design of complicated patterns for religious purposes or systems of reckoning for business and commerce, then one will develop quite different emphases.

There is a curious psychological feature that one detects among mathematicians of the Platonic persuasion when it comes to the question of mathematical discovery. At root, the Platonist believes he is exploring and discovering the structure of some other 'world' of truths, yet somehow the Platonic view allows him to feel as though his creativity has free rein. This is at root why Brouwer's philosophy of mathematics was so distasteful.

An ardent Platonist like Cantor has this in mind when he claims that 'the essence of mathematics is its freedom'. It was not that mathematicians have any deep-seated resentment of the idea of construction as the essence of mathematical reasoning. If that idea had allowed them to construct, albeit more laboriously, all the mathematics that had already been created by the other means they were in the habit of using, then they would have been perfectly happy with it. No, the real objection was that it did not permit all the deductions arrived at by other means to be called 'mathematical truths': it was a restriction upon freedom of thought.

Platonism allows freedom of thought, but only in the sense that it is your fault if you want to think the wrong thoughts. Whereas the formalist is free to create any logical system he chooses, a Platonist like Gödel maintained that only one system of axioms captured the truths that existed in the Platonic world. Although a conjecture like the continuum hypothesis was undecidable from the axioms of standard set theory, Gödel believed it was either true or false in reality and this would be decided by adding appropriate axioms to those we were in the habit of using for set theory. He writes of the undecidability of Cantor's continuum hypothesis:

> Only someone who (like the intuitionist) denies that the concepts and axioms of classical set theory have any meaning ... could be satisfied with such a solution, not someone who believes them to describe some well-determined reality. For in reality Cantor's conjecture must be either true or false, and its undecidability from the axioms as known today can only mean that these axioms do not contain a complete description of reality.

This seems a strange view because there would always exist other undecidable statements no matter what additional axioms are prescribed.

We could regard the Platonic world of mathematical truths as containing every single mathematical concept and relation that is posable by anyone at any time, in which case the entire notion seems to lose something of its appeal as a way of explaining anything.* Alternatively, we could regard the Platonic world of mathematical forms as stocked only with a rather select collection of special mathematical ideas. Maybe only one variety of geometry lives there (only the one that applies to the space of our physical Universe perhaps). However, were that the case, we could use those Platonic blueprints to invent other analogous mathematical structures for ourselves.

* One might argue that inconsistent mathematical logics are perfectly viable candidates for Platonic worlds. They would be systems in which any statement that could be framed in the language of the system could be proved true. Whilst such systems could not admit of any particular representation in the material world if we assume the natural world is a logically consistent representation, there seems to be no reason why they cannot be universals devoid of particulars.

That is, if a limited form of Platonism were true, it could be expanded by human mathematicians to contain all the same entities as a universal Platonism by the application of formalistic or constructive principles.

In our discussion of the origin of counting, we saw how difficult it is to substantiate the common view of the Platonist that mathematical ideas arose spontaneously, with similar forms, in separate cultures. We do indeed find very basic but sophisticated mathematical ideas, like Pythagoras' theorem, in different cultures (thousands of years before Pythagoras), but it is very hard to unravel the possible influences of cultural diffusion and determine to what extent these cultures invented the ideas independently. A supporter of the inventionist view, like the mathematician Raymond Wilder, would argue that the simultaneous discovery of similar mathematical concepts is to be expected even if the cultures are independent. If mathematical ideas emerge initially as a result of social and practical pressure to solve certain types of problem or to administer finances then we expect the great thriving cultures each to pass through similar stages of general development, to encounter the same sorts of practical problems over time-keeping, surveying, building, monetary systems, trade, navigation, wonderment at the motions of the heavenly bodies, and so forth. Is it really surprising that they develop the same notions in response? Of course, the Platonist will not find this response altogether convincing because he will point to the common nature of the problems being faced by the different societies as evidence of a common quantitative aspect to things which derives from the universal mathematical nature of reality.

## FAR AWAY AND LONG AGO

> *The Universe thus consists of the continued vanishing of all sets of possibilities but one.* DAVID COTTINGHAM

One of the most transparent manifestations of the Platonic assumption of the universality of mathematics is in the search for extraterrestrial intelligence in the Universe. For many years terrestrial radio telescopes have broadcast signals in the direction of nearby star systems and solar system probes have carried artefacts and information stores which are supposed to give any extraterrestrial discoverer definitive information about ourselves and our vicinity.* The information that is sent in these transmissions

* I must confess, irrespective of the likelihood of finding any such extraterrestrial intelligence capable of detecting and responding to these signals, to being unconvinced that we ought to be broadcasting our presence in the Universe. All our human experience on Earth has taught us that ethical and moral values do not go hand in hand with advanced science and technology. Advanced life-forms on Earth, like ourselves, do not have a history of showing impressive sympathy to lesser life-forms. Maybe we should be expending our energies developing effective forms of camouflage?

assumes that mathematics is a universal language. It gives examples of our own mathematical knowledge and even broadcasts on special frequencies which have fundamental significance for anyone possessing a sophisticated (enough to detect radio signals) knowledge of physics. There are even science fiction stories based upon the ideas that intelligent extraterrestrials have sent messages to us already and they are embedded within complicated mathematical structures like the infinite and unpredictable decimal expansion of the number 'pi'. Whilst no one expects extraterrestrials to have the same representations of mathematical concepts as we do, the assumption is that they would be in possession of the same *concepts* nonetheless. This implicitly assumes that there is a universal mathematics that all extraterrestrials will eventually discover. If the Platonic philosophy of mathematics is not true and mathematics possesses a strong cultural element, or is an approximation to reality that our minds have invented in response to the evolutionary pressures created by our particular terrestrial environment, then extraterrestrial mathematics will not be like ours. The communications we have been making will not be in a language that they can correlate with their own intuitions and conceptions of quantity and structure. Their own environment will have presented them with different problems to solve, different conditions to adapt to; their physiologies will differ in many ways and we have seen how our own physiologies played a key role in dictating the direction in which our counting practices moved. The only aspect of the environment that we may share with them is in the realm of astronomy. They will contemplate celestial motions that have much in common with our view of the heavens. Some writers have suggested that the common features in the mythologies and religions of many ancient terrestrial peoples owe much to their common experience of celestial phenomena like the Moon and stars, comets and shooting stars. Similar experiences of the stars and the Milky Way shared by different civilizations in our own Galaxy could begin with religious notions which evolve into mathematical ones. Attractive as such a speculation is, it only needs some cloud cover or a dusty environment in space close to the central plane of the Milky Way to make naked-eye astronomy an impossible activity for an extraterrestrial civilization. Needless to say, the discovery of an extraterrestrial radio signal sent to us which provided evidence of a similar mathematics in that civilization to that employed on Earth would have enormous impact upon the philosophy of mathematics and might provide very persuasive evidence that mathematics is a universal feature which comes into the mind rather than out of it.

When studying the simultaneous discovery of mathematical concepts one must be careful not to generalize over the whole of mathematics. The motivations for ancient mathematical discoveries are rather different to those which inspire developments in modern times. The most primitive notions of geometry and counting are tied to practical applications in obvious ways and there was clear motivation for their adoption by less developed cultures who wanted to trade and converse with more sophisticated neighbours. But in modern times mathematics need not be tied to practical applications and one cannot argue that notions of pure mathematics are developed in response to social or utilitarian pressures. Yet, they may be the outcome of the pressures within the society of mathematicians. Modern communications mean that mathematicians and scientists are in effect a single intellectual society that transcends national and cultural borders. There may exist certain national styles in the way mathematics is done or which influence the range of problems selected; so, for example, if your country does not have a ready supply of cheap and powerful small computers you will tend not to become involved in those research problems that require their use. But it is hard for different mathematicians to be truly independent in the deepest sense. They grow up in the body of mathematical knowledge that is taught to mathematicians of every colour and creed. They attend conferences, or read reports of them, and sense the direction of the subject, they see the great unsolved problems being laid out and conjectures as to their solution being proposed. Earlier, we saw how Hilbert's famous roster of problems guided the work of generations of mathematicians. Even Brouwer's thesis work began as an attack on some of these problems. Thus it is very hard to maintain that contemporary mathematicians are truly independent. They may be independent in the sense that they have not talked to each other directly or through second parties about the same problem, but they have common predecessors, common inherited intuitions and methods. They are fish that swim in the same big pond.

Another intriguing aspect of mathematics that seems to distinguish it from the arts and humanities is the extent to which mathematicians, like scientists, collaborate in their work. A very large fraction of the articles published in the research journals of mathematics have multiple authorship. Such collaboration is rare in the arts unless it arises through a rather strict demarcation of contributions, as was the case with Gilbert and Sullivan's operettas, where one of the duo composed the music, the other the lyrics. There are also some instances of painters who specialize in producing background landscapes on which others paint human figures or animals. As one moves farther away from realist to abstract painting, this division of labour becomes less and less feasible. But in science and mathematics the collaborative process often goes much deeper to entwine the authors in a process of dialogue and mutual criticism in which they are able to produce a result that could not have been even half-reached by one of them. Perhaps we should regard this distinction between the arts and the sciences in the same way as we do the fact that the same scientific or mathematical discoveries are often made 'independently' by different individuals but artistic creations are unique. Is it evidence that the sciences and mathematics are dominated by a strong objective element that is independent of the investigator(s), whose role is primarily that of discovery? The artist or creative writer, by contrast, is offering an almost entirely subjective creation that emanates from the creative mind of the individual. Indeed, it is this non-objective element in the creative process that is so attractive. It is inherently unpredictable and unsystematic.

This contrast between the arts and mathematics can be pursued in an area where the divide has been questioned. Suppose that instead of spending this chapter discussing the nature of mathematics we had picked upon music. We could then have debated whether musicians invented new tunes and rhythms or whether they discovered them in some world of universal musical entities. Indeed, the comparison is rather interesting because the Pythagoreans regarded music and mathematics as closely linked because of the mathematical relations between what are now called musical intervals. The discovery of the simple numerical ratios linking the musical scale with the 'harmonic' progression of string lengths was probably one of the key motivations for the Pythagoreans to conclude that everything was imbued by number. The formalists and constructivists could lay down definite rules of harmony and composition, but there would always be musicians who wish to innovate and improvise and break the rules. The real difference between the structures of mathematics and music is that, whereas there is no basic condition which tells whether or not something is music, irrespective of whether we happen to like the sound of it, there is such a criterion for mathematics. Inconsistency is sufficient to disqualify a prospective mathematical structure. It is this appeal to something that feels objective that is responsible for the suspicions that there is an aspect to be discovered, regardless of how much that element may subsequently be embroidered by human constructions.

## THE PRESENCE OF THE PAST

> *The old men ask for more time, while the young waste it. And the philosopher smiles, knowing there is none there. But the hero stands sword drawn at the looking glass of his mind, aiming at that anonymous face over his shoulder.*
>
> R.S. THOMAS

There is one curious example of the effectiveness of mathematics that weighs heavily upon the side of those who would convince us that mathematics is discovered and existed before there were any such creatures as mathematicians.

When modern astronomers observe the structure of the distant Universe of stars and galaxies they are not determining the nature of the Universe's structure 'now'. They are seeing distant objects as they were far in our past. In fact, in many cases we are seeing them long before any form of life existed on the Earth. For light travels through space at a speed that can never exceed three hundred thousand kilometres per second and so when we observe a distant quasar billions of light years away we are seeing it as it was when the light was emitted billions of years ago. The fact that the mathematical structure of the object being observed coincides with that given by our mathematical analysis on Earth *here and now* witnesses to the fact that there is an intrinsic mathematical aspect to these objects that is observer-independent. Moreover, we can actually observe different quasars so widely separated on the sky that there has not been time for light to travel between them during the time since the expansion of the Universe began. They are independent of one another and cannot influence one another in any known way, yet we find that detailed aspects of the spectrum of light that they emit are identical. This gives us confidence in the existence of some universal substructure that is mathematical in character.

The most extreme example of this direct observation of the past arises from our understanding of what occurred in the Universe during the brief interval of time between one second and three minutes after it began its present state of expansion from some unknown state that we usually call the 'beginning of the Universe' (although it may have been nothing of the sort). Our mathematical theory allows us to determine the ambient conditions in the Universe during those first few minutes of cosmic history from what we observe at present, some fifteen billion years later. It shows that during those first three minutes the entire Universe should then be hot enough and dense enough to sustain a brief chain of nuclear reactions which would burn the material in the Universe into heavier elements of deuterium, two different isotopes of helium, and lithium. The percentages of the total mass of the Universe to be found in these elements now are predicted at 1/1000, 1/1000, 22, and 1/100 000 000 respectively; the remaining nuclei (roughly 78 per cent) will be hydrogen. One of the great successes of modern cosmology, upon which our confidence in our understanding of the expanding Universe back to these early times rests, is the fact that our astronomical observations confirm these detailed predictions. Helium, deuterium, and lithium are all found in the abundances predicted by the mathematical theory of the expanding Universe.

Clearly, this remarkable success is relevant for our present inquiry. It confirms that the mathematical notions that we employ here and now apply to the state of the Universe during the first few minutes of its expansion history at which time there existed no mathematicians. The residual abundances of the elements that were created at those early times are like time capsules which carry information about the nature of reality fifteen billion years ago when the Universe was too hot to permit the existence of any sentient beings. This offers strong support for the belief that the mathematical properties that are necessary to arrive at a detailed understanding of events during those first few minutes of the early Universe exist independently of the presence of minds to appreciate them. And they are the same properties that hold here and now.

Cosmology, the study of the overall structure of the Universe, is the most vivid expression of the Platonic doctrine. Einstein created a theory of gravitation that enables us to find mathematical descriptions of the entire Universe. Modern cosmologists manipulate mathematical equations whose solutions describe the entire physical Universe. Of course, in practice only very approximate solutions can be found and they incorporate sweeping simplifications—like assuming the Universe to be the same everywhere. Yet, the assumption behind such a methodology is that mathematics is something larger than the physical Universe. Studies have begun to be made of the process by which the Universe of space and time might have come into being. Regardless of the correctness of such ideas, one should notice that they are all mathematical theories which assume that mathematics is transcendental: that it exists as a form of logic which governs how everything else should be. The inventionist has a hard job studying such cosmological problems with a clear conscience. If mathematics is not something larger than the Universe what tools should one use to describe the extraterrestrial world?

## THE UNREASONABLE EFFECTIVENESS OF MATHEMATICS

*The mathematician may be compared to a designer of garments, who is utterly oblivious of the creatures whom his garments may fit. To be sure, his art originated in the necessity for clothing such creatures, but this was long ago; to this day a shape will occasionally appear which will fit into the garment as if the garment had been made for it. Then there is no end of surprise and of delight!*

TOBIAS DANTZIG

When it comes to the problem of why mathematics is so successful an instrument for understanding the physical world, the Platonist feels unassailable. Whereas the other philosophies of mathematics we have explored have found this remarkable aspect of mathematics used by their critics to undermine the universality and persuasiveness of their stance, the Platonist alone cites it as part of the case for the defence. If there exists a world of mathematical abstractions of which the appearances are a reflection then we should not be at all surprised to find that mathematics works. The world simply *is* mathematical. But if we dig a little deeper we find nothing behind such a view. There is no explanation as to why the world of forms is stocked up with mathematical things rather than any other sort. All we learn is that mathematics is the most primary essence of things.

The most remarkable examples of the effectiveness of mathematics in describing the physical world are those which are to be found in the realm of the most elementary particles of matter and in the astronomical sciences. The description of the most elementary particles of matter is founded upon an experience of symmetry in the structure of the natural world and the recognition that laws of cause and effect can be replaced by equivalent statements that guarantee some pattern remains unchanged. Thus laws of Nature become equivalent to the maintenance of certain abstract patterns in the fabric of reality and the definition of all possible patterns of this sort is a mathematical problem that was solved long ago. When elementary particle physicists wish to explore a new law of Nature they can choose one of these mathematical patterns and explore the consequences that flow from it. Similar principles of mathematical symmetry lie behind the successful theory of gravitation that Einstein developed to supersede that of Newton. What is most remarkable about the success of mathematics in these areas of science is that they are the most remote from human experience. If we sought to explain the effectiveness of mathematics by appealing to the human propensity to create mathematical tools to fit the purposes required then one would expect to find the best applications of mathematics in the 'in-between world' of everyday dimensions, the mastery of which was essential for our evolutionary success. Instead, we find just the opposite.

The success of these esoteric parts of the scientific quest to understand the Universe around us points to another curiously confident methodology that can be traced to the unconscious influence of the Platonic viewpoint. Hundreds of years ago the route to understanding aspects of the world was to start by making a number of observations of local events, from which, by a process of trial and error, one builds up a larger and more general picture of the laws and constraints upon what happens in the part of the world under examination. Beginning with the local, one builds up, step by step, like a map-maker, a global picture of the world. Beginning with the observation of particular events, scientists pieced together the universal laws of Nature. Modern particle physicists do the opposite. Convinced of the mathematical nature of the Universe, they begin by examining the consequences of various plausible alternatives for the form of the universal laws. They go from the universals to deduce the particular outcomes of those laws.

## DIFFICULTIES WITH PLATONIC RELATIONSHIPS

> *There are the sets; beautiful (at least to some), imperishable, multitudinous, intricately connected. They toil not, neither do they spin. Nor, and this is the rub, do they interact with us in any way. So how are we supposed to have epistemological access to them? To answer, 'by intuition', is hardly satisfactory. We need some account of how we can have knowledge of these beasties.*
>
> PAUL BENACERRAF AND HILARY PUTNAM

The advantages of the Platonic picture of mathematics are obvious. It is simple. It makes the effectiveness of mathematics a thoroughly predictable fact of life. It removes human mathematicians from the centre of the mathematical universe where otherwise they must cast themselves in the role of Creators; and it explains how we can be so successful in arriving at mathematical descriptions of those aspects of the physical world that are furthest removed from direct human experience. So soothing do these features appear that it is easy to ignore the difficulties that such a philosophy of mathematics creates. It sheds no further light on why mathematical truths prove so useful to us. First, one should not necessarily expect a correct account of mathematical reality to answer this question; for we have seen that only particular varieties of computable mathematics are of great utility and our account might well not shed any light on why there are so many computable functions in the mathematical menagerie. Nonetheless, any good account of mathematical reality ought to cast some light upon the question of utility rather than bury it even deeper in mystery as Platonism seems to do. It is all very well to claim that ordinary objects are accounted for by another realm of abstract objects but why should any abstract realm have anything to do with any part of the mundane world.

Let us take a long hard look at what the Platonist is asking us to believe. We must have faith in another 'world' stocked with mathematical objects. Where is this other world and how do we make contact with it? How is it possible for our mind to have an interaction with the Platonic realm so that our brain state is altered by that experience? Many mathematicians of the Platonic persuasion are strongly influenced by the fact of their own and others' intuition. They have experience of just 'seeing' that certain mathematical theorems are true which makes it feel that they have suddenly come upon mathematical truth by a faculty of 'intuition' that is tantamount to discovery. This non-sensory awareness of abstract mathematical structures is a faculty that varies widely, even amongst mathematicians, and so the Platonist must regard the best mathematicians as possessing a means of making contact with the Platonic world more often and more clearly than other individuals.

## SEANCE OR SCIENCE?

*What man does not know*
*Or has not thought of*
*Wanders in the night*
*Through the labyrinth of the mind.*
GOETHE

The first reaction of the sceptical inquirer into the Platonic position is to draw attention to its frustrating vagueness. Where do all these mathematical objects live? What are they? How do we make contact with them? And why are they mindful of us? In an attempt to amplify this objection more clearly we should distinguish between different varieties of Platonism that can be found amongst mathematicians and philosophers. *Ontological Platonism* is the stance that mathematics is about real objects; although these objects presumably cannot be material things like tables and chairs because they are potentially infinitely more numerous than all the things in the the physical Universe. The alternative view, *epistemological Platonism*, enshrines the idea that we have a means of directly perceiving mathematical objects. We see at once that if the ontological Platonist is right and there do exist immaterial mathematical entities then the epistemological Platonist is in deep trouble. He finds himself trying to defend a form of extrasensory perception of 'another' mathematical world. Cognitive scientist Aaron Sloman summarizes the sceptical reaction to this position when discussing Roger Penrose's defence of epistemological Platonism:

> All he is claiming is that mathematical truths and concepts exist independently of mathematicians, and that they are discovered not invented. This, I believe, deprives platonism of any content ... Although there are many who contest platonism as somehow mystical, or anti-scientific, just as hotly as Penrose defends it, such disagreements are really empty. It makes not a whit of difference to anything whether [mathematical objects] exist prior to our discovering them, or not. The dispute, like so many in philosophy, depends on the mistaken assumption that there is a clearly defined concept (in this case 'existence of mathematical objects') that can be used to formulate a question with a definite answer. We all know what it means to say that unicorns exist...Quite different procedures are

involved in checking the equally intelligible question whether there exists a prime number between two given integers ... But there is no reason to assume that any clear content is expressed by the question whether *all* the integers do or do not 'really' exist, or exist independently of whether we study them or not ... Arguing [whether] certain things do or do not *exist* is pointless when the relevant notion of existence in question is so ill-defined. I conclude that the question whether platonism is true is just one of those essentially empty questions, that has an aura of profundity, like 'Where exactly is the Universe?' or 'How fast does time really flow?'

The vagueness of the Platonic position boils down to its fudging of the issue of how we gain access to this world of mathematical ideas—what is the source of mathematical intuition? It is one thing to maintain that there exists another eternal Platonic world of mathematical forms but quite another to maintain that we can dip into it through some special mental effort. But without the possibility of such contact, mathematical truths must be regarded as essentially unknowable and our theories of sets and numbers cannot really be about the mathematical entities themselves.

If we cannot have this contact with the Platonic realm, then Platonism is not worth considering any further. The mathematics that *we* do would have to be founded upon some principle other than discovery. The question of whether we have some particular type of mental connection with such a realm is one which seems to have been little considered by psychologists and neurophysiologists (presumably) because they do not take such a possibility very seriously. We would be forced to believe that there exists a way of tapping into the Platonic realm that barely exists for ordinary people, but which is well developed* amongst those particular individuals who are called 'mathematicians'. Moreover, we might ask whether they need to make use of this special connection only when engaged in mathematical discovery, so that mathematical ideas can be explained to other non-mathematicians without them needing access to the Platonic realm as well. Things become more complicated still when we encounter mathematicians who make new mathematical 'discoveries' by putting together ideas that have been learned directly from other mathematicians. Is this type of mathematical deduction inferior to that arrived at by 'pure thought' wherein some contact is made with the Platonic realm? Many mathematicians might tend to think so, regarding originality as a measure of one's independence of other mathematicians.

* The non-mathematical reader might at this point become concerned by the fact that so many of the greatest mathematicians who have played a role in our story (Cantor, Brouwer, Gödel) suffered from serious mental and psychological problems at various times. 'Great wits are sure to madness near allied' as Dryden cautioned.

The difference between first- and second-class mathematicians now seems to reduce to their ability to make contact with the other world of mathematical Ideas. Applied mathematicians seem doomed to be second-class because they might never feel the need to tap this well-spring of Ideas, finding intuition enough in the world of particulars they see around them. Clearly, this line of argument is becoming incredible if we are unconvinced that mathematicians possess these extrasensory powers, notwithstanding the fact that they *do* possess some special mental powers not shared by the population at large—that's why they are mathematicians, after all. Some mathematicians have believed that their discoveries were made possible by a special form of revelation. Cantor, for instance, believed that his vision of the never-ending hierarchy of infinities was revealed to him by God. The psychologist Jean Piaget, who had deep interests in the nature of mathematical intuition, felt that if the Platonic philosophy were correct then some mathematicians should see mathematical truths that were only dimly understood and which would later yield up their full meaning:

Platonism would doubtless compel universal recognition if one day an exceptional subject were to 'see' and describe in detail ideal entities which neither he nor his contemporaries were capable of understanding, and if his 'visions' duly recorded and put in protocol form gave rise 50 or 100 years later to explanatory works which would elucidate their full meaning. But this has never happened.

It might, of course, be argued by the defence that this has happened; that mathematicians often discover new structures which later turn out to possess unsuspected properties and interconnections with other parts of the subject, indicating that the process of mathematical discovery unveils something independent of all minds.

Platonists often cite Poincaré's account of his discovery of Fuchsian groups as recorded in Jacques Hadamard's book *The Psychology of Invention in the Mathematical Field* as evidence of the nature of mathematical invention. Poincaré described periods of intensive intellectual effort in his search for a type of mathematical structure that he would eventually term 'Fuchsian functions'. But he had failed to make progress. Then, he recalls,

I left Caen, where I was living, to go on a geologic excursion under the auspices of the School of Mines. The incidents of the travel made me forget my mathematical work. Having reached Coutances, we entered an omnibus to go to some place or other. At the moment when I put my foot on the step, the idea came to me, without anything in my former thoughts seeming to have paved the way for it, ... I did not verify the idea; I should not have had time, as upon taking my seat in the omnibus, I went on with a conversation already commenced, but I felt a perfect certainty. On my return to Caen, for convenience' sake, I verified the result at my leisure.

Much stress is usually laid upon the fact that the idea came to him whilst his conscious thoughts were elsewhere. However, I believe there is little significance in this story. For every example of such unconscious intuition one can find another mathematician who makes discoveries by laborious computational thinking. A nice example is that of John von Neumann whose terrifying mental quickness had peculiar consequences. A colleague once set him a little problem to see how quickly he would solve it. Two cars start sixty feet apart and travel towards one another, each moving at fifteen feet per second, and a bee flies back and forth between the cars at thirty feet per second as they approach one another. How far does the bee fly before the cars collide? Von Neumann gave the answer almost before his colleague had finished the question—sixty feet. 'How did you solve it?' he was asked. Did you just notice that the cars collide after 60 divided by 15 plus 15 = 2 seconds; in which time the bee flies 30×2 = 60 feet? Not a bit of it; von Neumann formulated the infinite sum of distances travelled by the bee in each of its trips between the approaching cars and then summed the series—just like that! Who needs insight when you can think that fast?

## REVEL WITHOUT A CAUSE

*A reality completely independent of the spirit that conceives it, sees it, or feels it, is an impossibility. A world so external as that, even if it existed, would be for ever inaccessible to us.* HENRI POINCARÉ

If we look at the intuitions that mathematicians do have, then we have the problem of deciding how we know that they are truly about some external reality. The fact that, in Gödel's words, they 'force themselves upon us' as being true, is hardly persuasive. Might this just mean that they arise as the result of mental pressures or some subtle form of social conditioning or as an inevitable by-product of the evolutionary process? But let us explore how we might come to 'know' what is in the Platonic world. If its inhabitants are not material objects, how can they be known? How exactly do they interact with us?

There seems to be no way in which mathematical truths can interact with our minds to create mathematical knowledge of them. At first one might think this problem is no different from that of believing in the existence of any type of physical reality external to our minds. Why is the mathematician in deeper water in this respect than the scientist? How do we 'know' that there exist atoms or magnetic fields? But there is a real difference. The objects of the material world can act as causes of our sense perceptions and thoughts about them. The fact that different individuals are induced to have similar thoughts about them indicates that they do indeed have such an influence upon us. This is called the 'causal theory of knowledge': it maintains that if any of our beliefs is to be credited as knowledge (as opposed to speculation or wishful thinking) then what makes the belief true must in some way be causally responsible for me holding that belief. But mathematical entities do not seem to permit such knowledge. Triangles and numbers do not seem to be able to create sense-perceptions which support our belief in their existence: mathematical entities are causally inert because they are not part of the physical Universe. This type of objection does not necessarily assume that mathematical entities are 'objects'; they could equally well be structures or properties whose character we would also be unable to change through our interaction with them. So it appears that if Platonism is true and abstract mathematical entities exist then we can have no mathematical knowledge. And if it is maintained that we do indeed possess such knowledge then Platonism must be a false philosophy of mathematics.

This type of objection—that to know something, we must causally interact with it, and this we cannot do with abstract objects—is by no means a new one. In the second century Sextus Empiricus used a very similar argument against the Stoics' concept of a logical 'proposition'. The argument was a particularly telling one because the Stoics were materialists who would not countenance any mystical qualities being introduced into their world-view in order to explain our apprehension of abstract concepts like 'propositions'. Here is Sextus' argument; it is almost identical to that raised against mathematical Platonism:

> Let it be supposed ... that 'expressions' are 'in existence', although the battle regarding them remains unending. If, then, they exist, the Stoics will declare that they are either embodied or abstract. Now they will not say they are embodied; and if they are abstract, either—according to them—they effect something, or they effect nothing. Now they will not claim that they effect anything; for according to them, the abstract is not of a nature either to effect or to be affected. And since they effect nothing, they will not even indicate and make evident the thing of which they are signs; for to indicate anything and to make it evident is to effect something. But it is absurd that the sign should neither indicate nor make evident anything; therefore the sign is not an intelligible thing, nor yet a proposition.

There are two ways to avoid being impaled upon this logical fork whilst still retaining the Platonic approach to the interpretation of mathematics. We can deny the traditional causal theory of how we acquire knowledge, so that we can come to 'know' these causally inert mathematical objects in some extrasensory fashion; or we can deny that mathematical objects are causally inert and so grant that we can come to know them in a manner that is concordant with the causal theory of knowledge-acquisition.

The first of these options is the more radical. It moves the notion of mathematical intuition very close to familiar religious beliefs which permit means of 'spiritual' contact between human beings and God or His intermediaries. Alternatively, it might be seen as the close cousin of the desire many people feel for some means of contact with 'other realities' by extrasensory means. When described in such stark terms, this route to making sense of Platonism would probably be anathema to almost every mathematician. But nonetheless it is at root a viewpoint that one finds being advocated, albeit in a sanitized form, by highly respected figures. Kurt Gödel, as we have already seen, was one such advocate of some other kind of relationship between ourselves and reality which would permit mathematical entities, like sets and numbers, to influence the human mind. He claims:

> It by no means follows, however, that [intuitions], because they cannot be associated with actions of certain things upon our sense organs, are something purely subjective ... Rather, they too may represent an aspect of objective reality, but, as opposed to the sensations, their presence in us may be due to another kind of relationship between ourselves and reality.

This requires us to contemplate a situation in which the brain initially contains no state which encodes a particular piece of mathematical knowledge, but then at some future time this information is encoded in a brain state but no interaction has occurred between the brain and some external source of mathematical information. This is not unlike our idea of a 'vision'. The Oxford philosopher Michael Dummett (the most notable current supporter of the intuitionists' constructivist philosophy of mathematics as part of his wider interests in the philosophy of 'meaning') detects in it 'the ring of philosophical superstition'. Gödel's strongest critic, the American philosopher Charles Chihara, finds his claims of no value because

> the 'explanation' offered is so vague and imprecise as to be practically worthless: all we are told about how the 'external objects' explain the phenomena is that mathematicians are 'in some kind of contact' with these objects. What empirical scientists would be impressed by an explanation this flabby? ... It is like appealing to experiences vaguely described as 'mystical experiences' to justify belief in the existence of God.

He believes that the unanimity of different mathematicians about so many aspects of mathematical intuition, which lies at the root of many independent discoveries of similar concepts, arise simply because 'mathematicians, regarded as biological organisms, are basically quite similar'. But our common biological heritage does not seem to prevent us having a wide spectrum of differing opinions about almost everything else under the sun; nor are our mathematical experiences terribly similar to mystical religious experiences. We can all do simple sums and get the same answer without having to conjure up a contact with the spiritual world of numbers.

Roger Penrose's defence of Platonism in *The Emperor's New Mind* seems to imply a similar form of extrasensory awareness of causally inert mathematical entities:

> I imagine that whenever the mind perceives a mathematical idea, it makes contact with Plato's world of mathematical concepts ... When mathematicians communicate, this is made possible by each one having a direct route to truth, the consciousness of each being in a position to perceive mathematical truths directly, through this process of 'seeing'.

This process of 'seeing' is still vague and amounts to a notion that mathematicians are in communion with the great ocean of truth:

> Because of the fact that mathematical truths are necessary truths, no actual 'information', in the technical sense, passes to the discoverer. All the information was there all the time. It was just a matter of putting things together and 'seeing' the answer! This is very much in accordance with Plato's own idea that ... discovering is just a form of remembering.

However, a mathematical physicist like Penrose is not much swayed by the causal theory of knowledge, because he knows of situations in which quantum theory holds sway where one can have events without causes and derive knowledge about a system without any causal interaction taking place. The quantum physicist familiar with the Einstein–Podolsky–Rosen non-locality, and the experiments of Alain Aspect which display it, feels no qualms about maintaining that the causal theory of knowledge is simply wrong. This is why Penrose looks to quantum interactions in the brain as the interface between the Platonic world of all possible mathematical concepts (enshrined somehow in the unobservable 'wave function') and the physical world of particulars (or 'observables' in physicists' parlance), although as yet we have no evidence that quantum uncertainty does play any decisive role in the brain's overall operation.

The great Hungarian mathematician, Paul Erdös, who is famous for his eccentric ways as well as his remarkable intuition regarding what will eventually be proved true or false in many areas of mathematics (so much so that he offers attractive cash prizes for proofs of his conjectures), talks of good mathematical ideas and proofs coming 'straight from the book'.

The Deity is fondly imagined to possess 'a transfinite book of theorems in which the best proofs are written. And if he is well-intentioned, he gives us the book for a moment.'*

All these leanings toward a contact with a higher realm of mathematical truth seem to make mathematics a rather exact form of spiritualism practised by those mediums, called 'mathematicians', who are able to make sure contact with the other realm. The teaching of mathematics has more in common with a seance than a science. Of course, the language used by Erdös and others is hyperbole, but if one probed to ask what it was believed to represent one would be inexorably pushed into defending one of these more extreme positions.

Penrose looks to the possibility of an intrinsically quantum-mechanical ingredient in the reception and processing of information in the human brain which, through the 'collapse' of the wave function of all possibilities into the one perceived outcome, produces contact with Plato's world. But, at present, there is no reason to believe that there are any intrinsically quantum-mechanical aspects of human neurophysiology.

## A COMPUTER ONTOLOGICAL ARGUMENT

*Ask not what's inside your head but what your head's inside of.*

WILLIAM MACE

If there is another way to overcome the problem of how we gain access into the Platonic realm, it would need to do away with the conventional assumption that mathematical entities are causally inert. How can we do this? It seems that it would require us to relinquish the idea that there are pure abstractions which act as the blueprints for the particular concepts we experience. But suppose that we dispose of the divide between the Platonic world and our material world. The inventionist achieves this by doing away with the Platonic world; what happens if we explore the alternative? To do this, we take a lesson from the development of computer simulations of physical events, which have become the mainstay of so many scientific studies of complex problems.

Suppose that an astronomer wanted to understand how galaxies of stars formed and evolved into the exotic configurations we see in space today. One way to study the problem is to create a computer simulation of the galaxy-formation process. We program the computer with the laws of Nature which govern the way gravity acts between particles of matter and

* Such a book is reminiscent of Borges's 'Book of Sand' with its infinite number of pages. It contains everything but scanning is not advisable; if you lose your place you can never find it again.

how gas and dust respond to gravitational forces, how heating and cooling take place, and so forth. Then we follow the computations of the computer as it provides us with an unfolding history of events which would develop from a particular, perhaps random, starting distribution of particles. At present this type of computer simulation can be run for days on end and can produce a rudimentary picture of how the overall structure of a gigantic galaxy is built up from the individual motions of billions of pieces of moving matter. But, let us now extrapolate this research project into the far future and imagine the exquisite detail which the simulations of the future might be able to produce. Suppose we tell the computer all the laws of Nature that we know. The computer can now do much more. Instead of merely telling us how pieces of material move about and cluster, it will be able to follow the condensation of that material into objects that we recognize as real stars. With greater resolution and computational speed we could then follow the creation of planets around some of those stars. With even more power and programming, that includes other laws of Nature, the simulation would begin to reveal the evolution of simple molecules on the surfaces of some of those planets. Later, these would give way to the production and replication of complex biological molecules. Next, with finer resolution still, the simulation should reveal the development of living things, appearing and dying on the accelerated time-scale which the computer hardware dictates. Ultimately, the simulation could give rise to states of such complexity that they exhibit rudimentary aspects of that phenomenon we call 'consciousness'. At this stage some parts of the computer simulation will be able to communicate with other parts of the simulation; they will be 'aware' of their own structure; they will be able to make observations to ascertain the overall structure of the simulation they are embedded within. This process they will call 'science' and it will enable them gradually to piece together the laws of Nature which we have programmed in to determine how the simulation at one time is connected to its past state. They will not, of course, be able to determine that they are part of someone else's simulation, but it is very likely that some self-conscious parts of the simulation would indulge in 'theological' speculations about the origin of the 'world' about them, the nature of its initial state and what lay before it, and whether these considerations point to some Initiator of Everything.

Let us now examine this state of affairs a little more closely. The computer simulation is just a means of carrying out very long sequences of intricate mathematical deductions according to particular rules commencing from a given starting state. The print-out of our results, either on paper or as a movie, is just a way of visualizing the consequences of the mathematical rules that we have laid down. If we remove the prop of the computer hardware, then we see that the entire sequence of events that unfold in the simulation—the stars, the planets, the molecules, and the 'people'—are all just mathematical states. So if we think of the simulation as a vast web of mathematical deductions spanning out from the starting state, then, as we search through this network of mathematical possibilities, we will eventually come across the mathematical structures that correspond to the self-conscious beings. Both they and the possible communications they can make with other parts of the mathematical structure are parts of the structure. Processes like those that we call 'thinking' are just particular types of a very complex interrelationship. When we reach this stage we see that we really have no need for the computer hardware we started with; indeed, its particular identity is really irrelevant. We could have run our program on all manner of different types of computer architecture. But surely, if we are of the Platonic viewpoint, we need not have run the program on any hardware at all. This means that we can think of the mathematical formalism as containing self-conscious states—'minds'—within it.

This speculative line of reasoning turns the Platonic position inside out. We no longer need to think of mathematical entities as abstractions that our material minds are battling to make contact with in some peculiar way. We exist in the Platonic realm itself. We are mathematical blueprints.

This allows us to take the second path we introduced above, because mathematical entities can no longer be regarded as causally inert. They can affect us in just the same way as any familiar material object because we, and those objects together with their means of physical interaction with us, are nothing more nor less than very complicated mathematical relationships.

This approach has all sorts of interesting ramifications. It means that anything that can happen—anything that is a possible consistent statement in the language of mathematics—does happen in every possible sense of the word.* It also allows us to explore the parts of mathematics that can and cannot give rise to consciousness in a new way.

* This echoes an intriguing correspondence that went on between David Hilbert, in the midst of prosecuting his Formalist programme of reducing mathematics to rules and axioms, and the fervent Platonist, Gottlob Frege. For Frege the intuitive truth of axioms guaranteed the absence of any contradictions between them; Hilbert believed the opposite, that absence of contradiction 'is the criterion of truth and existence'. Frege was worried that if there were some way of proving the consistency of a set of axioms then Hilbert's stance would permit a provable version of the ontological proof of the existence of God. For, if it could be shown that the three conditions 'X is an intelligent being', 'X is omnipresent', and 'X is almighty' are consistent then, without finding a particular X that satisfies them, it would follow that an almighty, omnipresent, intelligent Being exists. Frege concludes that 'I cannot accept such a manner of inferring from consistency to truth'.

The picture we have sketched in which consciousness is viewed as a form of software which requires no reference to any hardware invites us to think of the Platonist picture of mathematics slightly differently. One of the problems with the Platonic viewpoint which we have highlighted is that of saying 'where' mathematical objects are and how we interact with them. Ever since the Pythagoreans, the mainstream of thinkers who use and develop mathematics have moved away from the Hermetic notion that the numbers and symbols of mathematics possess some special occult meaning. Rather, we place significance upon the relationships between numbers or symbols. Thus we see that the practice of mathematics witnesses to the fact that its interest is in structures or patterns and interrelationships. We recall that the entire Bourbaki programme of formalizing the decidable part of mathematics was built upon the belief that one should classify and expose the common structures underlying superficially dissimilar parts of mathematics. This 'structuralist' perspective is interesting because it stresses that the Platonist's mathematical objects, whether they are 'real' or 'abstract', are just not like other things. One pattern may occur within itself, but a thing cannot exist within itself.

This emphasis upon the essence of mathematics as the interrelationship between things rather than in the things themselves was one which we used to characterize the transformation of numerology into mathematics. By looking at it a little more carefully, we can find a further difficulty for the simple realist view of a Platonic world full of 'things' like numbers and sets. There have been many attempts to define what numbers are in terms of sets. Russell and Whitehead did this in their *Principia Mathematica*. For example, one could define the number 1 as the set whose only member is the empty set, then the number 2 as the set whose only members are the empty set and 1, as just defined, and so on. The sets that result have all the properties of the natural numbers. However, it turns out that it is possible to carry out this defining of numbers by sets in other, different ways as well. So the problem is: Which of the sets defining the number 1 *really* is the number 1? If we think that it is some definite object that lives in the Platonic realm, then it should be only one of the set-theoretic constructions. The critic would argue that none of these constructions seems superior to any of the others; there is no good reason for saying that a number is a particular set; so numbers are surely not sets at all. One should not think of numbers as objects or sets at all. If one is presented with one in isolation, it is meaningless. It has no meaning save as part of a sequence linked by definite rules. It makes no sense to think of numbers as objects, because this divorces them from the abstract structure, without which they are nothing at all. Since numbers are so central to mathematics, this argument that they cannot be uniquely ascribed an existence as sets or objects in the usual sense is a telling one. And, of course, even if it had been possible to define numbers as sets in only one way, we would still have to deal with the fact that there are many different varieties of set theory. We are in no position to explore the Platonic realm to discover which, if any, of them are the true furniture of Heaven. If there is one and only one correct set theory, then the only way in which we might hope to find it is through its manifestation in the physical world. This is a remarkable conclusion because it means that the past emphasis upon applying mathematics *to* sciences like physics is reversed. We should study the physical world in order to determine the nature of the most basic mathematical structures. In fact, to some extent, this trend can be seen in the recent development of 'superstring' theory by physicists. Although motivated by imperatives of physics, this logical structure has pointed mathematicians in new directions and revealed deeper interconnections between pieces of disparate mathematics already known to them. And, perhaps most telling, these powerful abstract structures seem to be manifested in the unobservable part of physical theory; that is, in the *laws* of Nature, rather than in the outcomes of those laws which constitute the observable world.

## A SPECULATIVE ANTHROPIC INTERPRETATION OF MATHEMATICS

> *Why, for example, should a group of simple, stable compounds of carbon, hydrogen, oxygen and nitrogen struggle for billions of years to organize themselves into a professor of chemistry? What's the motive?*
>
> ROBERT PIRSIG

In our exploration of the formalist approach to mathematics, we saw how very simple axiomatic systems, like Euclidean geometry, which were not elaborate enough to include the whole of arithmetic, possess the property of completeness. But keep adding further axioms one by one to enlarge the scope of the system and a critical threshold is passed when the system grows large enough to include arithmetic. Gödel's theorem then applies and there necessarily exist statements whose truth cannot be decided within the system. If we pursue the image of the deductions and statements of mathematics laid out in a great web of interconnections, this theorem teaches us that some parts of the web are disconnected from others and cannot be reached from them. An interesting example is provided by John Conway's 'Game of Life' which we described in the last chapter. Despite its simple rules, it is rich enough to be equivalent to arithmetic and so manifests Gödel incompleteness. This is displayed by the fact that there exist some configurations of the game which cannot be reached from other configurations by any sequence of moves made using the rules of the game. An example is shown in Fig. 6.2.

Let us now imagine how we could set about creating a sequence of ever larger and richer axiomatic systems. We begin with the simplest systems of axioms which are complete, then enlarge them by adding a new axiom until they are rich enough to possess incompleteness. Then, we create new systems from the old ones by finding undecidable statements (like the 'continuum hypothesis' that we discussed earlier) so that the new offspring can be made in pairs by adding, in one case, one of these undecidable statements as a new axiom, and in the other by adding its negation. There are other ways of creating a never-ending sequence of (different) axiomatic mathematical structures; this example serves merely to fix ideas.

We can now examine the array of mathematical systems in the light of our recognition that some formalisms ultimately contain self-conscious 'observers' of their internal structure. We see that some very simple axiomatic systems will not possess a spectrum of consequences rich enough to sustain such complex states. Only when a certain level of complexity is attained will such 'observers' arise. This leaves us with the problem of deciding just what critical level of complexity is necessary to allow the mathematical formalism to contain self-awareness. This question is too difficult to answer; but we can show how it helps us understand other aspects of mathematics and human intelligence in interesting ways.

An interesting case study to consider is the widely publicized claim of Roger Penrose that human thinking is intrinsically non-algorithmic. This claim is based on the truth of Gödel's theorem and the ability of human mathematicians to see that it is true. It amounts to the claim that the process by which mathematicians 'see' that a theorem is true, or a proof is valid, cannot itself be a mathematical theorem because otherwise its Gödel number could be calculated and that would itself be a mathematical truth. Penrose concludes on the basis of a rigorous *reductio ad absurdum* argument* (based on one originally given by Alan Turing) that *human mathematicians are not using a knowably sound algorithm to ascertain mathematical truth*. A more detailed explanation of the argument, which dates back to Turing and Gödel himself, goes like this: If we have any formal system, call it *S*, rich enough to contain arithmetic, then by the process of Gödel numbering (associating a prime number with each logical ingredient of a statement *about* mathematics and hence associating a unique number with the whole statement by multiplying together the

* Hence, if the constructivist philosophy of mathematics were adopted this argument would no longer hold good.

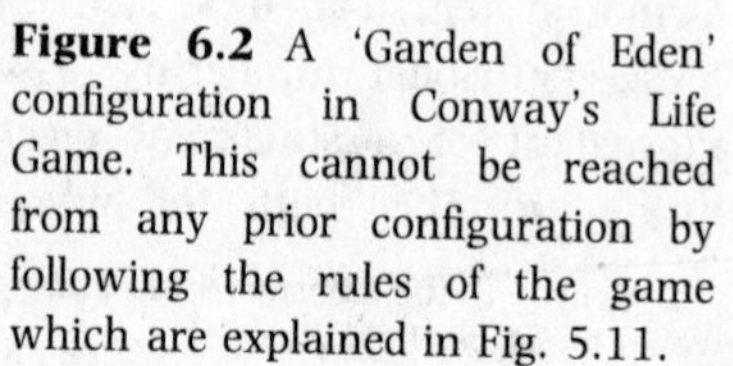

**Figure 6.2** A 'Garden of Eden' configuration in Conway's Life Game. This cannot be reached from any prior configuration by following the rules of the game which are explained in Fig. 5.11.

primes which define the logical ingredients) we can construct a series of arithmetical formulae of which the $k$th is $F_k(n)$. It is defined by the stipulation that the procedure $F_k(n)$ is true of the number $n$ if and only if there is no number $N$ for which the $N$th possible proof in the formal system $S$ is a proof whose Gödel number is $n$. This allows us to consider the formula $F_k(k)$ where $k$ is the Gödel number of the formula $F_k(k)$. The content of this valid construction is that there cannot be any way of deriving $F_k(k)$ or its opposite within $S$ because the unprovability of itself within the system is precisely what $F_k(k)$ asserts.

Penrose points to this as an example of something that we can see is true but which cannot be derived within the system $S$, despite the fact that $S$ (according to the 'Artificial Intelligence' enthusiasts) is supposed to be the formal system they are seeking which defines how all our minds work. The conclusion is that no formal system can capture all the insights and judgements that he can evidently make, so there can be no single algorithm that is equivalent to his mind. As with the original proof of Gödel's theorem, from which the argument is derived, we can always make a particular $F_k(k)$, which we find to be underivable, derivable by adding some new axioms to our formal system S, but this will always generate a new unprovable formula. Penrose's argument rests upon the following claim:

We *see* the validity of the Gödel proposition $F_k(k)$ though we cannot derive it from the axioms. The type of 'seeing' that is involved in a reflection principle requires a mathematical insight that is not the result of the purely algorithmic operations that could be coded into some mathematical formal system ... [so] ... If the workings of the mathematician's mind are entirely algorithmic, then the algorithm (or formal system) that he actually uses to form his judgements is not capable of dealing with the proposition $F_k(k)$ constructed from his personal algorithm. Nevertheless, *we* can see (in principle) that $F_k(k)$ is actually *true*! This would seem to provide *him* with a contradiction, since *he* ought to be able to see that also. Perhaps this indicates that the mathematician was *not* using an algorithm at all.

The emphasis here upon the centrality of a mathematician being able to 'see' the truth of Gödel's construction is rather worrying. If consciousness rests upon this ability, then we seem to be implying that we must regard young children and non-mathematicians as lacking a fundamental aspect of what it means to be conscious. If I do not see the truth of the Gödel sentence, does this mean I am not as conscious as someone who does, or that my mind could be simulated by a single algorithm whilst theirs could not? Consciousness is a universal trait amongst humans; it would be very odd if its essential element was the grasp of such a difficult and esoteric notion.

One can construct some interesting heuristic examples which show how we are able to discern the truth or falsity of undecidable statements. Suppose we take one of the great unsolved problems of mathematics; Fermat's 'last theorem' is a conjecture (which Fermat claimed he had proved although no proof was ever found), that remains unproved to this day. Fermat's conjecture is that there are no natural numbers *n*, *A*, *B*, and *C* for which

$$A^n + B^n = C^n \qquad (1)$$

is true when *n* is larger than 2.* Now suppose it was proved that Fermat's conjecture is *undecidable*; that is, it can neither be proved true nor be proved false using the axioms of arithmetic. If that were to happen, we would be able to conclude that Fermat's conjecture is *true*! This we can do because if Fermat's Conjecture is false then there exists some triplet of positive numbers *A**, *B**, and *C** which satisfy the equation (*) for some *n** bigger than 2. So, if we programmed a computer, or instructed some students, to search through all possible choices of the four numbers (*A*,*B*,*C*,*n*), they must eventually (perhaps only after billions of years) come across the set (*A**,*B**,*C**,*n**) which does the trick. But this means that Fermat's Conjecture would have been rendered decidable: and it is false. The only way this is compatible with it being undecidable is if no such counterexample can arise in the search process even if it goes on forever. But if no counterexample can exist then Fermat's conjecture *must be true*. If one is a Platonist who believes that mathematical statements are either true or false, then one would have to conclude that Fermat's Conjecture cannot be shown to be undecidable.

Other unsolved problems lead to similar conclusions. For example, there exists a famous unconfirmed conjecture made by the German-born mathematician Christian Goldbach in a letter to Leonhard Euler in 1742, that *every* even number greater than 2 is the sum of two prime numbers (for example, 4=2+2, 6=3+3, 8=5+3, 10=7+3, 12=5+7, etc.). No exception is known and no proof has ever been found.† Suppose that this were proved to be an undecidable statement. In order to perform the same trick as for Fermat's conjecture, we would need to set up a procedure that would search systematically through all even numbers to discover one

* For $n = 2$, as the Pythagoreans were fascinated to discover, there are a limitless number of examples: for example $A$=3, $B$=4, and $C$=5.

† To gauge the difficulty one should consider that no progress was made with the problem until 1931 when a Russian mathematician, Schnirelmann, showed that every number can be written as a sum of not more than 300,000 primes. Later, Vinogradoff proved that there is some number, N, such that every larger number can be expressed as the sum of either two, three or four primes, but no one knows how large the number N is.

that cannot be expressed as the sum of two primes. This can be done. We simply list all the even numbers and then systematically compare each one of them with the list of sums of all possible pairs of smaller prime numbers. If Goldbach's conjecture were 'undecidable' then this process could not ever find a counterexample. Hence, it must be true.

A third example, not of this sort, is the 'twin-prime conjecture' that there are an infinite number of consecutive odd numbers that are primes (like 3 and 5, 11 and 13, 101 and 103). Suppose this is undecidable, then if we search systematically through all the odd numbers we must never be able to come up with a counterexample. We might have a candidate to be the largest (and hence the last) pair of odd numbers which are both primes but we could not prove there were not bigger pairs in the infinite list of odd numbers that still remained to be checked out.

In general, the undecidability of some statement *S* implies the truth of whatever statement would be rendered false by the discovery of a counterexample that can be found by systematic search.

A curiosity of the argument Penrose gives is the way it goes against our intuitive expectations of what Gödel's incompleteness proofs would tell us about the relationship between the human mind and mathematical truth. For the Platonist, mathematical statements are independent of the human mind and so should be either true or false; we should be able to look over God's shoulder in the Platonic heaven and see. The proof of the 'existence' of undecidable propositions therefore seems to imply that mathematics and its formalisms are products of the human mind's inventiveness. Thus mathematics tells us something deep about the nature of the human mind. Penrose's argument uses Gödel's mental constructions to prove that because mathematics is not just in the mind, no mathematical algorithm can capture all those aspects of it that the mind seems able to.

The anti-Platonist might cite Gödel's theorem as evidence that there cannot exist an external world of mathematical objects because such a world should have a definite, complete, and consistent existence. But the Platonist should respond that Gödel's incompleteness has nothing to tell us about the existence or non-existence of the Platonic realm of mathematics, sets, and numbers; it simply displays the limitations of *symbolism* as a representation of its nature. No system of symbols is rich enough to capture all the possible interrelationships between numbers.

The earliest interpretation of Gödel's theorem as distinguishing the potential of a machine from the human mind seems to be Emil Post's (1941) statement about creativity and logic in which he writes,

> It makes of the mathematician much more than a kind of clever being who can do quickly what a machine could do ultimately. We see that a machine would never

give a complete logic; for once the machine is made *we* could prove a theorem that it does not prove.

A similar argument was made by John Lucas in 1961 in a discussion of the relative capabilities of minds and machines and has been a constant source of dispute ever since (see for example Douglas Hofstadter's book *Gödel, Escher, Bach*).

We have already seen in earlier chapters that mathematical truth itself is not algorithmic. Turing's discovery of non-computable mathematical functions established that over fifty years ago; what is being claimed here is that our 'understanding' of mathematics is not algorithmic. This is presented as a challenge to the goal of the 'Artificial Intelligensia' to produce an algorithm with the same information-processing capability as the human mind. Such a goal is often said to rest upon the assumption that the human mind performs only computable operations that can be simulated by a Turing machine. However, this seems to be an over-simplified view of the situations envisaged by the pursuers of Artificial Intelligence. The idea of 'mathematical computability', that is, equivalence to one of Turing's idealized machines, seems to be neither necessary nor sufficient to produce what we would be willing to call a mind.* There is no reason to believe that the mental processes of simple living creatures, like budgerigars, or even of unassisted human beings, possess all of the processing powers of a Turing machine. Nor would we expect a single Turing machine to be able to simulate the behaviour of a complex system which is linked to its chaotically unpredictable environment in all sorts of complicated (or even random) ways that produce non-computable responses. Nor need this system do anything that we would characterize as 'intelligent'. An example might be nothing more than a roulette wheel with various sensors which tie its behaviour sensitively to chaotic variations in the local air temperature and pressure, both of which vary in complicated ways. It is fair to say that most workers in the field of Artificial Intelligence are well aware of these possibilities and they do not, as Penrose claims, contend 'that the mere enaction of an algorithm would evoke consciousness'.

There are a variety of alternative conclusions that could have been

* Gödel objected to Turing's attempts to capture the capability of the human mind by his idealized machine on the grounds that 'Turing completely disregards [that] Mind, in its use, is not static, but constantly developing'. Although at present the human mind is like a computer with only a finite number of internal states 'there is no reason why the number [of states] should not converge to infinity in the course of its development ... there may be exact systematic methods of accelerating, specializing, and uniquely determining this development, for example by asking the right questions on the basis of a mechanical procedure [but a] procedure of this kind would require a substantial deepening of our understanding of the basic operations of the mind'.

drawn from this argument of Penrose's, other than that human mathematical understanding is non-algorithmic. The most reasonable is to conclude, as Turing himself did, that we simply make use of a fallible algorithm. It works most of the time but all of us (even those who are mathematicians) make mistakes. For the most part, the algorithm that we employ to judge mathematical truths is reliable just as we would expect it to be. After all, it has evolved through a process of natural selection that would penalize defective algorithms if they resulted in misjudgements about everyday events whose correct appraisal was essential for survival. But the esoteric areas of pure mathematics that now sit at the extremities of human mathematical thinking appear to lie outside of the range of mathematical intuitions which could have had any past survival value for an embryonic intelligence. There are even good examples of fallible 'algorithms' that we (and mathematicians) make frequent use of. If we are lost in a big city we could follow the logical algorithm that a mathematician would provide in order for us to find our way back home. It would tell us something like: head in a particular compass direction until the city limit is reached and then follow some procedure of walking up and down alternate streets so that every street will eventually be visited. Your home street will necessarily be one of them and you will be home at last. But this is a very inefficient procedure and no one would dream of adopting it in practice despite its foolproof nature. Instead, we would adopt a fallible exploratory heuristic. We might wander at random until a big street is found, then follow that in a particular direction of house numbers towards the bigger buildings; keep going until we recognize some familiar site, then adopt a new strategy of local searching, and so on. Here, a number of different simple procedures come together to create an overall response to the problem that is very complex, will usually solve it, but is not guaranteed to do so. We also carry out all manner of unwarranted generalizations, assuming that something will always hold if we have seen it to hold in a finite number of cases. This amounts to the employment of transfinite induction. It is unsystematic educated guesswork of this sort that lies behind many mathematical insights which enable us to 'see' solutions of problems (like the tiling of a plane with tiles of particular shapes) which are formally non-computable.

We have outlined Penrose's position in some detail because of its topicality and adherence to the Platonic philosophy. We have raised a number of difficulties that prevent its conclusion being entirely compelling. However, let us suppose its conclusion and supporting arguments are true. Then we see that it does pick out the critical level of structure that would be required of a mathematical system in order for it to possess the property of producing conscious states: it is simply the condition necessary for Gödel's incompleteness to arise; that is, the system must be rich enough to contain arithmetic. If all possible mathematical worlds exist, then only those rich enough to contain arithmetic and hence permit the derivation of Gödel's self-referential undecidable statement $F_k(k)$ permit consciousness in Penrose's sense. Only these structures can become self-aware and we necessarily find ourselves part of one of them. It is an intriguing question to ask whether or not there are other aspects of the axiomatic basis of mathematics which might play a role in determining whether or not self-conscious observers can arise or not.

As a postscript to these considerations we might return to the consideration of Post, one of the co-discoverers with Turing of non-computable operations, who said this about the divide between meaning and formalism in mathematics,

> *mathematical thinking is, and must be essentially creative.* It is to the writer's continuing amazement that ten years after Gödel's remarkable achievement current views on the nature of mathematics are thereby affected only to the point of seeing the need of many formal systems, instead of a universal one. Rather has it seemed to us inevitable that these developments will result in a reversal of the entire axiomatic trend of the late nineteenth and early twentieth centuries, with a return to meaning and truth. Postulation thinking will then remain as but one phase of mathematical thinking.

## MATHS AND MYSTICISM

> *Dear Toothfairy,*
> *Mummy lost my tooth please can you still leave me some money. Thank you.*
> ROGER BARROW

Throughout our investigation of the Platonic interpretation of mathematics there is the nagging feeling that one is being dragged into a rather refined field of exact mysticism, decked out with a litany of equations, but appealing to an unseen reality that seems to rival that of the tangible world around us. The American mathematician John Casti highlights the curious analogies that give new meaning to Sir James Jeans' famous remark that 'God is a mathematician',

> In the Reality Game religion has always been science's toughest opponent, perhaps because there are so many surface similarities between the actual practice of science and the practice of most major religions. Let's take mathematics as an example. Here we have a field that emphasizes detachment from worldly objects, a secret language comprehensible only to the initiates, a lengthy period of preparation for the 'priesthood', holy missions (famous unsolved problems) to

which members of the faith devote their entire lives, a rigid and somewhat arbitrary code to which all practitioners swear their allegiance, and so on.

Many sceptical philosophers would use the word 'mysticism' to describe human beliefs and intuitions which transcend physical experience. They would place science and mysticism at opposite poles of the spectrum of certain knowledge of the world. At first it appears that the Platonic picture of mathematics places it firmly at the mystic end of the spectrum. Although there exists a vast array of different forms of mysticism, it is useful to highlight some of the hallmarks of traditional mystical thinking so that some contrast can be drawn with Platonic mathematics. Mysticism generally possesses several features which distinguish it from the practice of mathematics. Most noticably it is passive in its attitude to reality. It does not seek to improve, or rearrange the Universe. It is a contemplation rather than an exploration. Bertrand Russell highlighted four characteristics of what might be called philosophical mysticism: a belief in insight or intuition as a valid route to knowledge that is distinct from discursive or analytic intellectual processes; a denial of the reality of time and its passage in the ultimate scheme of things; a belief in the holistic unity of all things and a resistance to any fragmentation or division of our knowledge of the world into compartments; a belief that all evil is mere appearance. It is clear that one can set up some form of strained correspondence between these attributes and features of Platonic mathematics. It is forced to appeal to immaterial forms of reality and new ways of 'knowing' in order to explain human access to the world of mathematics; it focuses upon timeless truths in contrast to those sciences which investigate the nature of change in the physical world; it seeks the common structures that lie behind specific mathematical structures and regards the unification of disparate parts of mathematics into a single axiomatic framework as a virtue to be sought after; and it regards irrationality and paradox as a manifestation of a fallible human perspective on things which can be removed by seeking out a deeper level of structure to mathematical reality. Perhaps it is an awareness of the closeness of this analogy that makes many mathematicians and scientists wary of the Platonic perspective. It appears to force them to take on board a host of consequences to which they did not plan on subscribing when they embarked upon a mathematical education. But there are real differences between mathematics and mysticism. The language of mathematics is directed towards precision and the establishment of interconnections whereas, in complete contrast, mystical language is used to express a personal commitment or oneness with some state of ultimate being. Whereas the Platonic mathematician directs his efforts to discovering mathematical 'reality', the mystic aims to partake of mystic reality in a 'spiritual' way, not to understand it or analyse it in any intellectual way. The mystic leans towards celebration; the mathematician to cerebration. Finally, the essence of any mathematical procedure is its repeatability by the same individual and by others. There are no gnostic secrets.

## SUPERNATURAL NUMBERS?

> *If the flesh came for the sake of the spirit, it is a miracle. But if the spirit for the sake of the flesh—it is a miracle of miracles.*
>
> THE GOSPEL OF THOMAS

More than anything else, mathematics impresses upon us that the deepest ideas are hallmarked by the manner in which an assumption too mundane for anyone to question results in a conclusion so paradoxical that no one can believe it. Human counting began in the performance of practical and ceremonial acts that from afar seem to bring together the mundane and the ridiculous. Yet from these natural numbers has sprung a tree of knowledge that makes mathematics seem like a living thing. It continues to deal with particulars, to describe the world around us, yet it has grown to become larger than the world in which it was conceived. And out of this growth it has shed the chrysalis of earthly application to emerge as something entirely different: more beautiful, more extensive, less encumbered than ever it was in its youth. In such a form it becomes the closest we have to a 'secret of the Universe'. Yet it is a secret with two sides. It is at once both the key which unlocks for us the unknown structure of the Universe, and the hidden kernel of reality that the Universe guards most impenetrably.

There have been many attempts to find that key and demythologize the power and significance of mathematics. We have explored many of their facets and failings. The formalist saw his desire to capture the meaning of mathematics in symbols and rules melt away in the face of the complexity that resides in arithmetic. No formal system can ensnare it. The inventionist strove to explain all from experience in his quest to reduce mathematics to something that comes out of the mind instead of something that enters it from outside. But how can this be? Our minds and the world of tangible things have not sprung ready-made into existence. They have evolved by a gradual process in which the persistent and the stable survive over long periods of time. These persistent structures are manifestations of the mathematical structure of the world long before they could become re-creators of it. At first the mathematical nature of reality is instantiated upon the patterns of matter and energy, but beyond some critical level of complexity the ability to construct sub-programs becomes manifest. Subsequently, rapid development of organized complexity proceeds by the creation of accurate representations of the true nature of the external mathematical reality of the world. For to fail to produce faithful representations of its nature would remove the very possibility of persistence and survival.

The constructivist seeks the 'Machine in the Ghost'. He sees mathematics as a human activity, not because of any desire for self-aggrandizement but because we have sound knowledge of no other activity. He is worried by the existence of things that cannot be realized in a finite number of definite operations and sees this restriction as a sure security against the infiltration of absurdity in the guise of abstraction. But the mathematics that results is a curious thing: finite, shorn of many truths that we had liked, divested of so many devices that were as much a part of human intuition as counting, and divorced from the study of the physical world. An ever-changing concomitant of human minds, constructivist mathematics is at risk of becoming merely a branch of psychology that lost first its meaning and then its mind when the computer appeared as a rival basis for the performance of finite deduction. But just as mathematics is larger than the scope of formal rules and symbols, so it proves to be larger than the encompassment of any computer. Yet the utility of mathematics for our understanding of the world is a reflection of the ubiquity of computable mathematical operations. They can be simulated by natural or unnatural sequences of events in the physical world, whether they be the motions of neurones or the swinging of pendulums, and so the description of such events by mathematics is a reflection of the prevalence of computable mathematical functions. They permit a gradual improvement of our description of the world through a sequence of small refinements that converge upon the fullest mathematical description of some aspect of the world. Constructivism may be 'true'. The physicists' search for a 'Theory of Everything' brings us face to face with a deep question concerning the bedrock of space and time in which we move and have our being. The physicist takes this to be a smooth continuum, infinitely divisible, which acts as the cradle of the laws of Nature. Such a picture requires the admission of all manner of notions that are neither finite nor constructible. Yet in opposition to this picture of the world stands a new paradigm, as yet immature and naked of the full clothing of its consequences, in which the world is not at root a continuum. It is discrete and bitty, and its ultimate legislation is not to be found in the patterns and symmetries of the elementary particle physicist but in the essence of the computational process. 'Bits' of information are the true reality; the laws of Nature we infer are just a software program that runs on the hardware of matter.

To decide between these radically different alternatives, we need to discover whether the laws of physics are prior to, in the sense of constraining, the possibilities of computation, or whether the laws of physics are themselves consequences of some deeper, simpler rules of step-by-step computation. If the latter is closer to the truth, then we may find that the constructivists' doctrines are what distinguishes mathematical physics from mathematics.

Finally, we entered the strangest, and yet the most familiar, of the possible mathematical worlds. For the Platonists mathematics is not of this world and the work of mathematicians is the exploration of a realm of non-spatial, non-mental, timeless entities whose acquaintance we can make with an intuitive facility that owes nothing to our five senses. Mathematics exists apart from mathematicians. Whether it be a world of structures, or a world of things, somehow we are in contact with this immaterial reality in mysterious ways. We have seen the problems with this doctrine. It is surely metaphysical and it is surely sufficient to account for the existence and utility of mathematics. But is it necessary? If we are ever to arrive at mathematical descriptions of the Universe as a whole, such as the cosmologist strives to achieve, perhaps it is? And we have seen how we are led to see ourselves not as parts of the material world that are mysteriously governed by mathematics but, ultimately, as parts of mathematics ourselves, through a conception in which the material world is neither necessary nor sufficient to explain our conscious experience of it. We are part of the Platonic world not communicators with it.

Why do we find this Platonic world so 'familiar', so 'accommodating', despite its mystery and metaphysics? Surely, the reason is clear. We began with a scientific image of the world that was held by many in opposition to a religious view built upon unverifiable beliefs and intuitions about the ultimate nature of things. But we have found that at the roots of the scientific image of the world lies a mathematical foundation that is itself ultimately religious. All our surest statements about the nature of the world are mathematical statements, yet we do not know what mathematics 'is'; we know neither why it works nor where it works; if it fails or how it fails. Our most satisfactory pictures of its nature and meaning force us to accept the existence of an immaterial reality with which some can commune by means that none can tell. There are some who would apprehend the truths of mathematics by visions; there are others who look to the liturgy of the formalist and the constructivists. We apply it to discuss even how the Universe came into being. Nothing is known to which it cannot be applied, although there may be little to be gained from many such applications. And so we find that we have adopted a religion that is strikingly similar to many traditional faiths. Change 'mathematics' to 'God' and little else might seem to change. The problems of human contact with some spiritual realm, of timelessness, of our inability to capture all with language and symbol—all have their counterparts in the quest for the nature of Platonic mathematics. In many ways the Platonists' belief in the existence of a realm of mathematical objects is like belief in the existence of God because one is at liberty to disbelieve it if one chooses. But the object of the theologians' belief is quite different in nature to that of the Platonic mathematicians. Were a mathematician to show a particular mathematical idea to be inconsistent or incoherent, then this would be sufficient reason to say it does not exist, just as the demonstration of the inconsistent nature of a God would be a reason to disbelieve in His existence. But whereas the Platonic mathematician would regard the consistency and coherence of a mathematical conception as the ground for its existence, the coherence of the notion of God would not on its own be regarded as sufficient reason for His existence. Most theologians view God's existence as the precondition for everything to exist and so there could not be any other condition whose fulfilment would be enough to ensure the existence of God.

Many see mathematics and the scientific edifice that is built upon it as the antithesis of traditional immaterial conceptions of reality. Yet at root they are strikingly similar in the tantalizing nature of their incompletenesses. And this is why we find the Platonic world so strangely attractive—we have been there before. Our ability to create and apprehend mathematical structures in the world is merely a consequence of our own oneness with the world. We are the children as well as the mothers of invention.

# FURTHER READING

*And further, by these, my son be admonished: of making many books there is no end; and much study is a weariness of the flesh.* ECCLESIASTES

This bibliography contains a selection of books and articles at different levels which I found to be interesting while writing this book. They should form a solid foundation for any further study of the nature and origins of mathematics.

*Chapter 1*


Allman, G. J., *Greek Geometry from Thales to Euclid* (University Press, Dublin, 1889).

Ashton, J., 'Mathematicians and the Mysterious Universe', *Thought* **6**, 258–74 (1931).

Barrow, J. D., *The World Within the World* (OUP, Oxford, 1988).

Barrow, J. D., 'The Mathematical Universe', *World and I Magazine*, May 1989, pp. 306–11.

Birkhoff, G., 'The Mathematical Nature of Physical Theories', *American Scientist* **31**, 281–310 (1943).

Bonola, R., *Non-Euclidean Geometry: A Critical and Historical Study of its Developments*, trans. H. S. Carslaw (Dover, New York, 1955).

Boole, G., *An Investigation of the Laws of Thought* (Dover, New York, 1958, orig. publ. 1854).

Browder, F., 'Does Pure Mathematics have a Relation to the Sciences?', *American Scientist* **64**, 542 (1976).

Cartwright, M. L., *The Mathematical Mind* (OUP, Oxford 1955).

Davies, P. C. W., 'Why is the Universe Knowable?', in *Maths and Science*, ed. R. E. Mickens (OUP, New York, 1989).

Davis, P. J. and Hersh, R., *The Mathematical Experience* (Birkhäuser, Boston MA, 1981).

de Broglie, L., 'The Role of Mathematics in the Development of Contemporary Theoretical Physics', in *Great Currents of Mathematical Thought*, Vol. 2, ed. F. Le Lionnais (Dover, New York, 1971) pp. 78–93.

Dyson, F., 'Mathematics in the Physical Sciences', *Scientific American*, (Sept. 1964), pp. 129–46.

Gorman, P., *Pythagoras: A Life* (Routledge, London, 1979).

Hamming, R. W., 'The Unreasonable Effectiveness of Mathematics', *Amer. Math. Monthly* **87**, 81–90 (1980).

Hardy, G. H., *A Mathematician's Apology* (CUP, Cambridge, 1941).

Heath, T., *The Thirteen Books of Euclid*, 2nd edn. (Dover, New York, 1956).

Hempel, C. G., 'On the Nature of Mathematical Truth', in *Readings in the Philosophy of Science*, eds. H. Feigl and M. Brodbeck (Appleton-Century-Crofts, New York, 1953).

Henderson, L. D. *The Fourth Dimension and Non-Euclidean Geometry in Modern Art* (Princeton UP, Princeton NJ, 1983).

Kline, M., *Mathematics: A Cultural Approach* (Addison-Wesley, Reading MA, 1962).

Leacock, S., *The Penguin Stephen Leacock*, selected and introduced by R. Davies (Penguin, London, 1981).

Lukasiewicz, J., *Elements of Mathematical Logic*, trans. O. Wojtasiewicz (Macmillan, New York, 1963).

Menger, K., 'The New Logic', trans. H. Gottlieb and J. Senior, *Phil. Sci.* **4**, 299-336 (1937).

Mickens, R. E. (ed.) *Mathematics and Science* (World Scientific, Singapore, 1990).

Nagel, E., 'The Formation of the Modern Conceptions of Formal Logic in the Development of Geometry', *Osiris* **7**, 142–222 (1939).

Philip, J. A., *Pythagoras* (Univ. Toronto Press, Toronto, 1966).

Purcell, E. A., *The Crisis of Democratic Theory* (Univ. Kentucky Press, Lexington KY, 1973).

Richards, J., 'The Reception of a Mathematical Theory: Non-Euclidean Geometry in England 1868–1883', in *Natural Order: Historical Studies of Scientific Culture*, ed. B. Barnes and S. Shapin (Sage Publications, Beverly Hills, 1979).

Rosser, B., 'On the Many-valued Logics', *Amer. J. Phys.* **9**, 207–12 (1941).

Russell, B., *An Essay on the Foundations of Geometry* (Dover, New York, 1956).

Schaaf, W. L., *Mathematics, Our Great Heritage: Essays on the Nature and Cultural Significance of Mathematics* (Harper & Row, New York, 1948).

Weyl, H., *Philosophy of Mathematics and Natural Science*, rev. edn., trans. O. Helmer (Princeton UP, Princeton NJ, 1949).

Wigner, E., 'The Unreasonable Effectiveness of Mathematics in the Natural Sciences', *Commun. Pure Appl. Math.* **13**, 1 (1960).

*Chapter 2*

Bag, A. K., *Mathematics in Ancient and Medieval India* (Chauhambha Orientalia, Varanasi, 1979).

Bennett, W. C., 'Lore and Learning, Numbers, Measures, Weights and Calendars', in *Comparative Ethnology of the S.A. Indians*, Vol. 5 of *Handbook of S.A. Indians*, Smithsonian Inst. Bur. Amer. Ethn. Bull. 143 (Washington DC, 1949).

Boas, F., 'The Kwakiutl of Vancouver Island, *Mem. Amer. Mus. Nat. Hist. (NY)* **8**, 2 (1908).

Bochner, S., *The Role of Mathematics in the Rise of Science* (Princeton UP, Princeton NJ, 1966).

Boden, M., *The Creative Mind: Myths and Mechanisms* (Weidenfeld & Nicholson, London, 1990).

Bogoshi, J., Naidoo, K., and Webb, J., 'The Oldest Mathematical Artefact', *Mathematical Gazette* **71**, 294 (1987).

Bolton, N. J. and Macleod, D. N., 'The Geometry of *Sriyantra', Religion* **7**, 66 (1977).

Bowman, M. E., *Romance in Arithmetic: A History of our Currency, Weights and Measures, and Calendar* (Univ. London Press, London, 1950).

Boyer, C. B., *A History of Mathematics* (Wiley, New York, 1968).

Boyer, C. B., 'Fundamental Steps in the Development of Numeration', *Isis* **35**, 157–8 (1944).

Brice, W. C., 'The Writing and System of Proto-Elamite Account Tables', *Bull. John Rylands Library* **45**, 15–39 (1962).

Cajori, F., *A History of Mathematical Notation*, Vol. 1 (Open Court, Chicago, 1928).

Closs, M. P., *Native American Mathematics* (Univ. Texas Press, Austin TX, 1986).

Conant, L. L., *The Number Concept* (Macmillan, London, 1923, orig. publ. 1910).

Crump, T., *The Anthropology of Numbers* (CUP, Cambridge, 1990).

Dantzig, T., *Number, The Language of Science* (Macmillan, New York, 1937).

Datta, B. and Singh, A. N., *History of Hindu Mathematics* (Asia Publ. House, Bombay and Motilah Banarshi Das, Lahore, 1935 & 1938).

de Heinzelin, J., 'Ishango', *Scientific American*, June 1962, pp. 105–16.

de Villiers, M., *The Numeral Words, Their Origin, Meaning, History and Lesson* (H. F. & G. Witherby, London, 1923).

Flegg, G., *Numbers: Their History and Meaning* (André Deutsch, London, 1983).

Frazer, J. G., *Folklore in the Old Testament* (Macmillan, London, 1918).

Frazer, J. G., *The Golden Bough* (Macmillan, New York, 1940).

Friberg, J., 'Numbers and Measures in the Earliest Written Records', *Scientific American*, Feb 1984, pp. 78–85.

Gallenkamp, C., *Maya: The Riddle and Rediscovery of a Lost Civilization* (D. McKay, New York 1959).

Gay, J. and Cole, M., *The New Mathematics in an Old Culture. A Study of Learning among the Kpelle of Liberia* (Holt, Rinehart & Winston, New York, 1967).

Gillings, R., *Mathematics in the Time of the Pharaohs* (Dover, New York, 1972).

Groom, A., *How We Weigh and Measure* (Routledge, London, 1960).

Halstead, G. B., *On the Foundation and Technic of Arithmetic* (Open Court, London, 1912).

Hill, G. F., *Arabic Numerals in Europe* (OUP, Oxford 1915).

Huntley, H. E., *The Divine Proportion: A Study in Mathematical Beauty* (Dover, New York, 1970).

Ifrah, G., *From One to Zero* (Viking, London, 1985).

Ivimy, J., *The Sphinx and the Megaliths* (Turnstone, London, 1974).

Joseph, G. G., *The Crest of the Peacock: Non-European Roots of Mathematics* (Tauris, London, 1991).

Kline, M., *Mathematics in Western Culture* (OUP, New York, 1953).

Kulaichev, A. P., '*Sriyantra* and its Mathematical Properties', *Indian J. Hist. Sci.* **19**, 279 (1984).

Kulkarni, R. P., 'Geometry as Known to the People of the Indus Civilization', *Indian J. Hist. Sci.* **13**, 117–124 (1971).

Kulkarni, R. P., *Geometry According to Sulba Sutra* (Vaidika Samsodhana Mandala, Pune, 1983).

Lévy-Brühl, L., *How Natives Think*, transl. L. A. Clare (Knopf, New York, 1926).

Libbrecht, U., *Chinese Mathematics in the 13th Century* (MIT Press, Cambridge MA, 1973).

Li Yan and Du Shiran, *Chinese Mathematics: A Concise History*, trans. J. N. Crossley and A. W.–C. Lun (OUP, Oxford, 1987).

Marshack, A., *The Roots of Civilization* (Weidenfeld & Nicholson, London, 1972).

McLeish, J., *Number* (Bloomsbury, London, 1991).

Menninger, K., *Number Words and Number Symbols*, trans. P. Broneer (MIT Press, Boston MA, 1970).

Morley, S. G., *The Ancient Maya* (Stanford UP, Stanford CA, 1946).

Nasr, S. H., *Science and Civilization in Islam* (Harvard UP, Cambridge MA, 1968).

Needham, J., *Science and Civilization in China*, Vol. 3 (CUP, Cambridge, 1970).

Neugebauer, O., 'Mathematical Methods in Ancient Astronomy', *Bull. Amer. Math. Soc.* **54**, 1013–41 (1948).

Neugebauer, O., *The Exact Sciences in Antiquity*, 2nd edn. (Harper, New York, 1962).

Neugebauer, O. and Sachs, A., *Mathematical Cuneiform Texts*, Amer. Oriental Series, Vol. 29 (Newhaven CT, 1946).

Nykl, A. R., 'The Quinary-Vigesmal System of Counting in Europe, Asia, and America' *Language* **2**, 165–73 (1926).

Ore, O., *Number Theory and its History* (McGraw-Hill, New York, 1948).

Peterson, F., *Ancient Mexico* (Capricorn Books, New York 1962).

Pullan, I. M., *History of the Abacus* (Ohio UP, Ohio, 1929).

Raum, O. F., *Arithmetic in Africa* (Evans Bros., New York, 1939).

Rawson, P., *Tantra: The Indian Cult of Ecstasy* (Thames & Hudson, London, 1973, reprinted 1989).

Richardson, L. J., 'Digital Reckoning Amongst the Ancients', *Amer. Math. Monthly* **23**, 7–13 (1916).

Sarasvati, T. A., *Geometry in Ancient and Medieval India* (Motilal Banarsidass, Delhi, 1979).

Schmandt-Besserat, D., 'The Envelopes that Bear the First Writing', *Technology and Culture* **21**, 357–85 (1980).

Scriba, C. J., *The Concept of Number* (B. Mannheim, Zurich, 1986).

Seidenberg, A., 'The Diffusion of Counting Practices', *University of California Publications in Mathematics* **3**, p. 215–99 (1960).

Seidenberg, A., 'The Ritual Origin of Geometry', *Arch. Hist. Exact Sci.* **1**, 488–527 (1960).

Seidenberg, A., 'The Ritual Origin of Counting', *Arch. Hist. Exact Sci.* **2**, 1–40 (1962).

Seidenberg, A., 'The Sixty System of Sumer', *Arch. Hist. Exact Sci.*, **2**, 436–40 (1962).

Smith, D. E. and Karpinski, L. C., *The Hindu–Arabic Numerals* (Ginn, London, 1911).

Srinivasiengar, C. N., *The History of Ancient Indian Mathematics* (The World Press, Calcutta, 1967).

Steele, R. (ed.) *The Earliest Arithmetics in England* (OUP, Oxford, 1922).

Thibaut, G., 'On the Sulvasutras', *J. Asiatic Soc. Bengal* **44**, 227–75 (1875).

Thomas, N. W., 'Bases of Numeration', *Man* **17**, 145–7 (1917).

Thompson, J. E. S., *Maya Hieroglyphs Without Tears* (British Museum, London, 1972).

Trumbull, J. H., 'On Numerals in the American Indian Languages', *Trans. Amer. Philo. Assn.* **5**, 41–76 (1874).

van der Waerden, B. L., *Geometry and Algebra in Ancient Civilizations* (Springer, New York, 1983).

van der Waerden, B. L., *Science Awakening* (OUP, Oxford 1961; orig. publ. Noordhoff, Groningen, 1954).

Wertheimer, M., 'Numbers and Numerical Concepts in Primitive Peoples', in *A Source Book of Gestalt Psychology*, ed. W. D. Ellis (Kegan Paul, London, 1950).

White, L., 'The Locus of Mathematical Reality: An Anthropological Footnote', *Phil. Sci.* **14**, 289–303 (1947).

Yeldham, F. A., *The Story of Reckoning in the Middle Ages* (Harrap, London, 1926).

Yushkevich, A., *A History of Mathematics in the Middle Ages* (CUP, Cambridge, 1985).

Zasvlasky, C., *Africa Counts: Number and Pattern in African Culture* (Prindle, Weber & Schmidt, Boston MA, 1973).

*Chapter 3*

Bourbaki, N., 'Foundations of Mathematics for the Working Mathematician', *J. Symbol. Logic* **14**, 1–8 (1949).

Bourbaki, N., 'The Architecture of Mathematics', trans. A. Dresden, *Amer. Math. Monthly* **57**, 221 (1950).

Candler, H., 'On the Symbolic Use of Number in the *Divina Commedia* and Elsewhere', *Trans. R. Soc. Literature* (Series 2) **30**, 1–29 (1910).

Casti, J., *Searching for Certainty: What Scientists Can Know About the Future* (Morrow, New York 1990).

Chaitin, G., *Algorithmic Complexity* (CUP, Cambridge, 1988).

Chaitin, G., 'Randomness in Arithmetic', *Scientific American*, July 1988, pp. 80–85.

Church, A., 'An Unsolvable Problem of Elementary Number Theory', *Amer. J. Math.* **58**, 345–63 (1936).

Cohen, P. J., 'The Independence of the Continuum Hypothesis: I and II', *Proc. Natl. Acad. Sci. USA* **50**, 1143–48 and **51**, 105–10 (1963).

Cohen, P. J., *Set Theory and the Continuum Hypothesis* (Benjamin, New York, 1966).

Curry, H., *Outlines of a Formalist Philosophy of Mathematics* (North–Holland, Amsterdam, 1951).

Davis, M. (ed.) *The Undecidable* (Raven, New York, 1965).

Dieudonné, J., 'L'École Française Moderne des Mathématiques', *Philosophia Mathematica* **1**, 97–106 (1964).

Dieudonné, J., 'Recent Developments in Mathematics', *Amer. Math. Monthly* **71** 239–48 (1964).

Eves, H. & Newsom, C. V., *An Introduction to the Foundations and Fundamental Concepts of Mathematics*, rev. edn. (Holt, Rinehart & Winston, New York, 1965).

Fang, J., *Bourbaki*, (Paideia Press, New York, 1970).

Fraenkel, A., and Bar-Hillel, Y., *Foundations of Set Theory* (North-Holland, Amsterdam, 1958).

Frege, G., *The Foundations of Arithmetic: A Logico-mathematical Enquiry into the Concept of Number*, trans. J. L. Austin (Blackwell, Oxford, 1980).

Frege, G., *Posthumous Writings* (Blackwell, Oxford, 1979).

Friedman, J., 'Some Set-theoretical Partition Theorems Suggested by the Structure of Spinoza's God', *Synthèse* **27**, 199–209 (1974).

Galileo Galilei, *Dialogues Concerning Two New Sciences*, trans. H. Crew and A. de Salvio (Macmillan, New York, 1914, orig. publ. 1632).

Gillies, D. A., *Frege, Dedekind, and Peano on the Foundations of Arithmetic* (Van Gorcum, Assen, 1982).

Gödel, K., *On Undecidable Propositions of Formal Mathematical Systems* (Princeton UP, Princeton NJ, 1943).

Gödel, K., 'What Is Cantor's Continuum Problem?', *Amer. Math. Monthly* **54**, 515 (1947). A later and expanded version of this paper can be found in the collection edited by Benacerraf and Putnam cited below under Chapter 6.

Grelling, K., 'The Logical Paradoxes', *Mind* **45**, 481–86 (1936).

Hilbert, D., *The Foundations of Geometry*, 10th edn., ed. P. Bernays, trans. L. Unger (Open Court, Chicago, 1971, orig. publ. 1899).

Hilbert, D., 'On the Foundations of Logic and Arithmetic', in *From Frege to Gödel: A Source Book in Mathematical Logic, 1879–1931*, ed. J. van Heijenoort (CUP, Cambridge, 1982).

Hilbert, D. and Cohn-Vessen, S., *Geometry and the Imagination*, trans. P. Nemenyi (Chelsea, New York, 1952).

Hughes, P. and Brecht, G., *Vicious Circles and Infinity* (Penguin, London, 1978; orig. publ. Doubleday, New York, 1975).

Kneebone, G. T., *Mathematical Logic and the Foundations of Mathematics: An Introductory Survey* (Van Nostrand, London, 1963).

Lakatos, I., *Proofs and Refutations: The Logic of Mathematical Disovery* (CUP, Cambridge, 1976).

Leibniz, G., *The Philosophical Works of Leibniz*, trans. G. Duncan (Tuttle, Morehouse & Taylor, New Haven CT, 1916).

McTaggert, J. E., 'Propositions Applicable to Themselves', *Mind* **32**, 462 (1923).

Moore, G. H., *Zermelo's Axiom of Choice* (Springer, Berlin, 1982).

Nagel, E. and Newman, J. R., *Gödel's Proof* (New York UP, New York, 1958).

Putnam, H., 'The Thesis that Mathematics is Logic', *Mathematics, Matter and Method, Philosophical Papers*, Vol. 1 (CUP, Cambridge, 1975).

Quine, W., 'Paradox', *Scientific American* **206**, 84–96 (1962).

Reid, C., *Hilbert* (Springer, New York, 1970).

Rota, G.-C. 'The Concept of Mathematical Truth', *Rev. Metaphys.* **44**, 483–94 (1991).

Rucker, R., *Infinity and the Mind* (Birkhäuser, Boston MA, 1982).
Russell, B., *Introduction to Mathematical Philosophy* (Macmillan, New York, 1919).
Ryll-Nardzewski, C., 'The Role of the Axiom of Induction in Elementary Arithmetic', *Fund. Math.* **39**, 239-63 (1953).
Shanker, S. G. *Gödel's Theorem in Focus* (Routledge, London, 1988).
Smullyan, R., *Forever Undecided: A Puzzle Guide to Gödel* (Knopf, New York, 1987).
Snapper, E., 'What Do We Do When We Do Mathematics?', *Mathematical Intelligencer* **10**(4), 53–8 (1988).
Tarski, A., *Logic, Semantics, Metamathematics* (OUP, New York, 1956).
Wang, H., *Reflections on Kurt Gödel* (MIT Press, Cambridge MA, 1988).
Weil, A., 'The Future of Mathematics', trans. A. Dresden, *Amer. Math. Monthly* **57**, 295–306 (1950).
Weyl, H., 'David Hilbert and his Mathematical Work', *Bull. Amer. Math. Soc.* **50**, 612–54 (1944).
Whitehead, A. N. and Russell, B., *Principia Mathematica* (CUP, Cambridge, 1925 & 1927).
Wilder, R. L., 'The Nature of Mathematical Proof', *Amer. Math. Monthly* **51**, 309–23 (1944).
Yates, F., *Giordano Bruno and the Hermetic Tradition* (Univ. Chicago Press, Chicago, 1964).

*Chapter 4*

Alder, A., 'Mathematics and Creativity', *The New Yorker*, 19 Feb. 1972, pp. 39–40.
Andreski, A., *Social Sciences as Sorcery* (St. Martin's Press, New York, 1972).
Beth, E. W. and Piaget, J., *Mathematical Epistemology and Psychology*, trans. W. Mays (Reidel, Dordrecht, 1966).
Black, M., 'The Relevance of Mathematical Philosophy to the Teaching of Mathematics', *Mathematical Gazette* **22**, 149–63 (1938).
Brainerd, C. J., 'The Origins of Number Concepts', *Scientific American*, March 1973, pp. 101–9.
Brainerd, C. J., 'Mathematical and Behavioural Foundations of Number', *J. Gen. Psychol.* **88**, 221–81 (1973).
Brooks, F. P., *The Mythical Man Month* (Addison-Wesley, New York, 1975).
Chomsky, N., 'The Linguistic Approach', in *Language and Learning* ed. M. Piatelli-Palmarini (Routledge, London, 1980).
Chomsky, N., *Language and the Problems of Knowledge* (MIT Press, Boston MA, 1988).
Colman, E., 'The Present Crisis in the Mathematical Sciences and General Outline for their Reconstruction', in *Historical Materialism*, ed. N. Bukharin (International Publ., New York, 1925).
Crump, T., 'Money and Number: The Trojan Horse of Language', *Man* (New Series) **13**, 503–18 (1978).
Crump, T., 'The Alternative Meanings of Number and Counting', in *Semantic Anthropology*, ed. D. Parkin (Academic Press, London, 1982).

Davis, C., 'Materialist Mathematics', in *For Dirk Struik*, ed. R. Cohen, J. Stachel, and M. Wartofsky (Reidel, Boston MA, 1974).

Dehoshe, G. and Seron, X., (eds.) *Mathematical Disabilities: A Cognitive Neurophysiological Perspective* (Erlbaum, Hillsdale NJ, 1988).

Fang, J., 'Kant and Modern Mathematics', *Philosophica Mathematica* **2**, 47–68 (1955).

Fang, J. and Takayama, K. P., *Sociology of Mathematics and Mathematicians* (Paideia Press, Hauppauge NY, 1975).

Field, H., *Science Without Numbers* (Blackwell, Oxford, 1980).

Field, H., *Realism, Mathematics and Modality* (Blackwell, Oxford, 1989).

Gay, J. and Cole, M., *The New Mathematics in an Old Culture* (Holt, Rinehart & Winston, New York, 1967).

Ghiselin, B. (ed.) *The Creative Process* (Univ. California Press, Berkeley, 1952).

Ginsburg, H. P., *Children's Arithmetic: The Learning Process* (Van Nostrand, New York, 1977).

Greenwood, T., 'Invention and Description in Mathematics', *Aristotelian Soc. Proc.* **30**, 88 (1929).

Gruber, H. and Vonéche, J. (eds.) *The Child's Understanding of Number* (Harvard UP, Cambridge MA, 1978).

Hadamard, J., *The Psychology of Invention in the Mathematical Field* (Princeton UP, Princeton NJ, 1945).

Harel, D., *Algorithmics: The Spirit of Computing* (Addison-Wesley, New York, 1987).

Jackson, D., 'The Human Significance of Mathematics', *Amer. Math. Monthly* **35**, 406–11 (1928).

Kemeny, J. G., 'Mathematics without Numbers', *Daedalus* **88**, 577–91 (1959).

Koehler, O., 'The Ability of Birds to Count', *Bull. Animal Behaviour* No. 9 (March 1951).

Marx, K., *The Mathematical Manuscripts of Karl Marx*, trans. C. Aronson, M. Meo, and R. A. Archer (New Park Publ., Clapham, 1983).

Miller, A. I., 'Scientific Creativity: A Comparative Study of Henri Poincaré and Albert Einstein', *Creativity Research Journal* (to be published) (1992).

Miller, G. A., 'The Magical Number Seven, Plus or Minus Two: Some Limits on Our Capacity for Processing Information', *Psychol. Rev.* **63**, 81–97 (1956).

Moulton, F. R., 'Laws of Nature and Laws of Man', *Bull. Amer. Association for the Advancement of Science* **2**, 25–26 (1943).

Piaget, J., *The Child's Conception of Number* (Routledge, London, 1952).

Piaget, J., 'How Children Form Mathematical Concepts', *Scientific American* **189**, 74–79 (1953).

Piaget, J., Inhelder, B. and Szeminska, A., *The Child's Conception of Geometry* (Basic Books, New York, 1960).

Piaget, J., *The Principles of Genetic Epistemology* (Routledge, London, 1972).

Piattelli-Palmarini, M., 'Evolution, Selection and Cognition: From 'Learning' to Parameter Setting in Biology and the Study of Language', *Cognition* **31**, 1–44 (1989).

Pinker, S. and Bloom, P., 'Natural Language and Natural Selection', *Behavioural and Brain Research* **13**, 707–84 (1990). Includes critical responses and replies by the authors.

Restivo, S., *The Social Relations of Physics, Mysticism and Mathematics* (Reidel, Dordrecht, 1983).

Restivo, S. and Collins, R., 'Mathematics and Civilization', *The Centennial Review* **26**(3), 277–301 (1982).

Ruelle, D., 'Is Our Mathematics Natural? The Case of Equilibrium Statistical Mechanics', *Bull. Amer. Math. Soc.* **19**, 259–68 (1988).

Spengler, O., *The Decline of the West*, Vol. 1 (Knopf, New York, 1926), Chap. 2.

Stein, S. K., *Mathematics, the Man-Made Universe* (Freeman, San Francisco, 1963).

Toulouse, E., *Henri Poincaré* (Flammarion, Paris, 1910).

Warrington, E. K., 'The Fractionation of Arithmetic Skills: A Single Case Study', *Quart. J. Exp. Psychol.* **34A**, 31–51 (1982).

Wilder, R. L., 'The Cultural Basis of Mathematics', *Proc. Int. Congr. Math.* **1**, 258–71 (1950).

Wilder, R. L., 'The Origin and Growth of Mathematical Concepts', *Bull. Amer. Math. Soc.* **59**, 423–48 (1953).

Wilder, R. L., 'Mathematics: A Cultural Phenomenon', in *Essays in the Science of Culture*, ed. G. E. Dale and R. L. Carneiro (T. Y. Crowell, New York, 1960).

Wilder, R. L., *Introduction to the Foundations of Mathematics*, 2nd edn. (Wiley, New York, 1967).

Wilder, R. L., *Evolution of Mathematical Concepts: An Elementary Study* (Wiley, New York, 1968).

*Chapter 5*

Andrews, G. E., 'An Introduction to Ramanujan's Lost Notebook', *Amer. Math. Monthly* **86**, 89–108 (1979).

Appel, K. and Haken, W., 'The Four-Color Problem', in *Mathematics Today*, ed. L. A. Steen, pp. 153–90, (Springer, New York, 1978).

Beeson, M. J., *Foundations of Constructive Mathematics* (Springer, Berlin, 1985).

Bennett, C. H., 'The Thermodynamic of Computation—A Review', *Int. J. Theoret. Phys.* **21**, 905 (1982).

Bennett, C. H. and Landauer, R., 'The Fundamental Physical Limits of Computation', *Scientific American* **253**, pp. 48 & 253 (issue 1) and p. 6 (issue 4) (1985).

Bishop, E., *Foundations of Constructive Analysis* (McGraw-Hill, New York, 1967). A new expanded edition was prepared under the authorship of D. Bridges and E. Bishop in 1985.

Bishop, E., 'The Crisis in Contemporary Mathematics', *Historia Mathematica* **2**, 507–17 (1975).

Bochner, M., 'Brouwer's Contribution to the Foundations of Mathematics', *Bull. Amer. Math. Soc.* **30**, 31–40 (1924).

Bridgman, P. W., 'A Physicist's Second Reaction to Mengenlehre', *Scripta Mathematica* **2**, 101–17 & 224–34 (1934).

Brouwer, L., 'Intuitionism and Formalism', trans. A. Dresden, *Bull. Amer. Math. Soc.* **30,** 81–96 (1913).

Brouwer, L., 'Historical Background, Principles and Methods of Intuitionism', *South African J. Sci.* **49**, 139–46 (1952).

Calder, A., 'Constructive Mathematics', *Scientific American*, Oct. 1979, pp. 134–43.

Cantor, G., *Contributions to the Founding of the Theory of Transfinite Numbers*, trans. P. Jourdain (Dover, New York, 1955).

Da Costa, N. C. A. and Doria, F. A. 'Undecidability and Incompleteness in Classical Mechanics' *Int. J. Theoret. Phys.* **30**, 1041–1073 (1991).

Dauben, J. W., *Georg Cantor* (Harvard UP, Cambridge MA, 1979).

Dawkins, R., *The Selfish Gene* (OUP, Oxford, 1976).

Deutsch, D., Quantum Theory, the Church–Turing Principle, and the Universal Quantum Computer, *Proc. R. Soc. (Lond.)* A **400**, 97 (1985).

Dummett, M., *Elements of Intuitionism* (OUP, Oxford, 1980).

Feynman, R., 'Simulating Physics with Computers', *Int. J. Theoret. Phys.* **21**, 467 (1982).

Freudenthal, H., *L. E. J. Brouwer's Collected Works* (North-Holland, Amsterdam, 1976).

Geroch, R. and Hartle, J., 'Computability and Physical Theories', in *Between Quantum and the Cosmos: Studies and Essays in Honor of John Archibald Wheeler* ed. W. H. Zurek, A. van der Merwe, and W. A. Miller, pp. 549–67 (Princeton UP, Princeton NJ, 1988).

Hardy, G. H., *Ramanujan* 3rd edn. (CUP, Cambridge, 1940).

Hays, J., 'The Battle of the Frog and the Mouse (from the *Fables of Aleph*)' *Mathematical Intelligencer* **6**, 77–80 (1984).

Herken, R. (ed.) *The Universal Turing Machine: A Half-Century Survey* (OUP, Oxford, 1988).

Heyting, A., *Intuitionism: An Introduction*, 3rd rev. edn. (North-Holland, Amsterdam, 1980, orig. publ. 1956).

Kangel, R., *The Man Who Knew Infinity: A Life of the Genius Ramanujan* (Scribners, New York, 1991).

Kronecker, L., *Leopold Kroneckers Werke*, ed. K. Hensel (Chelsea, New York, 1968, orig. publ. 1899).

Lam, C. W. H., 'How Reliable is a Computer-Based Proof?' *Mathematical Intelligencer* **12**, 8–12 (1990).

Landauer, R., 'Computation and Physics: Wheeler's Meaning Circuit', *Found. Phys.* **16**, 551 (1986). Reprinted in *Between Quantum and Cosmos: Studies and Essays in Honor of John Archibald Wheeler*, ed. W. H. Zurek, A. van der Merwe, and W. A. Miller, pp. 568–82 (Princeton UP, Princeton NJ, 1988).

Larguier, E. H., 'Brouwerian Philosophy of Mathematics', *Scripta Mathematica* **7**, 69–78 (1940).

Lloyd, S., 'The Calculus of Intricacy', *The Sciences*, Sep./Oct. 1990, pp. 38–44.

Poincaré, H., *Mathematics and Science: Last Essays*, trans. J. W. Boldue (Dover, New York, 1963).

Saaty, T. and Kainen, P., *The Four-Color Problem: Assaults and Conquest* (McGraw-Hill, New York, 1977).

Shannon, C. and Weaver, W., *A Mathematical Theory of Communication* (Univ. Illinois Press, Urbana IL, 1949).

Stewart, I., *Concepts of Modern Mathematics* (Penguin, London, 1981).

Turing, A., 'On Computable Numbers, With an Application to the *Entscheidungsproblem*', *Proc. Lond. Math. Soc.* **42**, 230–65 (1937), and erratum in **43**, 544–6 (1937).

Tymoczko, T., 'The Four-Colour Problem and its Philosophical Significance', *J. Phil.* **76**, 57–83 (1979).

van Dalen, D., 'The War of the Frogs and the Mice, or the Crisis of the *Mathematische Annalen*', *Mathematical Intelligencer* **12**, 17–31 (1990).

van Stigt, W. P., 'The Rejected Parts of Brouwer's Dissertation On the Foundations of Mathematics', *Historia Mathematica* **6**, 385–404 (1979).

van Stigt, W. P., *Brouwer's Intuitionism* (Elsevier North-Holland, Amsterdam, 1990).

von Neumann, J., *The Computer and the Brain* (Yale UP, New Haven CT, (1959).

Wavre, R., 'Is There a Crisis in Mathematics?', *Amer. Math. Monthly* **41**, 488–99 (1934).

Weyl, H., 'The Mathematical Way of Thinking', *Science* **92**, 437–46 (1940).

Weyl, H., 'Mathematics and Logic', *Amer. Math. Monthly* **53**, 2–13 (1946).

Weyl, H., 'Axiomatic versus Constructive Procedures in Mathematics', ed. T. Tonietti, *Mathematical Intelligencer* **7**, 10–17 (1985). Previously unpublished manuscript first written in about 1954.

Zurek, W. (ed.), *Complexity, Entropy and the Physics of Information* (Addison-Wesley, Redwood City, 1990).

*Chapter 6*

Barker, S. F., *Philosophy of Mathematics* (Prentice-Hall, Englewood Cliffs NJ, 1964).

Barrow, J. D., *Theories of Everything: The Quest for Ultimate Explanation* (OUP, Oxford, 1991).

Barrow, J. D., *The World Within the World* (OUP, Oxford, 1988).

Barrow, J. D. and Silk, J., *The Left Hand of Creation: The Origin and Evolution of the Universe* (Basic Books, New York, 1983).

Barrow, J. D. and Tipler, F. J., *The Anthropic Cosmological Principle* (OUP, Oxford, 1986).

Benacerraf, P., 'God, the Devil and Gödel', *The Monist* **51**, 9–32 (1967).

Benacerraf, P. and Putnam, H., (eds.), *Philosophy of Mathematics: Selected Readings* (Prentice-Hall, Englewood Cliffs NJ, 1964).

Boden, M., *The Creative Mind: Myths and Mechanisms* (Weidenfeld & Nicholson, London, 1990).

Brumbaugh, R., *Plato's Mathematical Imagination* (Indiana UP, Bloomington IN, 1954).

Buchanan, S., *Poetry and Mathematics* (John Day, New York, 1929).

Buckert, W., *Lore and Science in Ancient Pythagoreanism* (Harvard UP, Cambridge MA, 1972).
Campbell, A. D., 'A Note on the Sources of Mathematical Reality', *Amer. Math. Monthly* **34**, 263 (1927).
Davies, P. C. W., *The Mind of God: Science and the Search for Ultimate Meaning* (Simon & Schuster, London, 1992).
Dummett, M., *Frege: Philosophy of Mathematics* (Duckworth, London, 1991).
Dyck, M., *Novalis and Mathematics* (Univ. N. Carolina Press, Chapel Hill NC, 1960).
Gödel, K., 'Russell's Mathematical Logic', in *The Philosophy of Bertrand Russell*, ed. P. A. Schlipp (Northwestern Univ. Press, Evanston IL, 1944).
Gödel, K., 'What is Cantor's Continuum Problem?', *Amer. Math. Monthly* **54**, 515–25 (1947).
Gonseth, F., *Les mathématiques et la réalité* (F. Alcan, Paris, 1936).
Gonseth, F., *Philosophie mathématique* (Hermann, Paris, 1939).
Hardy, G. H., 'Mathematical Proof', *Mind* **30**, 1–25 (1929).
Heath, T., *A History of Greek Mathematics* (OUP, Oxford, 1927).
Hofstadter, D., *Gödel, Escher, Bach: The Eternal Golden Braid* (Harper & Row, New York, 1979).
Hofstadter, D. and Dennett, D., *The Mind's Eye: Fantasies and Reflections on Self and Soul* (Basic Books, New York, 1983).
Irvine, A. D., (ed.), *Physicalism in Mathematics* (Kluwer, Dordrecht, 1990).
Jeans, J., 'The Mathematical Aspect of the Universe', *Philosophy* **7**, 3–14 (1932).
Johnson, G., 'New Mind, No Clothes: Platonism and Quantum Physics Do Not Supplant Artificial Intelligence', *The Sciences*, Jul/Aug 1990, pp. 44–49.
Kitcher, P., *The Nature of Mathematical Knowledge* (OUP, Oxford, 1984).
Körner, S., *The Philosophy of Mathematics: An Introduction* (Harper & Row, New York, 1962).
Lassere, F., *The Birth of Mathematics in the Age of Plato* (Hutchinson, London, 1964).
Lehman, H., *Introduction to the Philosophy of Mathematics* (Blackwell, Oxford, 1979).
Lucas, J., 'Minds, Machines and Gödel', *Philosophy* **36**, 112 (1961).
Maddy, P., *Realism in Mathematics* (OUP, Oxford, 1990).
Mandlebrot, B., *The Fractal Geometry of Nature* (Freeman, San Francisco, 1982).
Masiarz, E. and Greenwood, T., *The Birth of Mathematics in the Age of Plato* (American Research Council, Washington, 1964).
Merlan, P., *From Platonism to Neoplatonism* (Nijhoff, The Hague, 1960).
Myhill, J., 'Philosophical Implications of Mathematical Logic', *Rev. Metaphys.* **6**, 165–98 (1952).
Penrose, R., *The Emperor's New Mind: Concerning Computers, Minds, and the Laws of Physics* (OUP, Oxford, 1989).
Penrose, R., 'Précis of *The Emperor's New Mind: Concerning Computers, Minds, and the Laws of Physics*' (together with responses by critics and a reply by the author), *Behavioural and Brain Sciences* **13**, 643–705 (1990).
Post, E. L., 'Absolutely unsolvable problems and relatively undecidable propositions: Account of an anticipation.' This article was originally submitted for

publication in 1941 and is printed in M. Davis, (ed.), *The Undecidable: Basic Papers on Undecidable Propositions, Unsolvable Problems and Computable Functions* (Raven Press, New York, 1965).

Russell, B., *An Introduction to Mathematical Philosophy* (Macmillan, New York, 1919).

Schaaf, W. L., 'Art and Mathematics: A Brief Guide to Source Materials', *Amer. Math. Monthly* **58**, 167–77 (1951).

Sextus Empiricus, 'Against the Logicians', in *Sextus Empiricus*, trans. R. G. Bury, (W. Heinemann, London, 1936) II, pp. 262–4.

Sloman, A., 'The Emperor's Real Mind', Unpublished Preprint, University of Birmingham, UK (1992).

Steiner, M., *Mathematical Knowledge* (Cornell UP, Ithaca NY, 1975).

Wedberg, A., *Plato's Philosophy of Mathematics* (Greenwood, Westport CT, 1977, orig. publ. 1955).

Weyl, H., *God and the Universe: The Open World* (Yale UP, New Haven CT, 1932).

Whitney, H., 'The Mathematics of Physical Quantities', *Amer. Math. Monthly* **75**, 115 & 227 (1968).

Yanin, Y., *Mathematics and Physics* (Birkhäuser, Boston MA, 1983).

# INDEX


abacists 101
aboriginals, Australian 34, 57
abstraction 23, 196
  growth of 249–51
aleph-nought 207, 211, 214
algebras 13, 167
algorithmists 101
algorithms
  concepts of 221
  fallible 291
  Markov 221
Al-Khowârizmî 100, 101
altars 73–5
animals, counting ability of 55, 168–9
Appel, Kenneth 230, 231, 306
Arabs 93
Aristotelian view of world 161, 163
Aristotle 15, 200, 243, 254
arithmetic 284, 285
  consistency of axioms of 114, 116, 121
  randomness in 137
  systems 143
  transfinite 206, 207, 210
*arrheton* 7
artificial intelligence 19, 247, 287, 290
arts 267, 268
astrology 106, 107
astronomy 86, 242, 266, 280
Augustine, St 256
automata, cellular 240, 241
axiomatic systems 12, 18, 141, 143, 144, 284
  building of 285
  consistency of 115, 120
axioms 17, 18, 126, 142, 143
axiom schemes 126
Aztecs 60, 61

Babylonians 54, 65, 66, 75, 80, 81, 244
  place-value system of 83, 84, 85, 91
  use of zero symbol by 85, 86
Barber paradox 110, 112
Barrow, Isaac 200
bases of counting systems 51
  base-2 (binary) 35, 51–6, 80–1
  base-5 58, 60–3
    5-10 system 61, 62, 63
    5-20 system 61, 62, 63
  base-10 (decimal), *see* decimal system
  base-12 60
  base-20 60, 62
  base-60 64–7, 85
Beaker People 78, 79
Bede, Venerable 49, 50
Bell, Eric T. 204
Ben Ezra, Rabbi 93
Bernays, Paul 258
binary arithmetic 104, 136
Bishop, Errett 220, 222, 223, 306
Boas, F. 39, 299
Bohr, Harald 220
Bohr, Niels 197
Boole, George 129, 298
Borel, Emile 259
Born, Max 219, 220
Bourbaki project 129–34
Bridgman, Percy 188, 262, 306
Brouwer, Luitzen 185, 190, 191, 192–6, 237–9, 245
  conflict with Hilbert 189–90, 195, 216–23
  philosophy of mathematics of 185, 196–8, 263
Bushmen, African 34, 51, 54

calculate, origin of word 42
Cantor, Georg 198, 201, 204, 217, 260, 264, 275
  conflict with Kronecker 199, 202–4, 213
  work on infinites 206–16
  works by 307
Cantor's Paradox 214
Carathéodory, Constantin 218, 219
cardinality
  of decimal numbers 213
  of natural numbers 206–7
  of real numbers 212
  of sets 206–7

cardinals 36
Carr, George 181
Casti, John 292, 302
Cayley, Arthur 227
Chaitin, Gregory 135, 136, 302
chaos 240, 241
Chihara, Charles 278
Chinese, numeral system of 88, 89, 91
Chomsky, Noam 171, 174, 304
Christianity 255–6
Church, Alonzo 238, 302
cipher, origin of word 90
civilization, cradles of 28–33
codes 151–4
  redundancy in 153, 154
Cohen, Paul 215, 302
collections 166, 167
commerce 157, 158, 250
complexity 159, 161–2, 163
compressions 24, 163–4, 247
computability 223, 238, 246, 290, 295
computer games 170–1
computer programming languages 222
computer programs 136, 138, 139, 239
  for four colour conjecture 230
  image of Universe as 248
computers 245–6
  idealized 246
  simple 254
  use of, in mathematics 222, 230, 234–6, 237–9, 240, 241
computer simulations 280–2
consciousness 148, 149, 169, 231, 244, 282–3, 287
conservation laws 160
consistency 114–17, 142
constructivism 188, 190, 194, 204, 220–2, 262, 295
continuum 211, 212, 214, 248
continuum hypothesis 214, 215, 264
Conway, John 239, 284
cosmology 262–3, 269, 270
counting 25
  human aptitude for 102
  neo-2 system of 56–9
  origin of 27, 31, 33, 81, 102–5, 244
  2-system 35, 51–6, 80–1
  use of fingers in 45–9, 50, 56
  *see also* bases of counting systems
Courant, Richard 220
cuneiform texts 65
cypher, origin of word 90

Dantzig, Tobias 168, 270, 300
Dawkins, Richard 244, 245 307
decimal systems 35, 51
  spread of 68–72
deductive reasoning 131
Descartes, René 150
Dieudonné, Jean 129, 130, 302
digits 38
Diophantine equations 135–7
Diophantus 135
dualism 150
Dummett, Michael 278, 307

Egyptians 80, 244
  hieroglyphic and hieratic symbols of 82–3, 84
Einstein, Albert 13, 117, 118, 119, 134, 188–9, 217
  and Hilbert–Brouwer conflict 216, 219, 220
  theories of relativity of 188, 262
  theory of gravitation of 270
ell 38
empiricism 257
Erdös, Paul 279, 280
Eskimos 61, 62
ethics 252
Euclid 10, 75, 115, 143
Euler, Leonhard 107, 288
evolution, human 244–5
extraterrestrial intelligence 265

Fang, Joong 175, 303, 305
Fermat's 'last theorem' 288
fingers, use of, in counting 45–9, 50, 56
finitists 215
formalism, *see under* mathematics
four-colour conjecture 227–34
fractions 190, 208–10
  concept of 38
  representation of 86
fraud 93

Frege, Gottlob 111, 192, 217, 282, 303
'frog-mice battle' 216, 220, 223
Fuchsian groups 275
functions
  computable 238, 247
  non-computable 238, 246, 247

Galileo Galilei 149–50, 205, 303
games 174–5
  algorithmic 239
  Game of Life 239, 240, 284–5, 286
gauge theories 160
Gentzen, Gerhard 121, 122
geometry 8, 204, 244, 253
  Euclidean 9–11, 13, 17, 198, 256
    consistency of 115
  non-Euclidean 12–14, 198, 242, 243
  ritual 73–81
Gerbert of Aurillac 93
God, arguments for the existence of 123, 162, 282, 297
Gödel, Kurt 18–19, 117–24, 125, 126, 142, 215, 285
  as exponent of Platonism 261, 278
  quotes by 261, 276, 278
  and Turing 290
  and Vienna circle 144
  works by 303, 309
Gödel numbers 124–5, 128, 285, 287
Gödel's theorems 119–20, 125–6, 135, 138, 139–40, 172, 284, 289
  impact on mathematics of 129–30
Goldbach, Christian 288
Goldbach's conjecture 179, 288
Gould, Stephen Jay 174
Greeks, early 157–8, 243, 249
Guthrie, Francis 227

Hadamard, Jacques 275, 305
Haken, Wolfgang 229, 230, 231, 306
Hamilton, William 13
handsbreadth 38
Hardy, G. H. 181, 182, 183, 184, 260
  works by 298, 307, 309
Hawking, Stephen 187
Hays, John 223, 307
Heawood, Percy 229
Hebrews, early 93, 158, 250
Heesch, Heinrich 229
Heinzelin, Jean de 32, 33
Heisenberg, Werner 18
Hermite, Charles 259, 260
Heyting, Arend 220, 307
Hilbert, David 113–17, 119, 121, 122, 125–6, 142, 143
  conflict with Brouwer 189–90, 195, 216–23
  early influence on Brouwer 192, 193
  and Frege, correspondence between 282
  quotes by 112, 213
  views on Gödel's theorems 120, 122
  works by 303
Hilbert's Hotel 207–8
Hindus 79
  use of zero symbol by 89
Hippasus 8
Hofstadter, Douglas 290, 309
Hume, David 191, 199
Hutchins, Edwin 174

idealism 255
impredicative definitions 112
Indian place-value system 91–2
Indo-Arab system of numbers 93
  evolution of number symbols 95–9
Indo-European languages
  distribution of 68–9
  propagation of 79
induction, mathematical 121, 126
  transfinite 291
Indus culture 89
infinites 205
  actual 203, 206, 216
  arithmetic of 206
  Cantor's work on 205–16
  concept of 185
  potential 201, 206, 215
information 135, 151, 154
*Interlingua* 129
intuition, nature of 196–8, 274, 276, 278
intuitionism 185, 191, 195, 217, 220–1, 248, 294
  neo-intuitionism 217
  primitive 200–1

intuitionism (*cont.*)
  and three-valued logic 186–7, 220
inventionism 155, 156, 165, 175, 176–7, 262, 280
Ishango bone 31–3
isomorphisms 125

Jaynes, Julian 169
Jeans, Sir James 292, 309

Kant, Immanuel 11, 164, 175, 193, 194, 197–8
  home town of 113
  scepticism of 147, 191
Kempe, Arthur 227
Kepler, Johannes 190
knowledge, causal theory of 227, 278, 279
Kolmogorov, A. N. 186
Korteweg, Diederik 192–3, 194
Kronecker, Leopold 189, 190–1, 195, 199–201
  conflict with Cantor 199, 202–4, 213
  works by 307
Kummer, Ernst 199

Lakatos, Imre 145–6, 303
language
  acquisition of 172
  evolution of 70, 102, 103, 173–4
  grammatical structure of 129
  hierarchy of languages 109
  universal 128
legal systems 174
Leibniz, Gottfried 127–9, 303
Lévy-Brühl, Lucien 27–8, 300
life sciences 162
logic 15–19, 134
  laws of 15, 257
    excluded middle 15, 16, 186
  symbolic 127, 129
  three-valued logics 16, 186, 220
  two-valued logics 15, 172, 186, 187
  *see also* logical systems
logical friction 140
logical systems
  completeness of 119
  self-consistency of 110, 119
  semantic incompleteness of 125
Lucas, John 19, 290, 309

Mannouri, Gerrit 192
'many', concept of 34
Markov, A. A. 221
Markov algorithm 221
Marx, Karl 157, 305
mathematical systems, basis of 17–18
mathematicians, characteristics of 274–5, 278–9
mathematics
  as an analogy 21
  applicability of 247, 248
  as an art 133
  categorical systems 125
  collaboration in 267
  completeness of 116, 122, 125
  consistency of 114, 115, 116, 122, 132
  and culture 243, 244, 245
  definitions of
    Bourbaki 132
    formalist 114, 126
    institutionist 196
  entities in
    causally inert 277, 280
    non-causally inert 282
  evolution of 244, 245
  experimental 179–81, 237, 239, 242
  extraterrestrial 266
  formalism in 117, 130, 131, 133, 176, 237, 292, 294
    case against 141–6
  foundation of 195
  future course of 236–43
  intuition in 189, 196, 272–3
    *see also* intuitionism
  as language of science 1
  Marxist 156–9
  meaning of 114, 124, 146, 188, 237, 296
  modern 104
  and music 268
  and mysticism 292–4
  natural structure to 104

mathematics (*cont.*)
nature of 173, 237, 296
Plato's philosophy of, *see* Platonic philosophy of mathematics
popular notion of 173
as psychology 165–71
and real world, relationship between 22–3, 134, 144, 256
as a science 258, 259
teaching of 133, 280
transhuman 234–6
truths in 234, 235, 236, 276
*Mathematische Annalen* 217–18, 219
Mayans 60, 61, 80
place-value system of 87, 91
use of zero symbol by 88, 89
mechanics 254
memes 244, 245
Merton, Robert 156
Mesopotamia 64
metalanguage 109, 110, 113
metamathematics 115, 190
Mill, John Stuart 258
mind, human
evolution of 147–9, 177
quantum mechanical aspects of 172, 280
simulation of 172, 290
structure of 172, 176
Mittag-Leffler, Gösta 203
models, mathematical 116, 155–6, 262
money 250
Morgenstern, Oskar 118
multiplication 57, 104
music 268
mysticism 292–4

Nature
forces of 160, 161
laws of 160–3, 165, 257, 272, 281, 284
neurophysiology 280
'new math' 133
Newton, Sir Isaac 10, 243
nine, origin of word 38
Noah 80–1
noise 154
non-euclideanism 14
nothing, notion of 89, 90
nought, origin of 86
Nozick, Robert 257
'number'
abstract concept of 4, 41, 105, 169, 190, 250
development of notion of, in childhood 166–70
sense 39–41
as symbolic language of Nature 20
numbers
as basic constituents of reality 251
complex 190
decimal 213
defined in terms of sets 283–4
even 207
irrational 7, 8, 190, 211, 212, 217
large 41
meanings of 5, 7
natural 126, 190, 191, 205, 294
cardinality of 206–7
in intuitionism 185, 188, 189, 190, 200–1
negative 90–1
odd and even 80
origins of 25, 26
prime 288–9
rational 7, 210
real 210, 211, 212
systems of 92, 93
evolution of 94–9
*see also* place-value systems
words for 44
*see also* zero
number theory 184
numerology 106–8, 283

observers, self-conscious 285, 292
ordinals 36

paradigms 243–5
paradoxes, logical 108–12
particle physics 159, 160, 162, 163, 271, 272
particulars 254, 255, 294
pattern recognition 164
Peano, Guiseppe 129
Penrose, Roger 19, 172, 285, 287, 289, 291

Penrose, Roger (*cont.*)
defence of Platonism by 261, 273, 279
and quantum mechanical aspects of human mind 172, 280
'singularity theorems' of 187
views on consciousness 290, 292
works by 309
perception, human 198
permanence 257
Persia, ancient 54
Philolaus of Croton 251
philosophy 24–5, 149, 193
of mathematics 25, 144–5, 194
of science 144
physicists 159, 160, 162, 163, 248, 255, 284
*see also* particle physics
Piaget, Jean 166, 168, 169, 170, 171, 172, 275
works by 305
place-value systems 83
Babylonian 83, 84, 85, 91
Indian 91–2
Mayan 87, 91
Plato 160, 252
Platonic philosophy of mathematics 252, 255, 258–65, 296–7
arguments for 268–72, 279
contrasted with mysticism 293–4
difficulties with 272–3, 277
manifestation of assumption of 265–6
vagueness of 273, 274
*see also* Platonism
Platonic view of world 160, 161, 163
Platonism 252–7
epistemological 273
mathematical 258
*see also* Platonic philosophy of mathematics
ontological 273
Plutarch 254
Poincaré, Henri 112, 189, 191, 193, 216, 259–60, 275–6
works by 307
Popper, Karl 144
populations, genetic closeness of 70–1
positivism, logical 144
Post, Emil 16, 121, 238, 289, 292, 309
predicate calculus 119
proof
notion of 179, 185, 230
process of 230–4
propositions 277
psychologists 169, 171
Pueblo Indians 43
Pythagoras 6–8, 75, 101, 251, 265
theorem of 75, 265
Pythagoreans 251, 268

quantity, notion of 103, 168, 250
quantum theory 279
quasars 269
quaternions 13

Ramanjuan, Srinvasa 181–4
randomicity 137
randomness 135, 137
Rao, Ramachandra 182
Rashomon effect 19–20
rationalism 257
realities, virtual 242
*reductio ad absurdum* 16, 186, 187, 201
relativity, Einstein's theories of 188, 262
religion 158, 278, 292, 296, 297
numerical 252
Riemann, Bernhard 12, 13
ritual, ancient 175
Roman counting board 93
Roman numerals 83, 93, 99
Russell, Bertrand 110–12, 114, 120, 192, 193, 283
paradoxes of 110–12, 114
and philosophical mysticism 293
project of founding mathematics on logic 110, 113, 120, 127, 185
quote by 188
works by 304, 310

Schwartz, Laurent 131, 133
science
collaboaration in 267–8
definition of 163
'hard' sciences 164
nature of 3, 24, 247

science (*cont.*)
  'soft' sciences 165
'scientific attitude' 2
Seidenberg, Abraham 53, 75, 79, 81, 301
self-discovery 176–7
sequences
  complexity of 138
  compressible 163
  incompressible 163
  random 138
sets 110, 112, 130, 261, 264, 284
  cardinality of 206–7
  definition of 206
  infinite 206
    countably 207, 208
    uncountably 211
  numbers as 283–4
  power 214
Sextus Empiricus 277, 310
Shannon, Claude 153, 154, 308
simplicity 159–61, 162
simulation 242
singularity theorems 187
Sloman, Aaron 273, 310
social sciences 155, 156, 159
Socrates 253
specialization 23
*Sriyantra* 77
Stoics 277
Stonehenge 78, 79
stones 38–9
subtraction 57
'Sulba-Sûtras' 73, 75
Sumerians 53, 64–5, 66, 67, 75
superstring theories 161, 284
symbolism 24, 70, 102, 106, 289
symbols
  manipulation of 167–8
  use of 20, 23–4
symmetry 160, 161, 162, 163, 248, 271
symmetry-breaking 162

taboos 72, 73
tallying 31, 32
  of large numbers 42–3
Tarski, Alfred 16, 109, 110, 125, 304
tautologies 112–13
Theory of Everything 140, 161, 163, 187, 248, 295
'threeness', notion of 249, 250
tokens 250
topology 194
toys 170
transfinite arithmetic 206, 207, 210
transfinitists 216
trap-door functions 151–3
truth, absolute 4, 13
Tsimshian language 39
Turing, Alan 235, 238, 246, 290, 291, 292
  works by 308
Turing machines 241, 246, 290
twenty, origin of word 47
2-system 35, 51–6, 80–1

undecidability 137, 138, 241, 288–9
universals 252, 254, 255
Universe
  origin of 187, 269–70
  paradigms of 243–4, 248, 295

visions 278
von Neumann, John 118, 120, 276, 308

Weierstrass, Karl 199, 200, 201
Weyl, Hermann 112, 195, 216, 217, 259
  works by 299, 304, 108, 310
Whitehead, A. N. 113, 114, 120, 127, 192, 283
  works by 304
Wilder, Raymond 175, 265, 304, 306
Wittgenstein, Ludwig 218

Yahweh 250
*yantras* 76, 77
year, division of 67

zero
  invention of 85–7, 88, 89
  origin of word 89–90